Advances in

CANCER RESEARCH

Volume 114

Guidance Molecules in Cancer and Tumor Angiogenesis

Advances in CANCER RESEARCH

Volume 114

Series Editors

Kenneth D. Tew

Professor and Chairman
Department of Cell and
Molecular Pharmacology
John C. West Chair of Cancer Research
Medical University of South Carolina
South Carolina, USA

Paul B. Fisher

Professor and Chairman
Department of Human & Molecular Genetics
Thelma Newmeyer Corman
Chair in Cancer Research
Director, VCU Institute of Molecular Medicine
Virginia Commonwealth University
School of Medicine
Richmond, Virginia, USA

Edited by

Ira O. Daar

Laboratory of Cell and
Developmental Signaling
National Cancer Institute
Frederick National Laboratory for
Cancer Research
Frederick, MD, USA

ELSEVIER

AMSTERDAM • BOSTON • HEIDELBERG • LONDON
NEW YORK • OXFORD • PARIS • SAN DIEGO
SAN FRANCISCO • SINGAPORE • SYDNEY • TOKYO
Academic Press is an imprint of Elsevier

Academic Press is an imprint of Elsevier
525 B Street, Suite 1900, San Diego, CA 92101-4495, USA
225 Wyman Street, Waltham, MA 02451, USA
32 Jamestown Road, London, NW1 7BY, UK
Linacre House, Jordan Hill, Oxford OX2 8DP, UK
Radarweg 29, PO Box 211, 1000 AE Amsterdam, The Netherlands

First edition 2012

ISBN: 978-0-12-386503-8
ISSN: 0065-230X

For information on all Academic Press publications visit our website at store.elsevier.com

Printed and bound by CPI Group (UK) Ltd, Croydon, CR0 4YY

Transferred to digital print 2012

Contents

Semaphorin Signals Tweaking the Tumor Microenvironment

Claudia Muratori and Luca Tamagnone

Guidance for Life, Cell Death, and Colorectal Neoplasia by Netrin Dependence Receptors

Christian Gespach

A Roundabout Way to Cancer

Mimmi S. Ballard and Lindsay Hinck

Role of the VEGF/VEGFR Axis in Cancer Biology and Therapy

Annamaria Rapisarda and Giovanni Melillo

Contributors

Numbers in parentheses indicate the pages on which the authors' contributions begin.

Mimmi S. Ballard, Department of Molecular, Cell and Developmental Biology, University of California, Santa Cruz, California, USA (187)

Jin Chen, VA Medical Center, Tennessee Valley Healthcare System; Department of Medicine; Department of Cancer Biology; Department of Cell and Developmental Biology and Vanderbilt-Ingram Cancer Center, Vanderbilt University School of Medicine, Nashville, Tennessee, USA (1)

Christian Gespach, INSERM U938, UMR_S938, Molecular and Clinical Oncology, UPMC Université Paris VI, Hôpital Saint-Antoine, Paris Cedex, France (87)

Lindsay Hinck, Department of Molecular, Cell and Developmental Biology, University of California, Santa Cruz, California, USA (187)

Giovanni Melillo, Discovery Medicine-Oncology, Bristol-Myers Squibb, Princeton, New Jersey, USA (237)

Claudia Muratori, University of Torino Medical School, Institute for Cancer Research (IRCC), Candiolo, Turin, Italy (59)

Annamaria Rapisarda, SAIC-Frederick, Inc., Frederick National Laboratory for Cancer Research, Frederick, Maryland, USA (237)

Ombretta Salvucci, Laboratory of Cellular Oncology, Center for Cancer Research, National Cancer Institute, Bethesda, Maryland, USA (21)

Luca Tamagnone, University of Torino Medical School, Institute for Cancer Research (IRCC), Candiolo, Turin, Italy (59)

Giovanna Tosato, Laboratory of Cellular Oncology, Center for Cancer Research, National Cancer Institute, Bethesda, Maryland, USA (21)

Preface

Progression of human cancer requires the coalescence of many processes that are also critical during development of the nervous system, such as cell proliferation, invasion, avoidance of apoptosis, and angiogenesis. Several families of neuronal guidance molecules, for example, semaphorins, ephrins, netrins, and slit, along with their cognate receptors (neuropilin/plexin, Eph, Unc/DCC, and Robo, respectively), have been identified that control guidance and morphogenetic functions in other tissues and the vascular system. One additional family, the vascular endothelial growth factors (VEGFs), may play a role in neurogenesis but are best known as critical regulators of vascular development. Several antiangiogenic therapies have been developed and used clinically that target VEGF or the cognate receptor, but the clinical outcomes have varying success. The role of the nervous system in tumorigenesis was originally thought to be that of a path by which the invading tumor cells travel or to act as a neoangiogenic supply line to tumors. However, now it has been suggested that the nervous system is functionally related to tumor progression by regulating a complex network of mediators (soluble factors, membrane-bound molecules, and receptors) that influence tumor progression. Emerging evidence indicates that there is also cross-talk between tumor cells and the nervous system, particularly between neural factors or guidance molecules and cancer cells. Understanding the relationship between guidance molecules (as well as neurotrophic factors and neurotransmitters), the tumor, and the tumor microenvironment may prove to be critical for devising appropriate therapies for intervention in cancer growth, invasion, and progression. This volume will focus on several guidance molecule systems including Eph/ephrin, semaphorins/plexins and neuropilins, slit/robo, netrins/unc and DCC, and VEGF/VEGFR and what is known (or implicated) about their role in cancer.

Ira O. Daar

Regulation of Tumor Initiation and Metastatic Progression by Eph Receptor Tyrosine Kinases

Jin Chen

VA Medical Center, Tennessee Valley Healthcare System, Nashville, Tennessee, USA
Department of Medicine, Vanderbilt University School of Medicine, Nashville, Tennessee, USA
Department of Cancer Biology, Vanderbilt University School of Medicine, Nashville, Tennessee, USA
Department of Cell and Developmental Biology, Vanderbilt University School of Medicine, Nashville, Tennessee, USA
Vanderbilt-Ingram Cancer Center, Vanderbilt University School of Medicine, Nashville, Tennessee, USA

In recent years, a growing body of evidence has indicated that signaling molecules previously implicated in axon guidance are important regulators of multistep tumorigenesis and progression. Eph receptors and ephrins belong to this special class of molecules that play important roles in both axon guidance and cancer. Tremendous progress has been made in the past few years in both understanding the role of Eph receptors and ephrins in cancer and designing therapeutic strategies for cancer therapy. This review will focus on new advances in elucidating the contribution of Eph/ephrin molecules to key processes in tumor initiation and metastatic progression, including cancer cell proliferation, invasion and metastasis, and tumor angiogenesis.

Advances in CANCER RESEARCH, Volume 114

0065-230X/12 $35.00
DOI: 10.1016/B978-0-12-386503-8.00001-6

I. INTRODUCTION

Cancer initiation and malignant progression are multistep processes that involve loss of growth control, evasion of apoptosis, sustained angiogenesis, tissue invasion, and metastasis (Hanahan and Weinberg, 2000, 2011). Emerging evidence has indicated that signaling molecules previously implicated in axon guidance are important regulators of multistep tumor initiation and progression (Adams and Eichmann, 2010; Eichmann *et al.*, 2005). These include Eph/ephrin, semaphorins/plexins, VEGF/VEGFR, chemokines/receptors, netrins/DCC UN5, Slit/Robo, and Notch/Delta. This review will focus on recent advances on dissecting the role of Eph/ephrin molecules in cancer and tumor angiogenesis. Due to space and scope constraints, we have limited our discussion in the text to the Eph receptors and ephrins for which the most data are available. Tumor and vascular phenotypes of Eph/ephrin knockout and transgenic animals are summarized in Tables I and II. More comprehensive reviews on Eph–ephrin signaling in physiology and disease can be found elsewhere (Kullander and Klein, 2002; Pasquale, 2005, 2008, 2010).

Table I Tumor Phenotypes of Eph/ephrin-Mutant Animals

Mouse strain	Tumor model	Phenotype	Reference
EphA2$^{-/-}$ knock-out mice	MMTV-Neu breast cancer	Decreased tumor incidence, tumor burden, and metastasis	Brantley-Sieders *et al.* (2008)
EphA2$^{i/i}$ gene trap mice	DMBA/TPA skin cancer	Increased tumor number, tumor burden, and invasiveness	Guo *et al.* (2006)
EphA2$^{i/i}$ gene trap mice	APC$^{Min/+}$ colon cancer	Decreased tumor number and size in small and large intestine	Bogan *et al.* (2009)
Villin-*ephrinA1* transgenic mice	APC$^{Min/+}$ colon cancer	Increased tumor numbers and invasiveness	Shi *et al.* (2008)
MMTV-*EphB4* transgenic mice	MMTV-Neu breast cancer	Accelerated tumor onset and increased metastasis	Munarini *et al.* (2002)
Villin-*EphB2*$^{\Delta C}$ *EphB3*$^{-/-}$ *Vil-Cre/ephrin-B1*$^{\Delta/\Delta}$	APC$^{Min/+}$ colon cancer	Increased tumor number and tumor invasion	Batlle *et al.* (2005), Cortina *et al.* (2007)

Table II Vascular Phenotypes of Eph/ephrin-Mutant Animals

Gene knockout/knockin	Phenotype	Reference
$ephrinA1^{-/-}$	Defects in heart valves; impaired cardiac function	Frieden *et al.* (2010)
$EphA3^{-/-}$	Perinatal lethality; hypoplastic heart valve	Stephen *et al.* (2006)
$EphA2^{-/-}$	Defects in tumor angiogenesis	Brantley-Sieders *et al.* (2005, 2008)
$EphB2^{-/-}EphB3^{-/-}$	Embryonic lethal, die at E10.5 (~30%); defective vessel remodeling, similar to those observed in $ephrinB2^{-/-}$	Adams *et al.* (1999)
$EphB4^{-/-}$	Embryonic lethal, die at E10.5; defective vessel remodeling, similar to those observed in $ephrinB2^{-/-}$	Gerety *et al.* (1999)
$ephrinB2^{-/-}$	Embryonic lethal, die at E10.5; defective vessel remodeling and sprouting	Wang *et al.* (1998)
$ephrinB2^{\Delta C/\Delta C}$ (deletion of cytoplasmic domain)	Defects in angiogenic remodeling similar to those observed in $ephrinB2^{-/-}$	Adams *et al.* (2001)
$ephrinB2^{\Delta V/\Delta V}$ (deletion of Val. residue in PDZ-binding motif)	Defects in angiogenic remodeling similar to those observed in $ephrinB2^{-/-}$; defects in tumor angiogenesis	Mäkinen *et al.* (2005), Sawamiphak *et al.* (2010)
$ephrinB2^{5Y/5Y}$ (knockin mutation in five tyrosine residues)	Mild lymphatic phenotype	Mäkinen *et al.* (2005), Sawamiphak *et al.* (2010)
$ephrinB2^{i\Delta EC}$ (VE-cadherin promoter-driven CreERT2, EC-specific knockout)	Defects in angiogenic sprouting in both blood vessels and lymphatic vessels	Wang *et al.* (2010)
$ephrinB2^{iGOF}$ (Tie2-rtTA driven tetO-ephrinB2 transgenic)	Increased angiogenic sprouting	Wang *et al.* (2010)
$ephrinB2^{\Delta PC/vSMC}$ (PDGFRβ promoter-driven Cre, mural cell-specific knockout)	Perinatal lethality; vascular defects in multiple organs; abnormal migration of smooth muscle cells to lymphatic capillaries	Foo *et al.* (2006)

II. DYSREGULATION OF EPH RECEPTORS IN CANCER

The Eph family of receptor tyrosine kinases (RTKs) is the largest identified in the vertebrate genome and is subdivided into class A and class B based on sequence homology and binding affinity for two distinct types of membrane-anchored ephrin ligands. In general, class A receptors interact with glycosyl-phosphatidylinositol-linked class A ephrins, while class B receptors bind to class B ephrins that are attached to the cell membrane by a transmembrane-spanning domain, although interclass binding does occur among certain family members (reviewed in Pasquale, 2005, 2008, 2010). Binding of ephrin to Eph receptor induces receptor clustering and activation. Recent structural studies of the Eph/ephrin interactions have not only identified ligand–receptor interaction sites but also discovered receptor–receptor interfaces that function in assembly of higher-order signaling clusters (Himanen *et al.*, 2010; Seiradake *et al.*, 2010). Because both ligand and receptor are membrane bound, engagement of ephrin with Eph receptor between adjacent cells induces bidirectional signaling through both ligand- and receptor-expressing cells (reviewed in Kullander and Klein, 2002; Pasquale, 2008). Originally characterized as axon guidance regulators, ephrins and Eph RTKs are subsequently recognized to regulate physiologic and pathologic processes during embryonic development, in normal tissue homeostasis, and in disease (reviewed in Pasquale, 2008, 2010).

Recent technological advances in analyzing the human cancer genome permit the study of gene copy number, expression level, and mutation status in tumor tissue. These studies demonstrated that Eph receptors are often dysregulated in cancer (reviewed in Brantley-Sieders, 2011; Brantley-Sieders *et al.*, 2011b; Pasquale, 2010). Gene expression studies by microarray analysis and immunohistochemistry on tumor tissue microarrays have correlated expression of some Eph receptors in tumor epithelium and/or vasculature with disease stage, metastasis, recurrence, and survival. For example, EphA2 expression is elevated in many types of cancer (Landen *et al.*, 2005b; Wykosky and Debinski, 2008) and high levels of EphA2 correlate with tumor malignancy and poor patient survival in breast cancer (Brantley-Sieders, 2011; Fournier *et al.*, 2006; Martin *et al.*, 2008; Zhuang *et al.*, 2010) and lung cancer (Brannan *et al.*, 2009; Faoro *et al.*, 2010; Kinch *et al.*, 2003). Likewise, levels of several EphB receptors are also elevated in different stages of colon cancer, lung cancer, or breast cancer (Batlle *et al.*, 2002; Brantley-Sieders, 2011; Ji *et al.*, 2011; Kumar *et al.*, 2009; Noren and Pasquale, 2007). However, decreased Eph receptor levels have also been reported in certain types of human cancer (Batlle *et al.*, 2005; Kumar *et al.*, 2009). Dysregulation of Eph receptor expression in tumor has been attributed to multiple mechanisms, including chromosomal abnormality, epigenetic regulation, mRNA stability, and transcriptional control (reviewed in Pasquale, 2010). Aside from changes in receptor level, somatic mutations have been found in nearly all Eph receptors. Notably, 11 somatic mutations in EphA3 receptor were identified in 5–10% of lung cancers, placing EphA3 among 27 most frequently mutated genes in human lung adenocarcinoma (Davies *et al.*, 2005; Ding *et al.*, 2008; Greenman *et al.*, 2007; Wood *et al.*, 2006). However, these mutations are scattered throughout the receptor, and it is unclear whether these nonrecurrent mutations are "driver" or biologically neutral "passenger" genetic alterations. Elucidating the effects of these mutations will greatly improve our understanding of how Eph receptors function in cancer.

III. ROLE OF EPH RECEPTORS IN TUMOR CELL PROLIFERATION

Sustained proliferative signals and evasion of growth suppressors are two hallmarks of cancer. However, Eph receptors were initially thought not to be involved in regulation of cell growth (Brambilla *et al.*, 1995; Bruce *et al.*, 1999; Lhotak and Pawson, 1993). More recent studies found that Eph receptors maintain tissue homeostasis by controlling the proliferation of stem and progenitor cells in adult. For example, in the intestinal stem cell niche, EphB signaling promotes cell-cycle reentry of progenitor cells through Abl and Cyclin D1 and accounts for approximately 50% of the mitogenic activity in the adult mouse small intestine and colon (Genander *et al.*, 2009). Likewise, Eph receptor signaling in neural progenitor cells in hippocampus also promotes proliferation (Chumley *et al.*, 2007). In contrast, Eph receptors can negatively regulate proliferation of hair follicle and epidermal progenitor cells in the adult mice (Genander *et al.*, 2010), as well as in neural stem/progenitor cells in the lateral ventricle (Conover *et al.*, 2000).

In cancer cells, the role of Eph receptors in regulation of tumor cell growth is similarly complex, and the effects of these receptors on cell proliferation are often dependent on ligand stimulation, signaling cross talk, or other contextual factors. This complexity is best illustrated in the case of EphA2 receptor. EphA2 receptor is highly expressed in a number of tumor types, including breast cancer, prostate cancer, ovarian cancer, lung cancer, and glioblastoma multiforme (Brantley-Sieders, 2011; Landen *et al.*, 2005b; Wykosky and Debinski, 2008). High levels of EphA2 are correlated with poor patient survival (Brannan *et al.*, 2009; Brantley-Sieders, 2011; Faoro *et al.*, 2010; Fournier *et al.*, 2006; Kinch *et al.*, 2003; Martin *et al.*, 2008; Zhuang *et al.*, 2010). Silencing of EphA2 in cancer cells by siRNA-mediated knockdown, antisense oligonucleotides, or targeted deletion of EphA2 in knockout mice inhibited tumor initiation and metastatic progression (Brantley-Sieders *et al.*, 2008; Duxbury *et al.*, 2004; Landen *et al.*, 2005a). Interestingly, the tumor promotion effects of EphA2 appear to be ligand independent and result from cross talk with HER2 RTK (Brantley-Sieders *et al.*, 2008). In support of this notion, overexpression of EphA2 in nontransformed MCF-10A breast epithelial cells induces anchorage-independent colony growth *in vitro* and tumor formation *in vivo* (Zelinski *et al.*, 2001). Further investigation revealed that high levels of EphA2 in these tumor cells are minimally phosphorylated, suggesting poor ligand-induced receptor forward signaling. Indeed, exogenous ligand stimulation of EphA2 suppressed colony formation on 2D culture and spheroid growth in 3D Matrigel (Miao *et al.*, 2001; Yang *et al.*, 2011). Treatment of mammary tumors with a ligand-mimetic activating antibody against EphA2 suppressed tumor growth *in vivo* (Brantley-Sieders *et al.*, 2008; Coffman *et al.*, 2003; Landen *et al.*, 2006), corroborating the notion that EphA2's effects in tumor promotion is ligand independent. Similarly, EphB4 is often overexpressed in human breast cancer (Wu *et al.*, 2004), and EphB4 knockdown inhibited tumor cell survival, migration, and invasion *in vitro*, and tumor growth *in vivo* (Kumar *et al.*, 2006). However, ephrin-B2-induced EphB4 forward signaling inhibits breast cancer cell viability and proliferation, as well as tumor volume in xenografts (Noren *et al.*, 2006). Taken together, these data suggest a model in which ephrin-induced Eph receptor forward signaling inhibits tumor cell proliferation, whereas in the absence of the ligand, cross talk between Eph receptors and other oncogenic proteins leads to enhanced cell proliferation and tumorigenesis, presumably independent of ephrin stimulation.

How does Eph receptor forward signaling inhibit tumor cell growth? Ephrin-B2-induced EphB4 receptor forward signaling activates an antioncogenic pathway, leading to Abl kinase activation and inactivation of Crk adaptor function through phosphorylation by Abl (Noren and Pasquale, 2007; Noren *et al.*, 2006). In the case of EphA2, several studies showed that stimulation of tumor cells with ephrin-A1 ligand inhibited the activation of both Ras–MAPK and Akt–mTOR pathways (Macrae *et al.*, 2005; Miao *et al.*, 2001; Yang *et al.*, 2011). However, in PC3 prostate cancer cells, MEK inhibitor does not appear to affect cell proliferation, whereas PI3K inhibitor and Rapamycin efficiently inhibited cell growth, suggesting that Akt–mTOR is the major pathway inhibited by EphA2 forward signaling (Yang *et al.*, 2011). It is interesting to note that EphA2 forward signaling does not change the activity of Akt upstream regulators such as Ras family GTPases, PI3 kinase, integrin, or the SHIP2 lipid phosphatase. Rather, EphA2 inactivates the Akt–mTORC1 oncogenic pathway through Akt dephosphorylation mediated by a PP1-like serine/threonine phosphatase (Yang *et al.*, 2011). Although inhibition of Akt–mTOR pathway by EphA2 forward signaling appears to be a major mechanism in regulation of cell growth in multiple tumor cell lines, the importance of EphA2-mediated suppression of Akt–mTOR pathway in tumor growth needs to be further validated in animal models *in vivo*.

Decades of research have demonstrated that "contact inhibition" is an important mechanism used by normal cells to suppress cell proliferation. The mechanism of contact inhibition appears to involve strengthening of cell–cell adhesion and maintaining cell polarity and tissue integrity. In MDCK cells, ligand-activated EphA2 signaling suppresses Arf6 GTPase activity, leading to cell compaction and polarization (Miura *et al.*, 2009), suggesting a role for EphA2 forward signaling in E-cadherin-based cell–cell adhesion and the apical–basal polarization of epithelial cells. However, overexpression of EphA2 receptor in MCF10A mammary epithelial cells destabilizes adherens junctions via a RhoA-dependent mechanism (Fang *et al.*, 2008b). The seemingly conflicting results in these two studies may be due to differences in signaling strength and composition of the signaling complex in the respective cell types, as illustrated in ephrin-B1 regulation of tight junctions (Lee and Daar, 2009; Lee *et al.*, 2008). Either overexpression or loss of ephrin-B1 can disrupt cell–cell contact and tight junctions. Overexpressed or unphosphorylated ephrin-B1 competes with active Cdc42 GTPase for binding to Par-6 and inhibits αPKC activation in the Par polarity complex, leading to tight junction disruption. Tyrosine phosphorylation of ephrin-B1 induces dissociation of ephrin-B1 and Par-6, which is now available to interact with Cdc42 and establish tight junction. However, ephrin-B1 is also associated with Par-6 at adherence junction. Loss of ephrin-B1 entirely may allow Par-6 at adherence junction and compete with Par-6 at tight junction for Cdc-42, resulting in disruption of tight junction. Interestingly, phosphorylation of ephrin-B1 can be EphB receptor independent and can be induced by association with tight junction protein claudin (Tanaka *et al.*, 2005). Together, these studies suggest that deregulation of ephrin/Eph signaling can affect cell–cell adhesion, but it remains to be determined whether dysregulation of cell–cell adhesion directly leads to changes in cellular proliferation. However, breaking down cell–cell adhesion is often a first step that leads to epithelial-to-mesenchymal transformation, a program that broadly regulates invasion and metastasis.

IV. ROLE OF EPH RECEPTORS IN INVASION AND METASTASIS

High endogenous levels of Eph receptors in cancer cells, or overexpression of Eph molecules, have been implicated in promoting tumor cell invasiveness and motility *in vitro* and distant metastasis *in vivo*. Targeted EphA2 receptor knockout significantly inhibited breast cancer lung metastasis in the MMTV-Neu transgenic mammary tumor model (Brantley-Sieders *et al.*, 2008). EphA2-deficient tumor cells isolated from the knockout mice displayed reduced cell invasion and migration in response to serum stimulation, indicating that EphA2 is required for tumor cell motility. High levels of EphA2 in glioblastoma cells were also shown to promote tumor cell invasiveness and motility (Miao *et al.*, 2009; Wykosky *et al.*, 2005). In both cases, EphA2-dependent cell motility appears to be mediated by ligand-independent signaling and cross talk with other oncogenic pathways. In breast cancer cells, EphA2 forms a complex with ErbB2 RTK and regulates the activity of Ras–MAPK and RhoA GTPase (Brantley-Sieders *et al.*, 2008). In glioblastoma cells, EphA2-mediated chemotactic cell migration required phosphorylation of EphA2 on Ser897 by Akt (Miao *et al.*, 2009). Ephrin-A1 stimulation of EphA2 negated Akt activation by growth factors and caused EphA2 dephosphorylation on Ser897. As noted above, inhibition of Akt activity by ligand-dependent EphA2 receptor signaling appears to be mediated by a PP1-like phosphatase, resulting in dephosphorylation of Akt at Thr308 and Ser374 sites (Yang *et al.*, 2011).

In colon cancer, EphB–ephrin-B interaction regulates both cell positioning and tumor metastasis (Merlos-Suárez and Batlle, 2008). In normal intestine, Wnt signaling drives EphB2 and EphB3 expression in crypts (Batlle *et al.*, 2002). EphB2 is expressed at highest levels in intestinal stem cells and its expression decreases in progenitor cells as they differentiate and migrate toward the lumen. EphB3 is localized at the bottom most positions of the crypt. Ephrin-B1 and B2 express complementarily in differentiated cells and their expression is negatively controlled by β-catenin/Tcf activity. In colon cancer, APC mutation activates the Wnt pathway and upregulates the expression of EphB2, B3, and B4 receptors. These tumor-initiating cells repopulate the crypts until they reach the surface epithelium where they encounter normal cells expressing ephrin-Bs. EphB–ephrin-B bidirectional signaling was proposed to restrict tumor spreading (Merlos-Suárez and Batlle, 2008). This model was supported by the observation that low- and medium-grade tumors were enriched in EphB-positive tumor cells, whereas high-grade tumors are often EphB negative, suggesting that silencing of EphB expression is associated with more malignant human tumors. Furthermore, loss of EphB3 or expression of a dominant-negative cytoplasmic deletion mutant of EphB2 accelerates tumorigenesis in the colon and rectum of $APC^{Min/+}$ mice and results in the formation of aggressive adenocarcinoma (Batlle *et al.*, 2005). The mechanism by which EphB receptors compartmentalize the expansion of colorectal tumor cells appears to be dependent on E-cadherin-mediated adhesion (Cortina *et al.*, 2007). Thus, constitutive Wnt signaling upregulates EphB expression at early stages of colon cancer.

However, EphB silencing is required for malignant tumor expansion at later stages, as clinical studies correlate loss of EphB expression with the transition from adenoma to adenocarcinoma (Batlle *et al.*, 2005).

Interaction between tumor cells and adjacent stromal cells greatly influences tumor cells' ability to move and invade into surrounding tissue. Most studies in the literature have focused on paracrine signaling mediated by secreted factors. Using a tumor cell-fibroblast coculture system, a recent report showed that a combinatorial code of Eph receptor activation dictates whether a cell moves or stops upon encountering another cell (Astin *et al.*, 2010). Prostate cancer cell line PC3 expresses high levels of ephrin-As and EphA2 and EphA4 receptors. Contact inhibition of locomotion is induced between PC3 cells by EphA forward signaling via RhoA activation and subsequent cell rounding. Knockdown of EphA2 and EphA4 abolished homotypic contact inhibition of motility in PC3 cells. In contrast, fibroblasts express high level of ephrin-B2, which activates EphB3 and EphB4 on PC3 cells to induce Cdc42 activation, lamellipodia formation, and cell migration. It is not known if the same combinatorial code of Eph molecules also functions in other tumor cell types. However, it is now clear that signals transduced upon direct cell–cell contact are also crucial for regulating contact inhibition of locomotion and invasiveness, possibly through Eph–ephrin signaling between adjacent cells.

V. REGULATION OF TUMOR ANGIOGENESIS BY EPH RECEPTORS

It is increasingly recognized that the tumor microenvironment plays crucial roles in tumor initiation and malignant progression. Angiogenesis, a process in which capillaries sprout from existing vessels, not only provides nutrients and oxygen for tumor growth but also allows tumor cells to intravasate and travel to a distant site to form metastatic lesions. Eph receptors have been long known to regulate angiogenesis. Early studies discovered that ephrin-A1, a prototypic ligand for the EphA2 receptor, induces corneal angiogenesis (Pandey *et al.*, 1995). Several years later, ephrin-B2 and EphB4 were shown to regulate angiogenic remodeling and arterial vein specification during embryonic development (Gerety *et al.*, 1999; Wang *et al.*, 1998). Subsequent studies revealed that ephrin–Eph molecules have diverse functions in regulating angiogenesis, including modulating endothelial cell motility and assembly, recruiting perivascular supporting cells, lymphangiogenesis, and, more recently, angiocrine signaling.

A. Ephrin-A1 and EphA2 in Tumor–Endothelium Interaction

In addition to dysregulation of Eph receptor in tumor cells, many Eph receptors and ephrins are also upregulated in the tumor vasculature. Elevated EphA2 receptor expression was found in a number of tumor xenografts and human breast cancer clinical specimens (Brantley-Sieders *et al.*, 2011a; Ogawa *et al.*, 2000). Host deficiency of EphA2 RTK results in reduced tumor angiogenesis and metastatic progression (Brantley-Sieders *et al.*, 2005). EphA2-deficient endothelial cells display impaired migration and assembly in response to either ephrin-A1 ligand or VEGF stimulation (Brantley-Sieders *et al.*, 2004; Chen *et al.*, 2006). The defective phenotypes resulted from, at least in part, dysregulation of PI3K, Vav guanine nucleotide exchange factor, and Rac1 GTPase signaling in EphA2 knockout endothelial cells (Brantley-Sieders *et al.*, 2004; Hunter *et al.*, 2006). Mapping of phosphorylated tyrosine residues of EphA2 revealed interaction sites between EphA2 and Vav GEF or p85 subunits of PI3K (Fang *et al.*, 2008a), suggesting a critical role for tyrosine phosphorylation in transducing EphA2 forward signaling in vascular endothelial cells. Indeed, knockout of Vav2 and Vav3 in mice resulted in similar vascular defects *in vitro* (Hunter *et al.*, 2006) and impaired tumor angiogenesis *in vivo* (Brantley-Sieders *et al.*, 2009), supporting the critical role of Vav–Rac signaling in EphA2-mediated neovascularization.

Expression analysis has correlated ephrin-A1 with tumor neovascularization and progression in several mouse models of cancer and in human samples. Ephrin-A1 expression is induced by TNF-α, VEGF, and HIF (Cheng *et al.*, 2002; Pandey *et al.*, 1995; Yamashita *et al.*, 2008). Targeted gene deletion of ephrin-A1 in mice results in defective heart valve and impaired cardiac function (Frieden *et al.*, 2008). In addition, ephrin-A1-null endothelial cells have reduced migratory and assembly response upon stimulation of VEGF *in vitro* and impaired angiogenesis in a sponge assay *in vivo* (V. Youngblood and J. Chen, unpublished data), suggesting a critical role of ephrin-A1 in angiogenesis. It is currently unknown whether ephrin-A1 functions in vascular endothelial cells through EphA2 forward signaling, ephrin-A1 reverse signaling, or both. Although ephrins lack a cytoplasmic domain, ephrin-A ligands can complex with transmembrane proteins to transduce their signals. Indeed, reverse signaling by ephrin-As upon binding EphAs controls axon guidance and mapping in the nervous system (Lim *et al.*, 2008), and cell attachment and motility in fibroblasts and tumor cells (Davy and Robbins, 2000; Davy *et al.*, 1999). One recent study showed that p75 neurotrophin receptor can form a complex required for Fyn phosphorylation upon binding EphAs, activating a signaling pathway leading to cytoskeletal changes (Lim *et al.*, 2008). It remains to be determined whether ephrin-A1 reverse signaling also plays significant roles in angiogenesis and which key transmembrane proteins act as coreceptors for ephrin-A1 in endothelial cells.

It is well known that tumor-derived proangiogenic factors induce neovascularization to facilitate tumor growth and malignant progression. Conversely, the concept of "angiocrine" signaling, in which signals produced by endothelial cells elicit tumor cell responses distinct from vessel function, has been proposed (Butler *et al.*, 2010). Recently, EphA2 receptor was found to regulate such "angiocrine" signaling (Brantley-Sieders *et al.*, 2011a). Cotransplantation of tumor cells with wild-type, but not EphA2-null, endothelial cells enhanced tumor growth before incorporation of exogenous endothelial cells into blood vessels, suggesting that endothelial-derived factors modulate tumor growth. Interestingly, EphA2 does not appear to upregulate angiogenic factors. Rather, loss of EphA2 results in elevated Slit2 production from endothelial cells, which inhibits tumor cells growth and motility. In human breast cancer, high levels of EphA2 are associated with low Slit2 expression in tumor endothelium, correlating with poor patient survival. Together, these data suggest that EphA2 RTK not only has a role in tumor-induced angiogenesis but also functions in angiocrine regulation of tumor growth and motility.

B. Ephrin-B2 and EphB4 in Tumor Angiogenesis

Ephrin-B2 and its receptor EphB4 were first discovered to play a role in arterial vein specification and angiogenic remodeling during embryonic development (Gerety *et al.*, 1999; Wang *et al.*, 1998). Knockout of either ephrin-B2 or EphB4 in mice resulted in similar defects of angiogenic remodeling. Because complete deletion of ephrin-B2 or EphB4 induces embryonic lethality, inducible endothelial-specific knockout animals or mice bearing specific mutations in the cytoplasmic tail of ephrin-B2 have been developed. Endothelial cell-specific deletion of ephrin-B2 affected retinal angiogenesis in newborn mice, as well as lymphatic vessel development in dermal skin (Wang *et al.*, 2010). These vascular defects appear to be caused by lack of ephrin-B2 reverse signaling through interaction with PDZ domain-containing proteins (Mäkinen *et al.*, 2005; Sawamiphak *et al.*, 2010). A single amino acid deletion in the PDZ-binding motif of the cytoplasmic tail of the ephrin-B2 leads to similar phenotypes as those in endothelial-specific knockout mice. In contrast, mutations in all five tyrosine residues of ephrin-B2 do not appear to affect ephrin-B2 function in angiogenesis, suggesting that ephrin-B2 reverse signaling in angiogenic remodeling is mediated by PDZ interaction (Mäkinen *et al.*, 2005; Sawamiphak *et al.*, 2010).

How does ephrin-B2 reverse signaling affects angiogenesis? Ephrin-B2 is expressed in sprouting capillaries, both in tip cells and stalk cells. One of the characteristics of angiogenic tip cells is that they send numerous fillopodia to sense the microenvironment in response to VEGF released from hypoxic tissues. These tip cells express high levels of VEGF receptors 2 and 3 and guide migration of the angiogenic sprout (Adams and Eichmann, 2010). Loss of ephrin-B2 or impaired ephrin-B2 reverse signaling significantly decreases the number of fillopodia extensions in angiogenic tip cells without affecting the proliferation of stalk cells (Sawamiphak *et al.*, 2010; Wang *et al.*, 2010). Because ephrin-B2 and VEGF receptor (VEGFR) are colocalized on fillopodia and the body of the tip cells, the cross talk between these two signaling pathways was investigated. Loss of ephrin-B2 or mutation in the PDZ-binding motif of the ephrin-B2 cytoplasmic tail inhibited VEGFR endocytosis and receptor phosphorylation. VEGFR endocytosis appears to be critical for its signaling, as inhibition of VEGFR internalization by Dynosore, an inhibitor of dynamin GTPase, reduced phosphorylation of VEGFR and Akt *in vitro* and fillopodia extension in angiogenic spouts. Together, these results elucidate how these two important signaling pathways cross talk in angiogenesis and lymphangiogenesis, although the molecular mechanism underlying how ephrin-B2 PDZ interaction affects VEGFR endocytosis awaits to be further investigated.

Aside from developmental and physiological angiogenesis, ephrin-B2 and EphB4 are also expressed in the tumor blood vessels in a variety of tumors (Erber *et al.*, 2006; Gale *et al.*, 2001; Shin *et al.*, 2001) and are induced under hypoxic conditions (Vihanto *et al.*, 2005), suggesting that this ligand–receptor pair may regulate tumor neovascularization. In support of this hypothesis, A375 melanomas form smaller, less vascularized tumors in the presence of the soluble, monomeric EphB4 extracellular domain *in vivo* (Martiny-Baron *et al.*, 2004). Soluble EphB4 may act, at least in part, by preventing binding of tumor cell EphB receptors to ephrin-B2-positive endothelium, thus disrupting tumor angiogenesis. Further support for this hypothesis is provided from studies in which overexpression of a truncated cytoplasmic deletion EphB4 receptor construct produced increased tumor growth and vascularity in mammary tumors, likely through ephrin-B2-mediated reverse signaling in host endothelium (Noren *et al.*, 2004). Indeed, transplantation of gliomas cells into the brain of mice with ephrin-B2 cytoplasmic mutation results in reduced tumor size and decreased tumor vessel density (Sawamiphak *et al.*, 2010), suggesting that ephrin-B2 reverse signaling is required for tumor angiogenesis. Taken together, these studies reveal a critical role for B class receptors and ligands in tumor progression and vascular recruitment for multiple types of human cancer.

C. Ephrin/Eph in Endothelial–Mural Cell Interaction

Perivascular supporting cells or mural cells, called pericytes in capillaries or vascular smooth muscle cells in larger vessels, are essential components of blood vessels and are critical in maintaining vessel stability. Alteration of pericyte density or the stable attachment of pericytes to the endothelium is associated with a number of human diseases including cancer. Paracrine signaling between endothelial and mural cells, such as PDGF-B/PDGFRβ and angiopoietin and Tie2 receptor, has been shown to play essential roles in recruitment of perivascular mesenchymal cells to differentiate into pericytes and maintaining tight interaction between endothelial cells and mural cells (reviewed in Gaengel *et al.*, 2009). More recently, ephrin–Eph receptor signaling, ephrin-B2 and EphB4 specifically, has also been implicated in blood-vessel-wall assembly.

During embryonic development, ephrin-B2 is expressed on arterial endothelial cells. However, as development proceeds, ephrin-B2 expression progressively extends through the capillary bed and to pericytes (Erber *et al.*, 2006; Gale *et al.*, 2001; Shin *et al.*, 2001), suggesting a role of ephrin-B2 in vessel wall formation. Mural cell-specific deletion of ephrin-B2 in mice results in perinatal lethality due to vascular defects in multiple organs and abnormal migration of smooth muscle cells to lymphatic capillaries (Foo *et al.*, 2006). Ephrin-B2-deficient pericytes have reduced contacts with endothelial cells and fail to envelope the endothelial tube (Foo *et al.*, 2006), resulting in dysfunction of microvessels and hemorrhage, indicating that ephrin-B2 is required for proper mural cell function. To determine whether the role of ephrin-B2 in mural cells is mediated by reverse signaling, Salvucci *et al.* mutated the five tyrosine residues of the ephrin-B2 cytoplasmic domain (Salvucci *et al.*, 2009). Interestingly, while mutation of all five tyrosine residues does not appear to affect angiogenic sprouting *in vivo* (Sawamiphak *et al.*, 2010; Wang *et al.*, 2010), the same mutant severely impairs the ability of pericyte and endothelial cell assembly into cordlike structures in a Matrigel plug assay (Salvucci *et al.*, 2009). Thus, ephrin-B2 reverse signaling via PDZ interaction appears to mediate angiogenic sprouting, whereas reverse signaling through phosphorylated tyrosine residues may function in vascular endothelial–pericyte interaction.

VI. CONCLUDING REMARKS

Both preclinical laboratory studies and data in human cancer clinical specimens provide compelling evidence that members of the Eph family of RTKs and their ephrin ligands regulate tumor growth, invasion and metastasis, and neovascularization. However, Eph receptor signaling is complex, influenced by differences in ligand-dependent versus ligand-independent signaling, forward versus reverse signaling, and kinase-dependent versus kinase-independent signaling. Some of the paradoxical functions of Eph also appear to be influenced by tissue type and oncogenic context. Given the complexity of signaling regulated by this RTK family, as well as extensive cross talk with other RTK families involved in cancer, future efforts should be aimed at understanding how Eph receptor expression and function is modulated in the context of relevant cancer pathways.

In spite of these challenges, Eph receptors are very attractive therapeutic targets. They are expressed in a broad range of human cancer types in both tumor and the surrounding host tissue. In addition, some of the Eph receptors are known to simultaneously regulate tumor growth and neovascularization, permitting molecularly targeted therapies to potentially disrupt at least two key processes in tumor progression with a single agent. Although the role of Eph receptors in tumor stem cells has not been investigated extensively, Eph receptors/ephrins are expressed in various other types of stem cells, such as in nervous system and the intestine (Genander and Frisén, 2010). Characterizing the Eph/ephrin in cancer stem cells will allow development of targeted therapies to this group of cells, preventing tumor recurrence and metastasis. Finally, because Eph family is the largest RTK family in the genome and its ligands and receptors often display overlapping expression patterns in both tumor cells and the surrounding host stroma, a key step in translating the biological and mechanistic data from the laboratory into the clinic will be to analyze the expression and mutations of individual Eph receptors in large datasets of human cancer. Together with mechanistic studies in cell lines and animal models, expression profiling and mutation analysis in human tissue will elucidate the mechanisms of tumor initiation and progression and lay foundations for development of new anticancer therapeutics.

ACKNOWLEDGMENTS

The author thanks Drs. Daar and Brantley-Sieders for helpful comments on the manuscript. Work in the author's laboratory is supported by a Merit Award from the Veterans Affairs Administration and grants from National Institutes of Health (CA95004 and CA114301).

REFERENCES

Adams, R. H., and Eichmann, A. (2010). Axon guidance molecules in vascular patterning. *Cold Spring Harb. Perspect. Biol.* **2**, a001875.

Adams, R. H., Wilkinson, G. A., Weiss, C., Diella, F., Gale, N. W., Deutsch, U., Risau, W., and Klein, R. (1999). Roles of ephrinB ligands and EphB receptors in cardiovascular development: demarcation of arterial/venous domains, vascular morphogenesis, and sprouting angiogenesis. *Genes Dev.* **3**, 295–306.

Adams, R. H., Diella, F., Hennig, S., Helmbacher, F., Deutsch, U., and Klein, R. (2001). The cytoplasmic domain of the ligand ephrinB2 is required for vascular morphogenesis but not cranial neural crest migration. *Cell* **104**, 57–69.

Astin, J. W., Batson, J., Kadir, S., Charlet, J., Persad, R. A., Gillatt, D., Oxley, J. D., and Nobes, C. D. (2010). Competition amongst Eph receptors regulates contact inhibition of locomotion and invasiveness in prostate cancer cells. *Nat. Cell Biol.* **12**, 1194–1204.

Batlle, E., Henderson, J. T., Beghtel, H., van den Born, M. M., Sancho, E., Huls, G., Meeldijk, J., Robertson, J., van de Wetering, M., Pawson, T., *et al.* (2002). Beta-catenin and TCF mediate cell positioning in the intestinal epithelium by controlling the expression of EphB/ephrinB. *Cell* **111**, 251–263.

Batlle, E., Bacani, J., Begthel, H., Jonkeer, S., Gregorieff, A., van de Born, M., Malats, N., Sancho, E., Boon, E., Pawson, T., *et al.* (2005). EphB receptor activity suppresses colorectal cancer progression. *Nature* **435**, 1126–1130.

Bogan, C., Chen, J., O'Sullivan, M. G., and Cormier, R. T. (2009). Loss of EphA2 receptor tyrosine kinase reduces ApcMin/+ tumorigenesis. *Int. J. Cancer* **124**, 1366–1371.

Brambilla, R., Schnapp, F., Casagranda, F., Labrador, J. P., Bergemann, A. D., Flanagan, J. G., Pasquale, E. B., and Klein, R. (1995). Membrane-bound LERK2 ligand can signal through three different Eph-related receptor tyrosine kinases. *EMBO J.* **14**, 3116–3126.

Brannan, J. M., Dong, W., Prudkin, L., Behrens, C., Lotan, R., Bekele, B. N., Wistuba, I., and Johnson, F. M. (2009). Expression of the receptor tyrosine kinase EphA2 is increased in smokers and predicts poor survival in non-small cell lung cancer. *Clin. Cancer Res.* **15**, 4423–4430.

Brantley-Sieders, D. (2011). Clinical relevance of Ephs and ephrins in cancer: lessons from breast, colorectal, and lung cancer profiling. *Semin. Cell Dev. Biol.* **23**, 102–108.

Brantley-Sieders, D., Caughron, J., Hicks, D., Pozzi, A., Ruiz, J. C., and Chen, J. (2004). EphA2 receptor tyrosine kinase regulates endothelial cell migration and assembly through phosphoinositide 3-kinase-mediated Rac1 GTPase activation. *J. Cell Sci.* **117**, 2037–2049.

Brantley-Sieders, D. M., Fang, W. B., Hicks, D., Koyama, T., Shyr, Y., and Chen, J. (2005). Impaired tumor microenvironment in EphA2-deficient mice inhibits tumor angiogenesis and metastatic progression. *FASEB J.* **19**, 1884–1886.

Brantley-Sieders, D. M., Zhuang, G., Hicks, D., Fang, W. B., Hwang, Y., Cates, J. M., Coffman, K., Jackson, D., Bruckheimer, E., Muraoka-Cook, R. S., *et al.* (2008). The receptor tyrosine kinase EphA2 promotes mammary adenocarcinoma tumorigenesis and metastatic progression in mice by amplifying ErbB2 signaling. *J. Clin. Invest.* **118**, 64–78.

Brantley-Sieders, D. M., Zhuang, G., Vaught, D., Freeman, T., Hwang, Y., Hicks, D., and Chen, J. (2009). Host deficiency in Vav2/3 guanine nucleotide exchange factors impairs tumor growth, survival, and angiogenesis in vivo. *Mol. Cancer Res.* **7**, 615–623.

Brantley-Sieders, D. M., Dunaway, C. M., Rao, M., Short, S., Hwang, Y., Gao, Y., Li, D., Jiang, A., Shyr, Y., Wu, J. Y., *et al.* (2011a). Angiocrine factors modulate tumor proliferation and motility through EphA2 repression of Slit2 tumor suppressor function in endothelium. *Cancer Res.* **71**, 976–987.

Brantley-Sieders, D. M., Jiang, A., Sarma, K., Badu-Nkansah, A., Walter, D. L., Shyr, Y., and Chen, J. (2011b). Eph/ephrin profiling in human breast cancer reveals significant associations between expression level and clinical outcome. *PLoS One* **6**, e24426.

Bruce, V., Olivieri, G., Eickelberg, O., and Miescher, G. C. (1999). Functional activation of EphA5 receptor does not promote cell proliferation in the aberrant EphA5 expressing human glioblastoma U-118 MG cell line. *Brain Res.* **821**, 169–176.

Butler, J. M., Kobayashi, H., and Rafii, S. (2010). Instructive role of the vascular niche in promoting tumour growth and tissue repair by angiocrine factors. *Nat. Rev. Cancer* **10**, 138–146.

Chen, J., Hicks, D., Brantley-Sieders, D., Cheng, N., McCollum, G. W., Qi-Werdich, X., and Penn, J. (2006). Inhibition of retinal neovascularization by soluble EphA2 receptor. *Exp. Eye Res.* **82**, 664–673.

Cheng, N., Brantley, D. M., Liu, H., Lin, Q., Enriquez, M., Gale, N. W., Yancopoulos, G., Cerretti, D. P., Daniel, T. O., and Chen, J. (2002). Blockade of EphA receptor tyrosine kinase activation inhibits VEGF-induced angiogenesis. *Mol. Cancer Res. (formerly Cell Growth and Differentiation)* **1**, 2–11.

Chumley, M. J., Catchpole, T., Silvany, R. E., Kernie, S. G., and Henkemeyer, M. (2007). EphB receptors regulate stem/progenitor cell proliferation, migration, and polarity during hippocampal neurogenesis. *J. Neurosci.* **27**, 13481–13490.

Coffman, K. T., Hu, M., Carles-Kinch, K., Tice, D., Donacki, N., Munyon, K., Kifle, G., Woods, R., Langermann, S., Kiener, P. A., *et al.* (2003). Differential EphA2 epitope display on normal versus malignant cells. *Cancer Res.* **63**, 7907–7912.

Conover, J. C., Doetsch, F., Garcia-Verdugo, J. M., Gale, N. W., Yancopoulos, G. D., and Alvarez-Buylla, A. (2000). Disruption of Eph/ephrin signaling affects migration and proliferation in the adult subventricular zone. *Nat. Neurosci.* **3**, 1091–1097.

Cortina, C., Palomo-Ponce, S., Iglesias, M., Fernandez-Masip, J. L., Vivancos, A., Whissell, G., Huma, M., Peiro, N., Gallego, L., Jonkheer, S., *et al.* (2007). EphB-ephrin-B interactions suppress colorectal cancer progression by compartmentalizing tumor cells. *Nat. Genet.* **39**, 1376–1383.

Davies, H., Hunter, C., Smith, R., Stephens, P., Greenman, C., Bignell, G., Teague, J., Butler, A., Edkins, S., Stevens, C., *et al.* (2005). Somatic mutations of the protein kinase gene family in human lung cancer. *Cancer Res.* **65**, 7591–7595.

Davy, A., and Robbins, S. M. (2000). Ephrin-A5 modulates cell adhesion and morphology in an integrin-dependent manner. *EMBO J.* **19**, 5396–5405.

Davy, A., Nicholas, W. G., Murray, E. W., Klinghoffer, R. A., Soriano, P., Feuerstein, C., and Robbins, S. M. (1999). Compartmentalized signaling by GPI-anchored ephrin-A5 requires the Fyn tyrosine kinase to regulate cellular adhesion. *Genes Dev.* **13**, 3125–3135.

Ding, L., Getz, G., Wheeler, D. A., Mardis, E. R., McLellan, M. D., Cibulskis, K., Sougnez, C., Greulich, H., Muzny, D. M., Morgan, M. B., *et al.* (2008). Somatic mutations affect key pathways in lung adenocarcinoma. *Nature* **455**, 1069–1075.

Duxbury, M. S., Ito, H., Zinner, M. J., Ashley, S. W., and Whang, E. E. (2004). EphA2: a determinant of malignant cellular behavior and a potential therapeutic target in pancreatic adenocarcinoma. *Oncogene* **23**, 1448–1456.

Eichmann, A., Makinen, T., and Alitalo, K. (2005). Neural guidance molecules regulate vascular remodeling and vessel navigation. *Genes Dev.* **19**, 1013–1021.

Erber, R., Eichelsbacher, U., Powajbo, V., Korn, T., Djonov, V., Lin, J., Hammes, H. P., Grobholz, R., Ullrich, A., and Vajkoczy, P. (2006). EphB4 controls blood vascular morphogenesis during postnatal angiogenesis. *EMBO J.* **25**, 628–641.

Fang, W. B., Brantley-Sieders, D. M., Hwang, Y., Ham, A. J., and Chen, J. (2008a). Identification and functional analysis of phosphorylated tyrosine residues within EphA2 receptor tyrosine kinase. *J. Biol. Chem.* **283**, 16017–16026.

Fang, W. B., Ireton, R. C., Zhuang, G., Takahashi, T., Reynolds, A., and Chen, J. (2008b). Overexpression of EPHA2 receptor destabilizes adherens junctions via a RhoA-dependent mechanism. *J. Cell Sci.* **121**, 358–368.

Faoro, L., Singleton, P. A., Cervantes, G. M., Lennon, F. E., Choong, N. W., Kanteti, R., Ferguson, B. D., Husain, A. N., Tretiakova, M. S., Ramnath, N., *et al.* (2010). EphA2 mutation in lung squamous cell carcinoma promotes increased cell survival, cell invasion, focal adhesions, and mTOR activation. *J. Biol. Chem.* **285**, 18575–18585.

Foo, S. S., Turner, C. J., Adams, S., Compagni, A., Aubyn, D., Kogata, N., Lindblom, P., Shani, M., Zicha, D., and Adams, R. H. (2006). Ephrin-B2 controls cell motility and adhesion during blood-vessel-wall assembly. *Cell* **124**, 161–173.

Fournier, M. V., Martin, K. J., Kenny, P. A., Xhaja, K., Bosch, I., Yaswen, P., and Bissell, M. J. (2006). Gene expression signature in organized and growth-arrested mammary acini predicts good outcome in breast cancer. *Cancer Res.* **66**, 7095–7102.

Frieden, L., Townsend, T. A., Vaught, D. B., Hwang, Y., Barnett, J. V., and Chen, C. (2008). Regulation of heart valve morphogenesis by ephrin-A1 in cardiac development in vivo. *Circ. Res.* **239,** 3226–3234.

Frieden, L., Townsend, T. A., Vaught, D. B., Hwang, Y., Barnett, J. V., and Chen, C. (2010). Regulation of heart valve morphogenesis by Eph receptor ligand, ephrin-A1. *Dev. Dyn.* **239,** 3226–3234.

Gaengel, K., Genove, G., Armulik, A., and Betsholtz, C. (2009). Endothelial-mural cell signaling in vascular development and angiogenesis. *Arterioscler. Thromb. Vasc. Biol.* **29,** 630–638.

Gale, N. W., Baluk, P., Pan, L., Kwan, M., Holash, J., DeChiara, T. M., McDonald, D. M., and Yancopoulos, G. D. (2001). Ephrin-B2 selectively marks arterial vessels and neovascularization sites in the adult, with expression in both endothelial and smooth-muscle cells. *Dev. Biol.* **230,** 151–160.

Genander, M., and Frisén, J. (2010). Ephrins and Eph receptors in stem cells and cancer. *Curr. Opin. Cell Biol.* **22,** 611–616.

Genander, M., Halford, M. M., Xu, N. J., Eriksson, M., Yu, Z., Qiu, Z., Martling, A., Greicius, G., Thakar, S., Catchpole, T., *et al.* (2009). Dissociation of EphB2 signaling pathways mediating progenitor cell proliferation and tumor suppression. *Cell* **139,** 679–692.

Genander, M., Holmberg, J., and Frisén, J. (2010). Ephrins negatively regulate cell proliferation in the epidermis and hair follicle. *Stem Cells* **28,** 1196–1205.

Gerety, S. S., Wang, H. U., Chen, Z. F., and Anderson, D. J. (1999). Symmetrical mutant phenotypes of the receptor EphB4 and its specific transmembrane ligand ephrin-B2 in cardiovascular development. *Mol. Cell* **4,** 403–414.

Greenman, C., Stephens, P., Smith, R., Dalgliesh, G. L., Hunter, C., Bignell, G., Davies, H., Teague, J., Butler, A., Stevens, C., *et al.* (2007). Patterns of somatic mutation in human cancer genomes. *Nature* **446,** 153–158.

Guo, H., Miao, H., Gerber, L., Singh, J., Denning, M. F., Gilliam, A. C., and Wang, B. (2006). Disruption of EphA2 receptor tyrosine kinase leads to increased susceptibility to carcinogenesis in mouse skin. *Cancer Res.* **66,** 7050–7058.

Hanahan, D., and Weinberg, R. A. (2000). The hallmarks of cancer. *Cell* **100,** 57–70.

Hanahan, D., and Weinberg, R. A. (2011). Hallmarks of cancer: the next generation. *Cell* **144,** 646–674.

Himanen, J. P., Yermekbayeva, L., Janes, P. W., Walker, J. R., Xu, K., Atapattu, L., Rajashankar, K. R., Mensinga, A., Lackmann, M., Nikolov, D. B., *et al.* (2010). Architecture of Eph receptor clusters. *Proc. Natl. Acad. Sci. USA* **107,** 10860–10865.

Hunter, S. G., Zhuang, G., Brantley-Sieders, D. M., Swatt, W., Cowan, C. W., and Chen, J. (2006). Essential role of Vav family guanine nucleotide exchange factors in EphA receptor-mediated angiogenesis. *Mol. Cell. Biol.* **26,** 4830–4842.

Ji, X. D., Li, G., Feng, Y. X., Zhao, J. S., Li, J. J., Sun, Z. J., Shi, S., Deng, Y. Z., Xu, J. F., Zhu, Y. Q., *et al.* (2011). EphB3 is overexpressed in non-small-cell lung cancer and promotes tumor metastasis by enhancing cell survival and migration. *Cancer Res.* **71,** 1156–1166.

Kinch, M. S., Moore, M. B., and Harpole, D. H. J. (2003). Predictive value of the EphA2 receptor tyrosine kinase in lung cancer recurrence and survival. *Clin. Cancer Res.* **9,** 613–618.

Kullander, K., and Klein, R. (2002). Mechanisms and functions of Eph and ephrin signaling. *Nat. Rev. Mol. Cell Biol.* **3,** 475.

Kumar, S. R., Singh, J., Xia, G., Krasnoperov, V., Hassanieh, L., Ley, E. J., Scehnet, J., Kumar, N. G., Hawes, D., Press, M. F., *et al.* (2006). Receptor tyrosine kinase EphB4 is a survival factor in breast cancer. *Am. J. Pathol.* **169,** 279–293.

Kumar, S. R., Scehnet, J. S., Ley, E. J., Singh, J., Krasnoperov, V., Liu, R., Manchanda, P. K., Ladner, R. D., Hawes, D., Weaver, F. A., *et al.* (2009). Preferential induction of EphB4 over EphB2 and its implication in colorectal cancer progression. *Cancer Res.* **69,** 3736–3745.

Landen, C. N., Jr., Chavez-Reyes, A., Bucana, C., Schmandt, R., Deavers, M. T., Lopez-Berestein, G., and Sood, A. K. (2005a). Therapeutic EphA2 gene targeting in vivo using neutral liposomal small interfering RNA delivery. *Cancer Res.* **65**, 6910–6918.

Landen, C. N., Kinch, M. S., and Sood, A. K. (2005b). EphA2 as a target for ovarian cancer therapy. *Expert Opin. Ther. Targets* **9**, 1179–1187.

Landen, C. N., Lu, C., Han, L. Y., Coffman, K. T., Bruckheimer, E., Halder, J., Mangala, L. S., Merritt, W. M., Lin, Y. G., Gao, C., *et al.* (2006). Efficacy and antivascular effects of EphA2 reduction with an agonistic antibody in ovarian cancer. *J. Natl. Cancer Inst.* **98**, 1558–1570.

Lee, H., and Daar, I. O. (2009). EphrinB reverse signaling in cell-cell adhesion. *Cell Adh. Migr.* **3**, 250–255.

Lee, H. S., Nishanian, T. G., Mood, K., Bong, Y. S., and Daar, I. O. (2008). EphrinB1 controls cell-cell junctions through the Par polarity complex. *Nat. Cell Biol.* **10**, 979–986.

Lhotak, V., and Pawson, T. (1993). Biological and biochemical activities of a chimeric epidermal growth factor–Elk receptor tyrosine kinase. *Mol. Cell. Biol.* **13**, 7071–7079.

Lim, Y. S., McLaughlin, T., Sung, T. C., Santiago, A., Lee, K. F., and O'Leary, D. D. (2008). p75 (NTR) mediates ephrin-A reverse signaling required for axon repulsion and mapping. *Neuron* **59**, 746–758.

Macrae, M., Neve, R. M., Rodriguez-Viciana, P., Haqq, C., Yeh, J., Chen, C., Gray, J. W., and McCormick, F. (2005). A conditional feedback loop regulates Ras activity through EphA2. *Cancer Cell* **8**, 111–118.

Mäkinen, T., Adams, R. H., Bailey, J., Lu, Q., Ziemiecki, A., Alitalo, K., Klein, R., and Wilkinson, G. A. (2005). PDZ interaction site in ephrinB2 is required for the remodeling of lymphatic vasculature. *Genes Dev.* **19**, 397–410.

Martin, K. J., Patrick, D. R., Bissell, M. J., and Fournier, M. V. (2008). Prognostic breast cancer signature identified from 3D culture model accurately predicts clinical outcome across independent datasets. *PLoS One* **3**, e2994.

Martiny-Baron, G., Korff, T., Schaffner, F., Esser, N., Eggstein, S., Marme, D., and Augustin, H. G. (2004). Inhibition of tumor growth and angiogenesis by soluble EphB4. *Neoplasia* **6**, 248–257.

Merlos-Suárez, A., and Batlle, E. (2008). Eph-ephrin signalling in adult tissues and cancer. *Curr. Opin. Cell Biol.* **20**, 194–200.

Miao, H., Wei, B.-R., Peehl, D. M., Li, Q., Alexandrou, T., Schelling, J. R., Rhim, J. S., Sedor, J. R., Burnett, E., and Wang, B. (2001). Activation of EphA receptor tyrosine kinase inhibits the Ras/MAPK pathway. *Nat. Cell Biol.* **13**, 527–530.

Miao, H., Li, D. Q., Mukherjee, A., Guo, H., Petty, A., Cutter, J., Basilion, J. P., Sedor, J., Wu, J., Danielpour, D., *et al.* (2009). EphA2 mediates ligand-dependent inhibition and ligand-independent promotion of cell migration and invasion via a reciprocal regulatory loop with Akt. *Cancer Cell* **16**, 9–20.

Miura, K., Nam, J. M., Kojima, C., Mochizuki, N., and Sabe, H. (2009). EphA2 engages Git1 to suppress Arf6 activity modulating epithelial cell-cell contacts. *Mol. Biol. Cell* **20**, 1949–1959.

Munarini, N., Jager, R., Abderhalden, S., Zuercher, G., Rohrbach, V., Loercher, S., Pfanner-Meyer, B., Andres, A. C., and Ziemiecki, A. (2002). Altered mammary epithelial development, pattern formation and involution in transgenic mice expressing the EphB4 receptor tyrosine kinase. *J. Cell Sci.* **115**, 25–37.

Noren, N. K., and Pasquale, E. B. (2007). Paradoxes of the EphB4 receptor in cancer. *Cancer Res.* **67**, 3994–3997.

Noren, N. K., Lu, M., Freeman, A. L., Koolpe, M., and Pasquale, E. B. (2004). Interplay between EphB4 on tumor cells and vascular ephrin-B2 regulates tumor growth. *Proc. Natl. Acad. Sci. USA* **101**, 5583–5588.

Noren, N. K., Foos, G., Hauser, C. A., and Pasquale, E. B. (2006). The EphB4 receptor suppresses breast cancer cell tumorigenicity through an Abl-Crk pathway. *Nat. Cell Biol.* **8**, 815–825.

Ogawa, K., Pasqualini, R., Lindberg, R. A., Kain, R., Freeman, A. L., and Pasquale, E. B. (2000). The ephrin-A1 ligand and its receptor, EphA2, are expressed during tumor neovascularization. *Oncogene* **19**, 6043–6052.

Pandey, A., Shao, H., Marks, R. M., Polverini, P. J., and Dixit, V. M. (1995). Role of B61, the ligand for the Eck receptor tyrosine kinase, in TNF-alpha-induced angiogenesis. *Science* **268**, 567–569.

Pasquale, E. B. (2005). Eph receptor signalling casts a wide net on cell behaviour. *Nat. Rev. Mol. Cell Biol.* **6**, 462–475.

Pasquale, E. B. (2008). Eph-ephrin bidirectional signaling in physiology and disease. *Cell* **133**, 38–52.

Pasquale, E. B. (2010). Eph receptors and ephrins in cancer: bidirectional signaling and beyond. *Nat. Rev. Cancer* **10**, 165–180.

Salvucci, O., Maric, D., Economopoulou, M., Sakakibara, S., Merlin, S., Follenzi, A., and Tosato, G. (2009). EphrinB reverse signaling contributes to endothelial and mural cell assembly into vascular structures. *Blood* **114**, 1707–1716.

Sawamiphak, S., Seidel, S., Essmann, C. L., Wilkinson, G. A., Pitulescu, M. E., Acker, T., and Acker-Palmer, A. (2010). Ephrin-B2 regulates VEGFR2 function in developmental and tumour angiogenesis. *Nature* **465**, 487–491.

Seiradake, E., Harlos, K., Sutton, G., Aricescu, A. R., and Jones, E. Y. (2010). An extracellular steric seeding mechanism for Eph-ephrin signaling platform assembly. *Nat. Struct. Mol. Biol.* **17**, 398–402.

Shi, L., Itoh, F., Itoh, S., Takahashi, S., Yamamoto, M., and Kato, M. (2008). Ephrin-A1 promotes the malignant progression of intestinal tumors in Apc(min/+) mice. *Oncogene* **27**, 3265–3273.

Shin, D., Garcia-Cardena, G., Hayashi, S., Gerety, S., Asahara, T., Stavrakis, G., Isner, J., Folkman, J., Gimbrone, M. A., Jr., and Anderson, D. J. (2001). Expression of ephrinB2 identifies a stable genetic difference between arterial and venous vascular smooth muscle as well as endothelial cells, and marks subsets of microvessels at sites of adult neovascularization. *Dev. Biol.* **230**, 139–150.

Stephen, L. J., Fawkes, A. L., Verhoeve, A., Lemke, G., and Brown, A. (2006). A critical role for the EphA3 receptor tyrosine kinase in heart development. *Dev. Biol.* **302**, 66–79.

Tanaka, M., Kamata, R., and Sakai, R. (2005). Phosphorylation of ephrin-B1 via the interaction with claudin following cell-cell contact formation. *EMBO J.* **24**, 3700–3711.

Vihanto, M. M., Plock, J., Erni, D., Frey, B. M., Frey, F. J., and Huynh-Do, U. (2005). Hypoxia up-regulates expression of Eph receptors and ephrins in mouse skin. *FASEB J.* **19**, 1689–1691.

Wang, H. U., Chen, Z. F., and Anderson, D. J. (1998). Molecular distinction and angiogenic interaction between embryonic arteries and veins revealed by ephrin-B2 and its receptor Eph-B4. *Cell* **93**, 741–753.

Wang, Y., Nakayama, M., Pitulescu, M. E., Schmidt, T. S., Bochenek, M. L., Sakakibara, A., Adams, S., Davy, A., Deutsch, U., Lüthi, U., *et al.* (2010). Ephrin-B2 controls VEGF-induced angiogenesis and lymphangiogenesis. *Nature* **465**, 483–486.

Wood, L. D., Calhoun, E. S., Silliman, N., Ptak, J., Szabo, S., Powell, S. M., Riggins, G. J., Wang, T. L., Yan, H., Gazdar, A., *et al.* (2006). Somatic mutations of GUCY2F, EPHA3, and NTRK3 in human cancers. *Hum. Mutat.* **27**, 1060–1061.

Wu, Q., Suo, Z., Risberg, B., Karlsson, M. G., Villman, K., and Nesland, J. M. (2004). Expression of Ephb2 and Ephb4 in breast carcinoma. *Pathol. Oncol. Res.* **10**, 26–33.

Wykosky, J., and Debinski, W. (2008). The EphA2 receptor and ephrinA1 ligand in solid tumors: function and therapeutic targeting. *Mol. Cancer Res.* **6**, 1795–1806.

Wykosky, J., Gibo, D. M., Stanton, C., and Debinski, W. (2005). EphA2 as a novel molecular marker and target in glioblastoma multiforme. *Mol. Cancer Res.* **3**, 541–551.

Yamashita, T., Ohneda, K., Nagano, M., Miyoshi, C., Kaneko, N., Miwa, Y., Yamamoto, M., Ohneda, O., and Fujii-Kuriyama, Y. (2008). Hypoxia-inducible transcription factor-2alpha in endothelial cells regulates tumor neovascularization through activation of ephrin A1. *J. Biol. Chem.* **283**, 18926–18936.

Yang, N. Y., Fernandez, C., Richter, M., Xiao, Z., Valencia, F., Tice, D. A., and Pasquale, E. B. (2011). Crosstalk of the EphA2 receptor with a serine/threonine phosphatase suppresses the Akt-mTORC1 pathway in cancer cells. *Cell. Signal.* **23**, 201–212.

Zelinski, D. P., Zantek, N. D., Stewart, J. C., Irizarry, A. R., and Kinch, M. S. (2001). EphA2 overexpression causes tumorigenesis of mammary epithelial cells. *Cancer Res.* **61**, 2301–2306.

Zhuang, G., Brantley-Sieders, D., Vaught, D., Yu, J., Xie, L., Wells, S., Jackson, D., Muroaka-Cook, R., Arteaga, C., and Chen, J. (2010). Elevation of receptor tyrosine kinase EphA2 mediates resistance to trastuzumab therapy. *Cancer Res.* **70**, 299–308.

Essential Roles of EphB Receptors and EphrinB Ligands in Endothelial Cell Function and Angiogenesis

Ombretta Salvucci and Giovanna Tosato

Laboratory of Cellular Oncology, Center for Cancer Research, National Cancer Institute, Bethesda, Maryland, USA

Eph receptor tyrosine kinases and their Ephrin ligands represent an important signaling system with widespread roles in cell physiology and disease. Receptors and ligands in this family are anchored to the cell surface; thus Eph/Ephrin interactions mainly occur at sites of cell-to-cell contact. EphB4 and EphrinB2 are the Eph/Ephrin molecules that play essential roles in vascular development and postnatal angiogenesis. Analysis of expression patterns and function has linked EphB4/EphrinB2 to endothelial cell growth, survival, migration, assembly, and angiogenesis. Signaling from these molecules is complex, with the potential for being bidirectional, emanating both from the Eph receptors (forward signaling) and from the Ephrin ligands (reverse signaling). In this review, we describe recent advances on the roles of EphB/EphrinB protein family in endothelial cell function and outline potential approaches to inhibit pathological angiogenesis based on this understanding.

I. INTRODUCTION

The Eph receptor family, which includes 14 members, constitutes the largest family of tyrosine kinase receptors in mammals. Eph receptors and their Ephrin (Eph receptor interacting) ligands form a system of cell

Advances in CANCER RESEARCH, Volume 114
2012, Published by Elsevier Inc.
0065-230X/12 $35.00
DOI: 10.1016/B978-0-12-386503-8.00002-8

communication with widespread roles in physiology and disease. In mammals, there are nine EphA (EphA1–8, and EphA10) receptors, which promiscuously bind five glycosylphosphatidylinositol (GPI)-linked EphrinA ligands and five EphB (EphB1–4 and EphB6) receptors, which promiscuously bind three transmembrane EphrinB ligands (Chrencik *et al.*, 2006; Egea and Klein, 2007; Kullander and Klein, 2002; Pasquale, 2010). There are some exceptions: EphA4 can bind both A-type and most B-type Ephrins, EphB2 can bind EphrinA5, and EphB4 essentially binds only EphrinB2 (Table I). Typically, Eph receptor–Ephrin ligand interactions occur at the cell's surface at sites of cell-to-cell contact and result in bidirectional signaling from the Eph receptor (forward signaling) and from the Ephrin ligand (reverse signaling) (Chrencik *et al.*, 2006; Egea and Klein, 2007; Pasquale, 2010).

EphrinB2 and its receptor EphB4 play critical roles in the development of the vascular system and contribute to the function of the adult vasculature (Pitulescu and Adams, 2010). Knockout mice lacking EphrinB2 or EphB4 expression and mice with deletion of EphrinB2 targeted to the endothelial cells display a severely compromised vascular system and die at midgestation (Adams *et al.*, 1999; Gerety and Anderson, 2002; Gerety *et al.*, 1999). The targeted deletion of EphrinB2 in mural cells (pericytes and smooth muscle cells) leads to diffuse tissue edema, hemorrhaging, and perinatal death of the mice

Table I Listing of EphA, EphrinA, EphB, and EphrinB Proteins

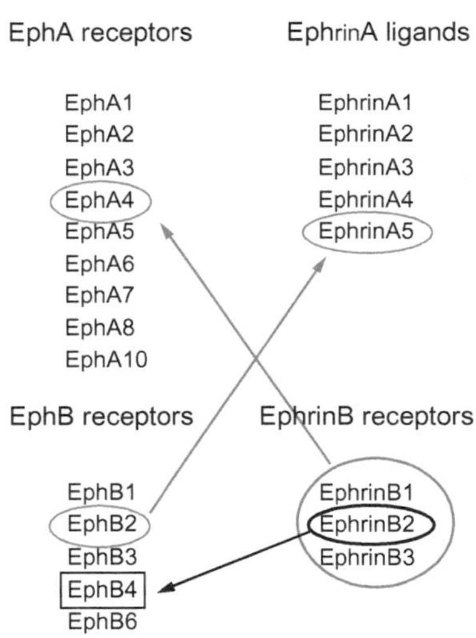

Although EphrinA ligands generally bind to EphA receptors, there are exceptions noted by the red arrows. Although EphrinB ligands generally bind all EphB receptors, EphrinB2 essentially binds only to EphB4. (For color version of this Table, the reader is referred to the Web version of this chapter.)

(Foo *et al.*, 2006). EphrinB is phosphorylated in angiogenic blood vessels, and inhibition of phosphorylation-dependent or PDZ-dependent signaling downstream of EphrinB ligands prevents endothelial cell sprouting and the proper assembly of endothelial cells with other endothelial cells and with pericytes (Salvucci *et al.*, 2006, 2009; Sawamiphak *et al.*, 2010b; Wang *et al.*, 2010b).

In this review, we will present current information on structural features, signaling mechanisms, and expression and function of the EphB/EphrinB proteins in the context of blood vessel physiology and pathology.

II. STRUCTURAL FEATURES AND INTERNALIZATION

The extracellular portion of Eph receptors (both A and B families) includes a globular, ligand-binding domain; a cysteine-rich EGF-like region; and two fibronectin-type III repeats (Himanen and Nikolov, 2003; Himanen *et al.*, 2001; Toth *et al.*, 2001). The intracytoplasmic portion consists of a short membrane-proximal region, a tyrosine kinase domain, a sterile alpha motif (SAM) protein–protein interaction domain, and a PDZ-binding C-terminal motif (Fig. 1A; Chrencik *et al.*, 2006). The B Ephrins display an extracellular Eph receptor-binding domain, a single-pass transmembrane region, a short intracellular domain with several sites for tyrosine and serine phosphorylation, and a C-terminal PDZ-binding motif (Fig. 1A; Chrencik *et al.*, 2006).

Initial high-affinity interaction between the Eph receptor globular domain and a protruding loop of EphrinB leads to conformational changes of receptor and ligand, and the formation of heterodimers of an Eph receptor with an Ephrin molecule (Himanen *et al.*, 2001; Toth *et al.*, 2001). Other residues in the Eph receptor contribute to a lower-affinity binding interface for much of the extracellular region of EphrinB2, leading to the association of two homodimers to form a tetrameric complex comprising two Eph receptors and two Ephrin molecules (Chrencik *et al.*, 2006; Himanen and Nikolov, 2003). Eph/Ephrin tetramers can aggregate into larger clusters through several low-affinity Eph–Eph and Ephrin–Ephrin interactions identified in the extracellular domains (Fig. 1B; Smith *et al.*, 2004).

One of the reported consequences of high-affinity interactions involving EphB and EphrinB expressed on adjacent cells is internalization of both molecules into Eph or Ephrin-expressing cells (Lauterbach and Klein, 2006; Mann *et al.*, 2003; Marston *et al.*, 2003; Zimmer *et al.*, 2003). This process provides a mechanism for removal of Eph receptors and ligands from the cell surface and termination of receptor/ligand adhesive interactions, which may explain the occurrence of cell repulsion (Egea and Klein, 2007). EphB–EphrinB complexes, detected in intracellular vesicles as full-length proteins, are believed to be the result of *trans*-endocytosis from one cell to the adjacent

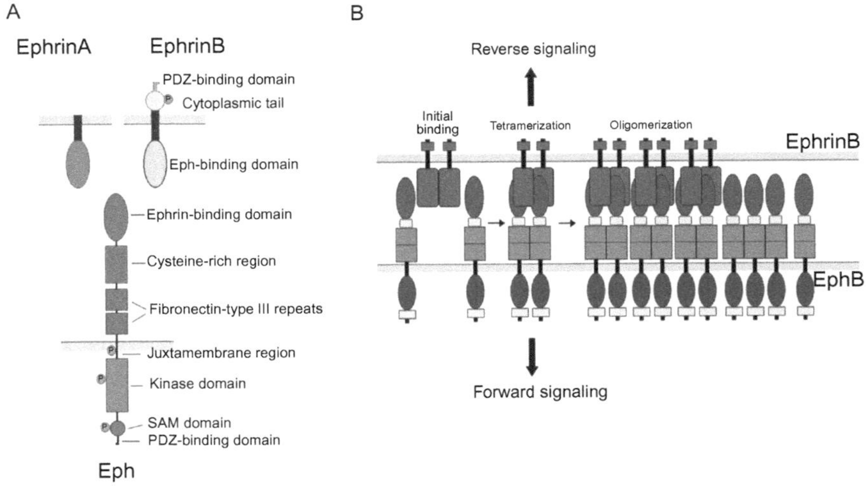

Fig. 1 (A) Schematic representation of the domain structure and binding interfaces of Ephrins and Eph receptors. EphrinA ligands are attached to the cell surface through a glycosylphosphatidylinositol (GPI)-anchor; the extracellular domain contains an Eph receptor-binding domain that is connected to the transmembrane segment. EphrinB ligands are transmembrane proteins with an extracellular Eph receptor-binding domain connected to a transmembrane segment, which is followed by a short intracellular domain. The Eph receptors include an extracellular domain composed of an Ephrin-binding domain, a cysteine-rich segment that contains an epidermal growth factor (EGF)-like motif, and two fibronectin-type III domains; and a cytoplasmic region that contains a juxtamembrane region, the kinase domain, a sterile a-motif (SAM), and a binding site for PDZ-containing proteins. (B) Representation of initial binding of cell surface Eph and Ephrin molecules to form heterotetramers, which initiate signaling, and subsequent oligomerization to form large receptor/ligand clusters that expand laterally through hemophilic interactions between Eph receptors. (See Page 1 in Color Section at the back of book.)

cell (Fig. 2). The direction of endocytosis seems to be dependent on the direction of signaling. For example, if a cell expressing EphB2 contacts a cell expressing EphrinB1 in which the C-terminal domain is truncated, endocytosis of the complex occurs preferentially into the EphB2-expressing cell (Zimmer *et al.*, 2003). By contrast, if EphB2 is signaling-deficient, the internalization occurs in the EphrinB1-expressing cell. If, however, both EphB and EphrinB are signaling-impaired, internalization of the Eph/Ephrin complex is not observed (Zimmer *et al.*, 2003). Although the biochemical basis for internalization of Eph/Ephrin is currently poorly defined, EphB1 was found to be associated with caveolin-1, suggesting the potential involvement of caveolae (Vihanto *et al.*, 2006), and EphrinB1 was found to be associated with clathrin-coated vesicles, suggesting a clathrin-dependent mechanism (Vihanto *et al.*, 2006). Once internalized, signaling (reverse or forward) has been shown to persist within the recipient cell (Marston *et al.*, 2003).

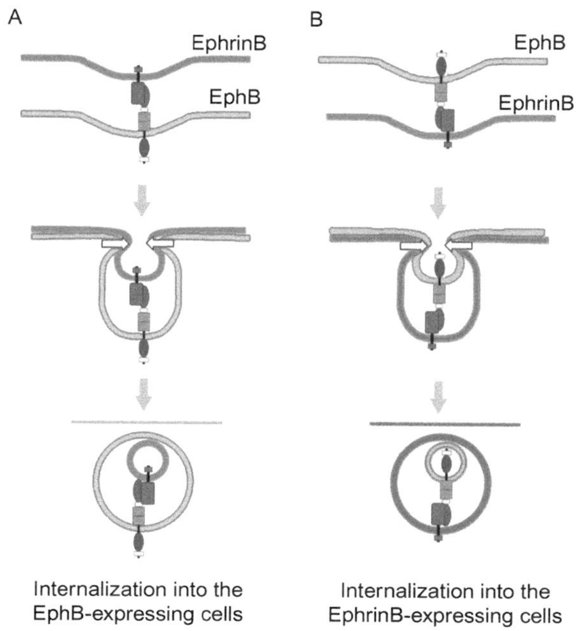

Fig. 2 Interaction of EphrinB with EphB can lead to (A) *trans*-endocytosis in the direction of the receptor (forward endocytosis) or (B) *trans*-endocytosis in the direction of the ligand (reverse endocytosis). Both processes lead to the internalization of full-length ligand and receptor. (For color version of this figure, the reader is referred to the Web version of this chapter.)

In addition to undergoing internalization of themselves, EphB and EphrinB can promote the internalization of the surrounding membrane and other proteins. For example, EphB signaling has been shown to modulate the clathrin-mediated endocytosis of amino-3-hydroxy-5-methyl-4-isoxazol-propionic acid-type glutamate receptors in neurons (Irie *et al.*, 2005), and EphrinB signaling has been reported to promote the internalization of VEGFR2 and VEGFR3 in endothelial cells (Sawamiphak *et al.*, 2010b; Wang *et al.*, 2010b).

III. BIDIRECTIONAL SIGNALING

Eph receptor signaling induced by Ephrin binding (forward signaling) is initiated by autophosphorylation and Src family kinases-mediated phosphorylation of the intracellular tyrosine residues, resulting in the activation of the tyrosine kinase catalytic domain (Kalo and Pasquale, 1999; Knoll and Drescher, 2004; Kullander and Klein, 2002; Murai and Pasquale, 2003).

Once the Eph receptors are phosphorylated, adaptor proteins containing Src-homology 2 (SH2) domains can bind and initiate phosphorylation of downstream substrates (Brantley-Sieders *et al.*, 2004; Kalo and Pasquale, 1999; Murai and Pasquale, 2003; Wybenga-Groot *et al.*, 2001; Zisch *et al.*, 2000). Activated Eph receptors can also mediate other protein–protein interactions via the SAM and PDZ-binding motifs, which contribute to signaling (Chrencik *et al.*, 2006; Kullander and Klein, 2002; Noren and Pasquale, 2004). Key components of Eph signaling are the Rho family of GTPases, including RhoA, Cdc42, and Rac, which are involved in the regulation of the actin cytoskeleton and cell shape, movement, and adhesion (Klein, 2009; Ogita *et al.*, 2003; Shamah *et al.*, 2001). Rho GTPases shuttle between an inactive (GDP-bound) and a signaling-active (GTP-bound) state, and this transition is regulated by guanidine nucleotide exchange factors (GEFs), which activate the Rho GTPases (Holland *et al.*, 1997). Phosphorylated Eph receptors have been shown to associate and activate a number of Rho GEFs, including Vav2, Tiam, Kalirin, and Intersectin (Cowan *et al.*, 2005; Klein, 2009; Murai and Pasquale, 2005; Sahin *et al.*, 2005). In addition, phosphorylated EphB has been shown to induce the ubiquitination and degradation of the GEF Ephexin 5, which binds to EphB and inhibits RhoA activity (Margolis *et al.*, 2010).

Unlike Eph receptors, B Ephrins do not possess intrinsic catalytic activity, and thus rely on the recruitment of signaling molecules to signal. EphrinB signaling upon receptor engagement and clustering (reverse signaling) is initiated by the recruitment and activation of Src family kinases, which phosphorylate specific tyrosine residues in the intracytoplasmic domain of B Ephrins (Palmer *et al.*, 2002; Fig. 3A). The activation of Src kinases is dependent upon metalloproteinase and presenilin-1-γ-secretase-mediated enzymatic cleavage of an intracellular peptide of EphrinB that binds to Src inhibiting its association with the inhibitory protein Csk (Georgakopoulos *et al.*, 2006). Phosphorylated EphrinB provides a docking site for the SH2 domain-containing adaptor protein Grb4 and the signal transducer and activator of transcription STAT3 (Bong *et al.*, 2007; Cowan and Henkemeyer, 2001; Segura *et al.*, 2007; Tanaka *et al.*, 2003; Xu and Henkemeyer, 2009). The EphrinB–Grb4 complex results in the activation of focal adhesion kinase (FAK) catalytic activity and recruitment of the G-protein-coupled receptor kinase interacting protein (GIT) 1 (Cowan and Henkemeyer, 2001; Segura *et al.*, 2007). Grab4 can also associate with other proteins involved in cytoskeleton regulation, including dynamin, Cbl-associated protein (CAP/ponsin), the Abl-interacting protein (Abi-1), and p21-activated kinase (PAK1; Cowan and Henkemeyer, 2001). Upon binding to EphrinB, STAT3 undergoes Jak-2-dependent phosphorylation and migrates to the nucleus, where it regulates a variety of target genes, thus linking membrane EphrinB to the nucleus (Bong *et al.*, 2007).

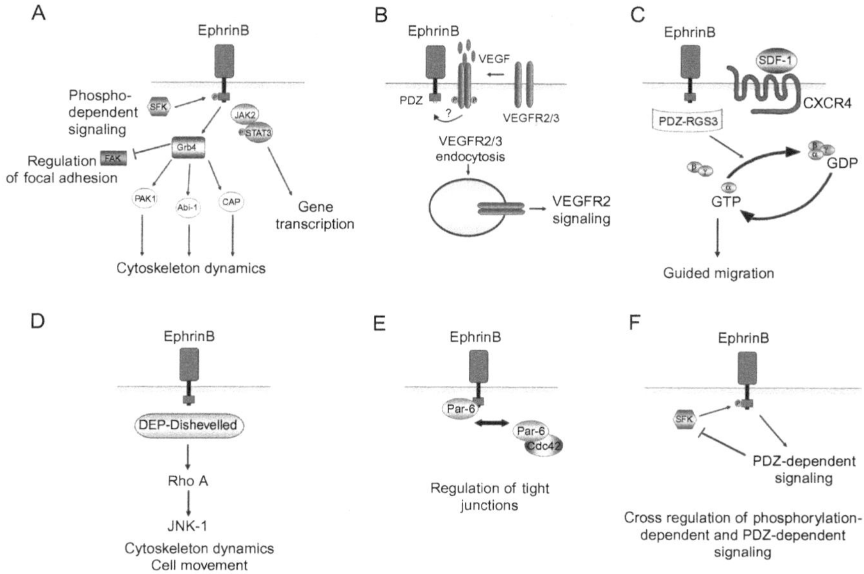

Fig. 3 Mechanisms of EphrinB reverse signaling. (A) EphrinB activation by EphB receptor leads to the recruitment of Src family kinases (SFKs) that phosphorylate EphrinB intracellular domain at tyrosine residues. The adaptor molecule Grb4, which contains a Src-homology-2 (SH2) domain, is recruited to the phosphorylated EphrinB and initiates a number of signaling events that regulate cytoskeleton dynamics and focal adhesions. The phosphorylated EphrinB can also recruit the Jak2/STAT3 complex; the phosphorylated STAT3 translocates to the nucleus regulating expression of target genes. (B) Activated EphrinB regulates the internalization of VEGFR2 and VEGFR3 through PDZ-mediated binding of yet unknown protein(s). After VEGF binding, VEGFR is phosphorylated; signaling from activated VEGFR requires internalization, which is positively regulated by EphrinB signaling. (C) RGS3 is a PDZ-containing protein that constitutively binds to EphrinB and links signaling from the G-protein-coupled receptor CXCR4 to EphrinB. CXCR4 signaling in response to the ligand SDF1 is induced by the dissociation of G$\beta\gamma$ and GTP-Gα subunits. PDZ-RGS3 can inhibit CXCR4 signaling by enhancing the GTPase activity of the Gα subunit, resulting in the reformation of the inactive heterodimeric CXCR4 receptor. (D) The scaffold protein Dishevelled binds EphrinB through its DEP domain and mediates signaling via the Rho small GTPase pathway. (E) The scaffold protein Par6 associates with EphrinB resulting in the loss of tight junctions. PAR6 forms a complex with activated apical protein kinase C (aPKC) and Cdc42-GTP; the complex localizes to the apical cell junctions where it regulates tight junctions. EphrinB1 can compete with Cdc42 for binding to Par6 and thus reduce tight junctions. (F) Cross-regulation between EphrinB phosphorylation- and PDZ-dependent signaling pathways. The phosphatase PTB-BL, recruited through its PDZ domain to EphrinB cytoplasmic domain, inactivates Src and dephosphorylates EphrinB. (For color version of this figure, the reader is referred to the Web version of this chapter.)

EphrinB can also become phosphorylated in the absence of receptor engagement. In the chicken retina, FGF can induce tyrosine phosphorylation of endogenous EphrinB, presumably through the coexpressed FGF receptor

(Chong *et al.*, 2000; Lee *et al.*, 2009). In fibroblasts, activation of the PDGF receptor induces EphrinB1 phosphorylation (Bruckner *et al.*, 1997), as does activation of Tie2 by angiopoietin-1 in endothelial cells (Adams *et al.*, 1999).

A number of studies have highlighted the importance of tyrosine phosphorylation-independent signaling by EphrinB originating from protein–protein interactions involving the C-terminal PDZ-binding motif (Egea and Klein, 2007; Lin *et al.*, 1999; Noren and Pasquale, 2004). Genetic studies have shown that knock-in mice expressing a mutant EphrinB2, in which the five tyrosines have been substituted to prevent phosphorylation, are only modestly affected (Makinen *et al.*, 2005). In contrast, knock-in mice expressing a mutant EphrinB2, in which the PDZ-interacting domain was disrupted, present major defects in lymphatic vascular (Makinen *et al.*, 2005) and retinal vessel development (Sawamiphak *et al.*, 2010b).

Biochemical studies have provided examples of phosphorylation-independent EphrinB signaling. In one such example, both the WT and phosphorylation-deficient EphrinB2 proteins were reported to similarly regulate the internalization of VEGFR2, whereas the PDZ-mutant EphrinB2 did not. This observation suggested that the binding of activated EphrinB2 to a PDZ-containing protein (yet to be identified) mediates the internalization of VEGFR2 or VEGFR3 (Sawamiphak *et al.*, 2010b; Fig. 3B). In another example, the EphrinB cytoplasmic domain was reported to bind to the PDZ-containing RGS3 (regulator of G-protein signaling) protein, which can regulate G-protein-coupled receptor signaling (Lu *et al.*, 2001; Fig. 3C). Another PDZ-dependent signaling pathway involves the interaction between EphrinB and PAR-3, a scaffold protein member of the Par (partitioning defective) complex (Lin *et al.*, 1999), which has been shown to stimulate Cdc42-induced Rac activation (Lin *et al.*, 1999; Nishimura *et al.*, 2005). Dishevelled, a scaffold protein that plays important roles in the Wnt signaling pathway, has been shown to bind to unphosphorylated EphrinB as well as to tyrosine-phosphorylated EphrinB, a binding which is mediated by Grb4 (Boutros and Mlodzik, 1999; Tanaka *et al.*, 2003). The phosphorylation-independent EphrinB1 binding of Dishevelled was found to require the presence of the DEP (Dishevelled, Egl–10, Pleckstrin), but not the PDZ domain, of Dishevelled (Lee *et al.*, 2006). Functionally, the binding of Dishevelled to EphrinB1 has been shown to mediate signals through the Rho small GTPase pathway (Lee *et al.*, 2006; Tanaka *et al.*, 2003; Fig. 3D). More recently, Par-6, another member of the Par protein family, was found to bind to EphrinB1 at the C-terminus through a non-PDZ-binding motif (Lee *et al.*, 2009). This Par-6/EphrinB1 complex was found to compete with the small GTPase Cdc42 for binding to Par-6, providing evidence that Par-6 is a mediator of EphrinB1 signaling (Lee *et al.*, 2009; Fig. 3E). Recent genetic experiments have further suggested the existence of phosphorylation and PDZ-independent EphrinB signaling, potentially mediated by Grb4 docking via the EphrinB SH3 domain or the association with Dishevelled (Dravis and Henkemeyer, 2011).

There is evidence of cross-regulation between phosphorylation and PDZ-dependent EphrinB signaling. For example, the tyrosine phosphatase PTB-BL, which contains a PDZ motif, is recruited to the active EphrinB and can negatively regulate EphrinB phosphorylation and Src activity (Palmer *et al.*, 2002; Fig. 3F).

Many of the signaling pathways initiated by EphB/EphrinB signaling described above have been identified in endothelial cells and have been linked to various endothelial cell functions (Chrencik *et al.*, 2006). EphB4 signaling has been shown to activate the Akt, PI3K (phosphatidylinositol 3-kinase), and the MAPK (mitogen activated protein kinases) pathways promoting endothelial cell proliferation and migration (Steinle *et al.*, 2002). EphB2 and EphB4 signaling was found to enhance SDF1/CXCL12-induced Akt phosphorylation (Salvucci *et al.*, 2006). EphrinB phosphorylation was reported to transiently activate Src family kinases, which are positive regulators of EphrinB phosphorylation, inducing endothelial cell sprouting (Palmer *et al.*, 2002; Salvucci *et al.*, 2009). Also, EphrinB phosphorylation was reported to induce Jak2-dependent STAT3 phosphorylation, contributing to endothelial cell assembly onto extracellular matrix (Salvucci *et al.*, 2009). EphrinB2 stimulation resulted in the activation of the PI3K and MAPK pathways in vascular endothelial cells promoting their proliferation (Steinle *et al.*, 2003) and migration (Maekawa *et al.*, 2003). EphrinB1 activation promoted JNK phosphorylation through interaction of the C-terminal domain with PDZ-containing proteins enhancing endothelial cell attachment and migration (Huynh-Do *et al.*, 2002).

Studies *in vivo* have revealed the importance of EphrinB2–PDZ interaction for reverse signaling underlying the development of lymphatic vessels (Makinen *et al.*, 2005), and the regulation of VEGFR2 and VEGFR3 internalization and signaling in blood vessels thereby modulating endothelial cell growth, survival, and migration (Sawamiphak *et al.*, 2010b). Although the biochemical basis for signaling from EphrinB/PDZ interactions has not been clarified in endothelial cells, the contribution of Rho family GTPases might be important in this context, analogous to the situation in neuronal cells and other cells (Egea and Klein, 2007; Lee *et al.*, 2006; Noren and Pasquale, 2004; Tanaka *et al.*, 2003, 2004).

IV. ROLE OF B-TYPE EPHS AND EPHRINS IN VASCULAR DEVELOPMENT

The formation of blood vessels starts early during development of the mouse embryo, and the presence of functional vessels becomes necessary around midgestation. Two processes contribute to the development of vascular structures: vasculogenesis, the process by which endothelial cell precursors (angioblasts) form endothelial tubules, which contributes to the formation of the primary vascular plexus in the yolk sac, and angiogenesis, the process by which an existing vasculature forms sprouts and generates new vessels, thus expanding and remodeling an existing vascular bed, which contributes to the formation of most vessels during development and after birth.

Genetic experiments in mice have demonstrated that the global deletion of EphB4 or EphrinB2 leads to embryonic death with marked defects in the "remodeling" of the primary vascular plexus, suggesting that these molecules are critical to early vascular development (Adams *et al.*, 1999; Gerety *et al.*, 1999; Wang *et al.*, 1998). Consistent with this interpretation, deletion of EphrinB2 targeted to the developing vasculature produced a similar phenotype (Gerety and Anderson, 2002). The phenotype of EphB4-null or EphrinB2-null mice was indistinguishable, suggesting that forward signaling, reverse signaling, or a combination of forward and reverse signaling is necessary for normal vascular development. Mutant mice, in which the carboxy-terminal cytoplasmic tail of EphrinB2 was truncated (residues 264–336), displayed a very similar vascular phenotype to that of EphrinB2-null mice (Adams *et al.*, 2001), suggesting that EphrinB2 reverse signaling is critical to early vascular development. A subsequent study using the same EphrinB2 C-terminal mutation in a different construct concluded that the cytoplasmic domain of EphrinB2 is not essential for vascular development as the mice were born live (Cowan *et al.*, 2004). This study suggested that the discordant results were attributable to differences in trafficking of the mutant EphrinB2 to the cell surface and that the defective vascular development was attributable to defective stimulation of Eph signaling (Cowan *et al.*, 2004). Nonetheless, cardiovascular development was compromised also in these mice, suggesting that EphrinB2 reverse signaling is critical at later stages of cardiovascular development (Cowan *et al.*, 2004). Other genetic experiments examined the relative signaling contribution of phosphorylation and the PDZ-binding domain of EphrinB2 to vascular development (Makinen *et al.*, 2005). Knock-in mice expressing a mutant EphrinB2, in which the conserved tyrosine residues were mutated to prevent phosphorylation, had no appreciable blood vascular defects (Makinen *et al.*, 2005). Similarly, knock-in mice expressing a PDZ-mutant EphrinB2 were born normally without apparent blood vascular defects (Makinen *et al.*, 2005). However, these EphrinB2 PDZ-mutant knock-in mice exhibited marked defects in lymphatic vessel development, whereas the phosphorylation-deficient knock-in mice had only minor defects in the lymphatic vessels (Makinen *et al.*, 2005). Additional experiments using EphrinB2 PDZ-mutant mice concluded that EphrinB2 signaling is required for the normal development of retinal vessels (Sawamiphak *et al.*, 2010b). Together, these results suggested that EphrinB2 reverse signaling mediated by the PDZ-binding domain is a critical contributor to lymphatic and retinal blood vessel development (Makinen *et al.*, 2005; Sawamiphak *et al.*, 2010b). EphrinB2-targeted deletion in pericytes and smooth muscle cells caused perinatal lethality associated with developmental defects in small-diameter blood vessels, which were not properly covered with smooth muscle cells/pericytes (Foo *et al.*, 2006). Similar to the observations in mutant mice with a targeted deletion of EphrinB2 in the endothelium, overexpression of EphrinB2 in the endothelium during early development caused marked vascular defects and premature death at midgestation, providing evidence for a dosage-dependent function of EphrinB2 during vascular development (Wang *et al.*, 2010b).

Several studies have revealed the importance of EphB4 and EphrinB2 in the determination of arterial–venous fate of endothelial cells (Adams *et al.*, 1999; Gerety *et al.*, 1999; Wang *et al.*, 1998; Zhong *et al.*, 2001; Fig. 4). Studies in zebrafish demonstrated that angioblasts migrating to the trunk (where they form the dorsal aorta and the cardinal vein) are predetermined to either an arterial or a venous fate based on their selective expression of the arterial marker EphrinB2 or venous marker EphB4. Initially, angioblasts assemble into a single precursor vessel from which the venous-fated, EphB4-expressing endothelial cells migrate and segregate to form the cardinal vein, whereas the EphrinB2-expressing cells do not move ventrally (Herbert *et al.*, 2009). Various approaches to limit EphrinB2 or EphB4 function have established that this process of cell segregation is regulated by Eph/Ephrin interactions (Herbert *et al.*, 2009). For example, if EphrinB2 expression is silenced by morpholinos or if EphrinB2 signaling is prevented by a C-terminal EphrinB2 deletion, the dorsal aorta contains very few cells. Conversely, if EphB4 is targeted with morpholinos, the cardinal vein is depleted of cells.

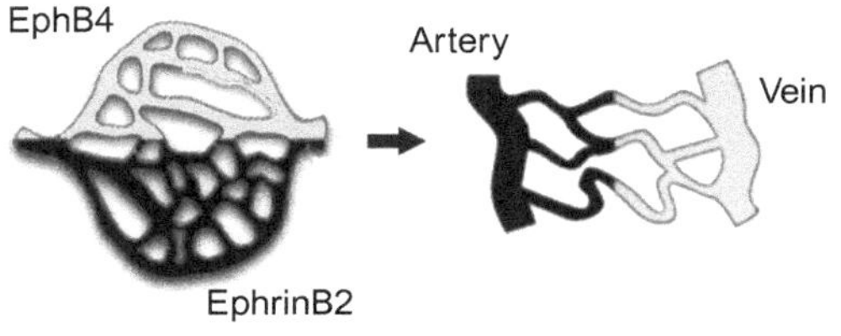

Fig. 4 A role for EphB4 and EphrinB2 in arterial and venous fate determination. In the primitive vascular plexus, presumptive arterial and venous territories are marked by the distinctive expression of EphrinB2 or EphB4, prior to the formation of EphrinB2^{+} arteries and EphB4^{+} veins. (For color version of this figure, the reader is referred to the Web version of this chapter.)

V. ANGIOGENESIS AND SPROUTING ANGIOGENESIS

Several mechanisms have been proposed to underlie the process of angiogenesis, that is, the formation of new vessels from existing vessels, including sprouting, elongation/widening, incorporation of circulating endothelial precursors, and formation of a lumen. There is evidence for a critical contribution of EphB4 and EphrinB2 function in several steps in angiogenesis.

EphrinB2 is expressed in the endothelium and mural cells of adult arteries, arterioles, and capillaries in many tissues, and levels of EphrinB2 expression increase in angiogenic vessels under physiological and pathological conditions (Gale *et al.*, 2001; Korff *et al.*, 2008; Shin *et al.*, 2001). EphrinB is broadly phosphorylated in angiogenic vessels of the retina, skin wounds, and tumor vessels, but not in the resting endothelium (Salvucci *et al.*, 2009). EphB4 is expressed most prominently in endothelial cells of venous derivation (Wang *et al.*, 1998). This pattern of segregated expression of EphrinB2 and EphB4 has raised questions about the sites and extent of receptor/ligand interactions, since they would be limited to arterial/vein boundaries. However, expression of EphrinB2 and EphB4 partially overlaps in retinal vessels (Wang *et al.*, 2010b), and endothelial cells derived from many sources, including the umbilical vein, human aorta, and dermal microvasculature, were found to express EphrinB2 and EphB4. This suggested that EphrinB2 and EphB4 have much broader avenues for interaction than previously appreciated (Gale *et al.*, 2001; Salvucci *et al.*, 2006; Shin *et al.*, 2001). In addition, certain tumor cells express EphB4 and other members of the B- and A-type receptors, providing an opportunity for functional interactions with angiogenic tumor vessels that express EphrinB2 (Fukai *et al.*, 2008; Noren *et al.*, 2006; Zhuang *et al.*, 2010).

Angiogenic sprouting is characterized by the appearance of a pioneering endothelial cell, identified as a "tip" cell, which forms filopodial extensions and is locally invasive and motile in response to surrounding VEGF-A gradients (Gerhardt *et al.*, 2003; Ferrara *et al.*, 2003; Ruhrberg *et al.*, 2002; Fig. 5). Other endothelial cells, named "stalk" cells, follow the tip cell and form the extending sprout at the base of the stalk (Gerhardt *et al.*, 2003). The Notch and VEGF signaling pathways play critical roles in determining the distinctive roles of tip and stalk cells. Tip and stalk cells have both been shown to be responsive to local VEGF-A gradients (Ferrara *et al.*, 2003), but this responsiveness appears to be modulated by opposing functions of the Notch ligands Dll4 and Jag1 (Benedito *et al.*, 2009). VEGF-A induces the expression of the Notch ligand Dll4 in tip cells, which interacts with Notch1/4 expressed on the adjacent stalk cells reducing their expression of VEGFR2 and thus limiting stalk cell responsiveness to VEGF-A (Lobov *et al.*, 2007; Suchting *et al.*, 2007). A second Notch ligand,

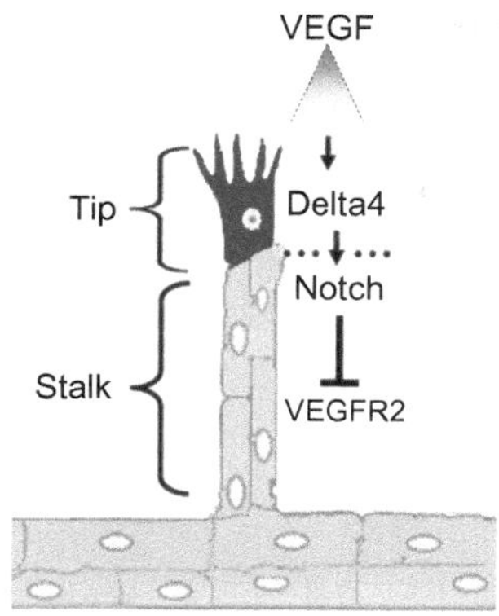

Fig. 5 Angiogenesis is often initiated by the sprouting of endothelial cells from existing vessels. The leading cell in the sprout, the so-called tip cell, is characterized by filopodia extensions; the cells at the base of the sprout, the so-called stalk cells, are proliferating. VEGF/VEGFR2, the Notch ligands Delta4 and Notch signaling participate in orchestrating sprouting angiogenesis. (For color version of this figure, the reader is referred to the Web version of this chapter.)

Jag1, is expressed by stalk cells and serves to stimulate Notch1/4 on tip cells contributing to their responsiveness to VEGF-A (Benedito *et al.*, 2009). Another potential regulator of tip and stalk cell function is soluble VEGFR1. Stalk cells produce soluble VEGFR1, which neutralizes VEGF-A by competing with receptor binding, and may thus further limit or modulate responses to VEGF-A in the growing sprout (Chappell *et al.*, 2009).

Recent studies have outlined an important role of EphrinB2 signaling in sprouting angiogenesis, particularly in the regulation of tip cell function. Developing retinal vessels grow by sprouting (Gariano and Gardner, 2005) and widely express EphrinB2, which is concentrated on the tip cells and their protrusions (Sawamiphak *et al.*, 2010b; Wang *et al.*, 2010b). EphrinB2 is diffusely phosphorylated in the developing retinal vessels, but phosphorylation subsides once the retinal vessels have fully developed (Salvucci *et al.*, 2009; Fig. 6). Expression of a PDZ-mutant EphrinB2 impaired the development of retinal vessels, providing evidence that EphrinB2 signaling is critical to retinal vessel development (Sawamiphak *et al.*, 2010b). In particular, the PDZ-mutant tip cells displayed a significantly reduced ability to form filopodia. This ability of EphrinB2 to promote filopodia extension was confirmed *in vitro* using primary endothelial cells stimulated with EphB4-Fc (Sawamiphak *et al.*, 2010b) and retinal explants *ex vivo* (Sawamiphak *et al.*, 2010a).

In the mouse retina, filopodia from adjacent tip cells have been shown to bridge with each other to form connections between developing sprouts, which is an initial step leading to the development of a lumen that permits

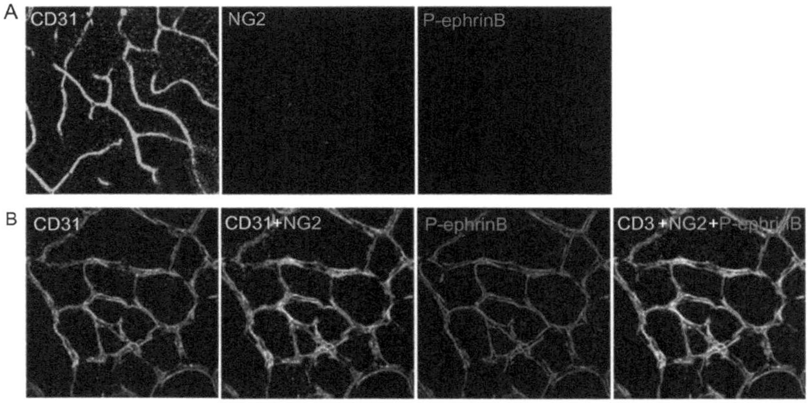

Fig. 6 EphrinB is phosphorylated in remodeling but not in fully developed retinal vessels. (A) EphrinB phosphorylation is not detected in fully developed mouse retinal vessels, including the CD31$^+$ endothelial cells and the NG2$^+$ pericytes. (B) EphrinB2 is widely phosphorylated in the remodeling retinal vessels from 6-day-old mice, both in the endothelial cells and in the pericytes. (For color version of this figure, the reader is referred to the Web version of this chapter.)

blood flow in the new vessel (Bentley *et al.*, 2009). The formation of such connecting bridges requires endothelial cell expression of EphrinB and signaling (Sawamiphak *et al.*, 2010b). A similar process of bridging between endothelial cells can be observed *in vitro* when primary endothelial cells are incubated onto extracellular matrix to form a characteristic network (Salvucci *et al.*, 2006, 2009). Endothelial-to-endothelial cell assembly onto extracellular matrix *in vitro* is characterized first by the appearance of needle-like extensions and protrusions from endothelial cells resulting in the joining or endothelial cells with each other (Fig. 7A). The appearance of these needle-like protrusions is followed by extensive changes in the cytoskeletal structure and shape of the joined cells, movement of the cell nuclei along the cytoplasm, and the thickening of connecting bridges (Salvucci *et al.*, 2006, 2009). The initial steps in this process of endothelial-to-endothelial cell assembly are associated with time-dependent phosphorylation of EphrinB2 (Salvucci *et al.*, 2009; Fig. 7B). This phosphorylation is likely induced by EphB/EphrinB interactions as specific peptide inhibitors of EphB prevent the formation of cord-like structures on Matrigel (Fig. 7C). A number of experiments provide evidence that such phosphorylation is critical to EphrinB2 function as a regulator of endothelial-to-endothelial cell assembly: if EphrinB2 is silenced in endothelial cells or if the endothelial cells express a signaling-deficient EphrinB2, the assembly of endothelial cells into networks is markedly reduced. Other studies have also shown that if

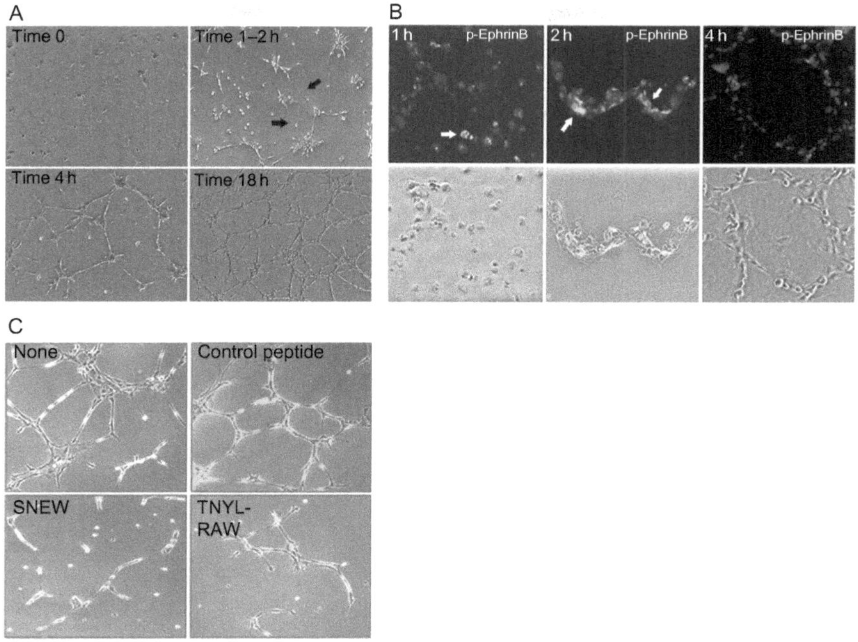

Fig. 7 Primary endothelial cells assemble into cord-like structures on extracellular matrix, a process that requires EphB/EphrinB interaction. (A) Time-dependent assembly of primary endothelial cells on Matrigel. Note the appearance after 1–2h of needle-like endothelial cell extensions that eventually form the initial bridging of endothelial cells with each other. (B) EphrinB2 is broadly and time-dependently phosphorylated in endothelial cells as they assemble into cord-like structures. (C) The EphB receptor inhibitors SNEW and TNYL-RAW peptides prevent the assembly of endothelial cells into cord-like structures. (For color version of this figure, the reader is referred to the Web version of this chapter.)

EphrinB2 is overexpressed, appropriate endothelial cell assembly is prevented, providing evidence that an optimal EphrinB2 expression level is required (Salvucci *et al.*, 2009; Sawamiphak *et al.*, 2010b; Wang *et al.*, 2010b).

An important question relates to the mechanism by which EphrinB2 is activated in sprouting endothelium, particularly in tip cells where it seems to play an essential role in the path-finding, probing, and extension function of tip cells. One of the possibilities is that, in the retina, astrocytes provide a source of cell-associated EphB4 that stimulates the developing retinal vasculature (Sawamiphak *et al.*, 2010b). Another possibility is that retinal endothelial cells express EphB4, which can stimulate EphrinB2 in interacting endothelial cells (Wang *et al.*, 2010b). Experiments *in vitro* using primary endothelial cells provide evidence that both EphB4 and EphrinB2 can be simultaneously expressed and that the assembly of vascular networks is dependent upon cell-to-cell interactions resulting in EphrinB signaling induced by endothelial cell-derived EphB4 (Salvucci *et al.*, 2006, 2009). Blocking such EphB/EphrinB cell-to-cell interactions prevents endothelial cell assembly into vascular networks (Salvucci *et al.*, 2006, 2009).

What is the mechanism by which EphrinB2 signaling contributes to sprouting angiogenesis, including "tip" cell function and endothelial cell assembly? Two recent studies in mouse and zebrafish have unveiled a role of EphrinB2 signaling in the regulation of VEGFR2 and VEGFR3 function. If EphrinB2 is silenced or if a signaling-deficient (PDZ-mutant) EphrinB2 is expressed in endothelial cells, VEGFR2 and VEGFR3 signaling is defective, as evidenced by reduced phosphorylation of the receptor following stimulation by the cognate ligand. In addition, activation of the downstream VEGFR signaling components Rac1, Erk1/2, and Akt is reduced (Sawamiphak *et al.*, 2010b; Wang *et al.*, 2010b). Recent studies have concluded that VEGFR trafficking away from the plasma membrane into the endocytic compartment contributes or is necessary for VEGFR function in response to appropriate VEGF ligands (Lampugnani *et al.*, 2006; Lanahan *et al.*, 2010; Sawamiphak *et al.*, 2010b; Wang *et al.*, 2010b). Thus, by promoting the internalization of VEGFR2 and VEGFR3, EphrinB2 signaling may critically regulate VEGFR function in response to the VEGF ligand. It was also observed that EphrinB2 signaling induces some degree of VEGFR2 internalization in the absence of exogenous VEGF (Sawamiphak *et al.*, 2010b). There is evidence for the existence of an endogenous pool of VEGF, which is not secreted outside the cells but contributes to endothelial cell function; it is possible that VEGFR internalized by EphrinB2 signaling is activated by autocrine endogenous VEGF (Lampugnani *et al.*, 2006; Lee *et al.*, 2007).

Besides serving as inducers of EphrinB2 reverse signaling, EphB receptors have also been reported to play a signaling role of their own in endothelial cells. Two studies have concluded that EphB4 forward signaling in endothelial cells represses endothelial cell migration, adhesion, and proliferation *in vitro*, and may serve to repel endothelial cells from each other and maintain arterial/venous boundaries in capillary beds (Fuller *et al.*, 2003; Kim *et al.*, 2002). Thus, in these studies, EphB4 forward signaling appeared to have opposite effects to those of EphrinB2 reverse signaling in the context of endothelial cell-sprouting angiogenesis. However, other studies have suggested that EphB4 forward signaling in endothelial cells promotes some degree of endothelial cell proliferation and angiogenesis (Stein *et al.*, 1997, 2002). A detailed analysis of EphB2/4 expression and signaling in the context of endothelial cell assembly into cord-like structures *in vitro* showed characteristic kinetic changes suggestive of a role of EphB forward signaling (Salvucci *et al.*, 2006). When first dispersed on extracellular matrix, primary endothelial cells broadly express surface EphB2/4, but as the cells contact each other to form cord structures, EphB2/4 are internalized and no longer detected on the cell surface (Salvucci *et al.*, 2006). In addition, EphB2/4 become phosphorylated with kinetics suggestive of their activation in association with the occurrence of cell-to-cell contact and receptor internalization (Salvucci *et al.*, 2006). EphB receptor signaling has been linked to the activation of secondary molecules that regulate cytoskeleton structure and cell movement, particularly small Rho family GTPases (Egea and Klein, 2007; Noren and Pasquale, 2004). Consistent with this ability of Eph signaling to regulate cell structure in response to cell-to-cell contact, a typical feature of endothelial cell assembly on extracellular matrix is the time-dependent activation of F (filamentous)-actin filaments in conjunction with cell-to-cell contact and changes in cell shape (Salvucci *et al.*, 2006). The *in vivo* relevance of these observations remains to be determined, and progress on the role of EphB4 forward signaling in endothelial cell-sprouting angiogenesis will likely benefit from use of mutant mice-expressing signaling-deficient forms of EphB targeted to the endothelial cells.

VI. VESSEL REMODELING AND STABILIZATION

Once formed through sprouting angiogenesis, newly formed vessels undergo a number of changes as a result of the establishment of blood flow, changes in tissue metabolism, endothelial growth factor availability, and other factors. Some of the newly formed vessels regress, whereas others mature through the establishment of a basal membrane composed of extracellular matrix proteins produced by endothelial cells and surrounding stromal cells. A critical step in this process of maturation is the recruitment of pericytes/smooth muscle cells to the nascent vessel, which stabilize the vessel wall and regulate endothelial cell survival, growth, and permeability (Armulik *et al.*, 2005; Hellstrom *et al.*, 2001). Electron microscopy studies have shown that pericytes are inserted into the basal membrane of vessels and come into direct contact with the endothelium through cytoplasmic extensions that penetrate the basement membrane and push deeply onto the endothelial cell surface membrane (Cuevas *et al.*, 1984). Pericytes are found around some blood capillaries, precapillary arterioles, postcapillary venules, and collecting venules, but pericyte coverage of vessels is partial and variable (ranges between 10% and 50%), depending on the tissue. The highest pericyte coverage is found in the central nervous system, where pericytes may regulate the blood–brain barrier functions (Armulik *et al.*, 2010); in the retina, the frequency of pericytes is similar to that of the endothelial cells. Although the derivation of pericytes/mural cells is still somewhat controversial, recent studies suggest that such cells are similar to mesenchymal stem cells, derive from the bone marrow, and display multilineage potential, which explains in part their morphological and phenotypic diversity (Crisan *et al.*, 2008).

Genetic mouse models have implicated several genes in the generation and proper assembly of pericytes, including the PDGF/PDGFR-1, Ang-1/Tie2, and the spingosine 1-phosphate (S1P/S1P1) genes. PDGFB and PDGF-Rβ-null mice die during development with microvascular aneurisms and lack of pericyte coverage on some vessels (Hellstrom *et al.*, 2001). Ang1- or Tie2-null mice die at midgestation with poorly organized basal membranes and reduced coverage/detachment of pericytes (Dumont *et al.*, 1994; Sato *et al.*, 1995; Suri *et al.*, 1996). Conversely, the overexpression of Ang1 promotes pericyte coverage and vessel leakage resistance (Suri *et al.*, 1998; Thurston *et al.*, 1999). Disruption of the s1p1 gene in mice causes prenatal death with vascular abnormalities characterized by defective pericyte coverage (Liu *et al.*, 2000).

EphrinB2 also plays an important role in pericyte function. The mural-cell-specific inactivation of EphrinB2, using PDGF-cre mice, caused the embryos to have edema and extensive hemorrhaging in a variety of tissues and was associated with perinatal death of the mutant mice (Foo *et al.*, 2006). Interestingly, the EphrinB2-null pericytes appeared morphologically normal in many of these mice, but they associated poorly with the vessels, showing a scattered distribution resulting in incomplete vessel coverage (Foo *et al.*, 2006).

During mouse development, EphrinB2 is expressed in the mural cells that cover arteries and veins (Foo *et al.*, 2006). In the adult mouse, EphrinB2 is expressed in a proportion of smooth muscle cells/pericytes surrounding arteries (Shin *et al.*, 2001). Importantly, levels of EphrinB2 expression in smooth muscle cells are comparable to those detected in the arterial endothelium (Shin *et al.*, 2001). Human smooth muscle cells located in the media of adult coronary arteries are partially positive for EphrinB2 (Korff *et al.*, 2008). Human bone marrow-derived mesenchymal stem cells, which either represent pericytes or are related to pericytes, express a functional EphrinB, which can be phosphorylated by EphB4-Fc (Salvucci *et al.*, 2009).

The silencing of EphrinB2 in pericytes/smooth muscle cells derived from the mouse aorta has remarkable effects on cell morphology: the cells become elongated, do not properly spread, and show numerous active lamellipodia protrusions, without evidence of polarization and without cell-to-cell contact (Foo *et al.*, 2006). Some of these defects appeared to be cell-contact-independent, suggesting that EphrinB2 may have cell-autonomous roles (Foo *et al.*, 2006). Functionally, EphrinB2-deficient smooth muscle cells displayed increased, but random, migration, which was associated with defective formation of focal adhesions. Such phenotype was reversed by reexpression of EphrinB2 or activation of Rho-like GTPases (Foo *et al.*, 2006). Consistent with EphrinB2 playing a role in mural cell function,

EphB4 (expressed by endothelial cells) induced activation of EphrinB2 on mural cells and enhanced mural/endothelial cell association within tumor blood vessels (Erber *et al.*, 2006).

In vitro, mesenchymal cells/pericytes, which express EphrinB2, characteristically make contact with endothelial cells on extracellular matrix, and when this contact occurs, EphrinB2 phosphorylation is clearly and transiently detected in the mesenchymal stem cells/pericytes (Salvucci *et al.*, 2009). The silencing of EphrinB2 in bone marrow-derived mesenchymal cells/pericytes prevented their proper assembly with endothelial cells (Salvucci *et al.*, 2009; Fig. 8), suggesting that phosphorylation-dependent EphrinB2 signaling is critical for pericytes to make proper contact with endothelial cells (Salvucci *et al.*, 2009).

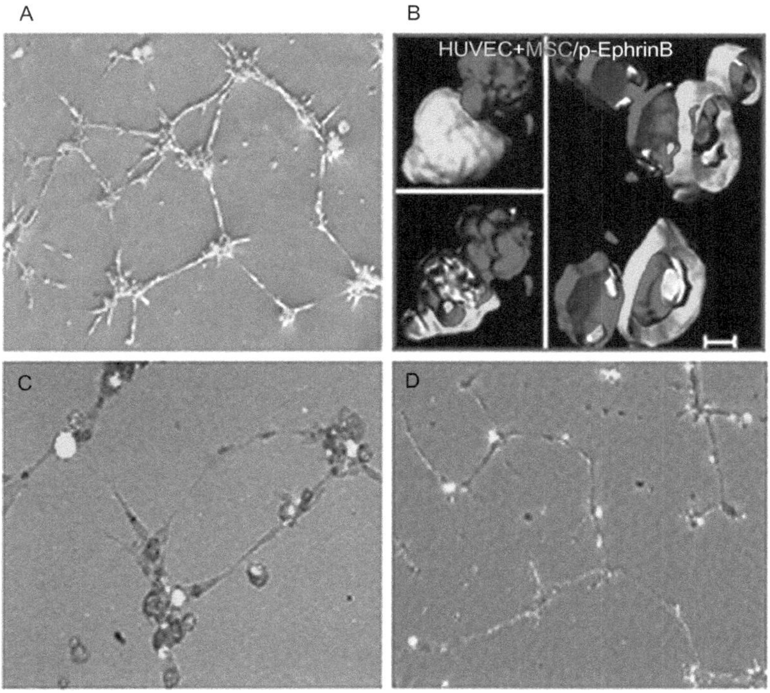

Fig. 8 Endothelial cells and mesenchymal stem cells/pericytes assemble in cord-like structures on Matrigel, with mesenchymal stem cells (green) in the nodes anchoring the cord structures. (B) EphrinB is phosphorylated (white) at points of contact between endothelial cells (HUVEC green) and mesenchymal stem cells (MSC red) forming cord-like structures. (C) The silencing of EphrinB2 in mesenchymal stem cells (green) prevents their assembly with endothelial cells (unstained). (D) The silencing of EphrinB2 in endothelial cells (unstained) prevents their assembly with MSC (green). (For interpretation of the references to color in this figure legend, the reader is referred to the Web version of this chapter.)

Little is known about the role of EphB receptors and their signaling in pericytes. Human mesenchymal stem cells express EphB2 and EphB4 (Salvucci *et al.*, 2009), and smooth muscle cells derived from human thymus also express EphB4 (Korff *et al.*, 2008). Activation of EphB4 in smooth muscle cells reduced their ability to migrate, suggesting that EphB4 signaling in pericytes may secure their attachment to the endothelial cells stabilizing the vessel wall (Korff *et al.*, 2008). Vascular smooth muscle cells also express EphA4 receptors, which can be activated by EphrinA1 and by many EphrinB ligands (Ogita *et al.*, 2003). When EphA4 is activated in smooth muscle cells, their ability to form lamellipodia and to spread is inhibited, a process that was linked to inhibition of the small GTPase Rac1 and the kinase Pak1 (Deroanne *et al.*, 2003).

VII. PHYSIOLOGICAL ANGIOGENESIS IN THE ADULT

Physiological angiogenesis in the adult is limited: it only occurs in conjunction with the menstrual cycle and pregnancy, in the skeletal bone during growth, and during wound repair.

The critical role of VEGF as a promoter of angiogenesis during the menstrual cycle, embryo implantation, and pregnancy are well recognized, but little is known about the potential roles of Eph/Ephrins in this context. Studies on corpora lutea have shown that ovulation was followed by a rapid increase in the expression of EphrinB1 and EphrinB2 in the luteinizing granulose (Egawa *et al.*, 2003; Xu *et al.*, 2006) and suggested the occurrence of EphB/EphrinB interactions between the granulosa cells of the corpus luteum and proximal angiogenic capillaries (Egawa *et al.*, 2003; Xu *et al.*, 2006).

Placental implantation of the embryo involves the aggressive invasion of maternal tissues by the embryonic/fetal cytotrophoblasts. This process ultimately results in the cytotrophoblast invasion of the maternal spiral arterioles that supply blood to the placenta, creating a chimeric vasculature composed of maternal and fetal cells (Fisher *et al.*, 1989; Librach *et al.*, 1991). The invasive behavior of the embryonic/fetal cytotrophoblasts is accompanied by remarkable changes in gene expression and phenotype; in particular, they acquire vascular-type markers (Damsky *et al.*, 1992; Zhou *et al.*, 1997) and produce proangiogenic factors (Zhou *et al.*, 2003). Important changes occurring in the cytotrophoblasts as they penetrate the uterine wall are the loss of EphB4 expression and an increase in EphrinB2 expression. Notch signaling is a critical regulator of the divergent expression of EphrinB2 and EphB4 in these cells (Hunkapiller *et al.*, 2011; Red-Horse *et al.*, 2005). This phenotypic change in embryonic/fetal cytotrophoblasts has been proposed to underlie the selective tropism of the cytotrophoblasts toward the maternal arterioles, thus limiting interactions with the maternal venules (Red-Horse *et al.*, 2005).

Angiogenesis in the context of skin wound healing occurs through endothelial cell sprouting from capillaries at the edge of the wound when the dermis is injured. The process is driven by VEGF-A (Bao *et al.*, 2009) that is first produced by platelets, which contribute to the formation of the blood clot that initially seals the wound. Inflammatory monocytes and neutrophils are additional sources of VEGF in the wound and represent a local defense system against pathogens. They are recruited by a variety of signals, including complement components, platelet degranulation products, and potentially bacterial particles (Banks *et al.*, 1998; Mohle *et al.*, 1997). Sprouting wound vessels are components of the "granulation tissue," which is a hallmark of wounds undergoing repair. Besides sprouting vessels, the granulation tissue contains a fibrin clot, fibroblasts, collagen, and inflammatory cells (Bao *et al.*, 2009; Murohara *et al.*, 1998). Granulation tissue undergoes extensive remodeling in repairing wounds, and the process is orchestrated by a series of enzymes including MMP1, MMP2, urokinase-type, and tissue-type plasminogen activators, and plasminogen activator inhibitor (Schultz and Wysocki, 2009).

A study reported that circulating mouse neutrophils express EphrinA2 and B2 (Schruefer *et al.*, 2006). Other studies have demonstrated that inflammatory macrophages express EphrinB2 and suggested a role for monocyte-derived EphrinB2 in the stimulation of venous endothelium that expresses EphB4 (Yuan *et al.*, 2000, 2004). Recent studies have confirmed the expression of EphrinB2 in the developing dermal vasculature of mice (Wang *et al.*, 2010b). In a murine model of wound healing in the skin, the sprouting endothelial cells from the margins of the wound expressed EphrinB2, which was prominently and time-dependently phosphorylated in these cells, indicative of active signaling (Salvucci *et al.*, 2009; Fig. 9A). In contrast, EphrinB2 was not phosphorylated in the surrounding normal vessels or in the new capillaries once sprouting had subsided and the granulation tissue was well developed (Salvucci *et al.*, 2009). Additionally, EphrinB2 expression and phosphorylation were transiently detected in wound-infiltrating inflammatory cells (Salvucci *et al.*, 2009).

VIII. PATHOLOGICAL ANGIOGENESIS

A. Neovascular Disorders of the Eye

Intraocular neovascularization is a pathological complication of many eye diseases, including diabetic retinopathy, retinopathy of prematurity, and age-related (neovascular, wet-type) macular degeneration (AMD). Since these conditions represent the major cause for blindness, control of

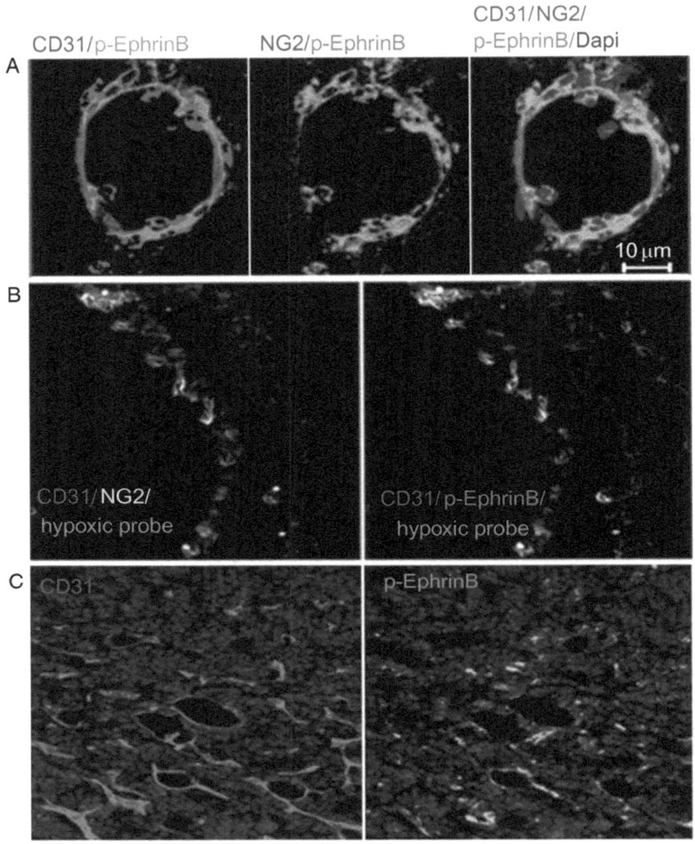

Fig. 9 (A) EphrinB is phosphorylated in angiogenic vessels associated with skin wound healing. (B) EphrinB is phosphorylated in angiogenic retinal vessels induced by hypoxia (areas of hypoxia visualized in green; ROP model). (C) EphrinB is phosphorylated in angiogenic tumor vessels. CD31 identifies endothelial cells; NG2 pericytes/mural cells. (For interpretation of the references to color in this figure legend, the reader is referred to the Web version of this chapter.)

pathological angiogenesis is an important therapeutic goal. Retinal ischemia and increased VEGF expression are often the principal causes of pathological retinal angiogenesis (Alon *et al.*, 1995; Fraisl *et al.*, 2009; Miller *et al.*, 1994). Oxygen levels and the levels of VEGF are tightly regulated in the normal retina (Stenzel *et al.*, 2011). Hypoxia regulates expression of VEGF largely through the activities of the transcription factors HIFs (hypoxia inducing factors) (Semenza, 2003; Tang *et al.*, 2011). The HIF-1α subunit is maintained at low levels during normoxia through an oxygen-dependent degradation process, which is mediated by the von Hippel Lindau (VHL) protein, a component of the E3 ubiquitin ligase complex (Semenza, 2011). Once the oxygen pressure drops below 5%, HIF-1α is not sufficiently degraded, resulting in the activation of genes involved in the maintenance of oxygen homeostatic levels. One of these target genes is VEGF, which promotes the growth of blood vessels toward hypoxic zones (Ferrara *et al.*, 2003; Gerhardt *et al.*, 2003). Consistent with its role in the degradation of HIF-1α, loss-of-function mutations of VHL result in increased tumor angiogenesis (Pugh and Ratcliffe, 2003).

Abnormally high levels of VEGF and pathological angiogenesis also characterize the wet-type AMD (Jager *et al.*, 2008). In this condition, choroidal neovascularization causes exudation and hemorrhaging, which impair the central portion of the retina that is responsible for high-resolution vision (Jager *et al.*, 2008). VEGF is detected at high levels in the choroidal membranes from patients with wet AMD (Kvanta *et al.*, 1996; Lopez *et al.*, 1996); various approaches to limit VEGF activity in the eye have demonstrated a marked clinical benefit to the patients.

A number of mouse models of ischemia-induced intraocular neovascularization have been developed, including the retinopathy of immaturity (ROP) model, which has been useful to study ischemia-induced retinal angiogenesis (Smith *et al.*, 1994). In this model, 7-day-old mice are exposed to high (75%) oxygen tension for a period of 5 days and then are exposed to room air. High oxygen tension causes physiological postnatal retinal angiogenesis to come to a stop and promotes degeneration of the retinal vessels that had recently sprouted. Once the retinas are reexposed to air oxygen, retinal neovascularization resumes with vigorous retinal vessel sprouting after 3 days (Smith *et al.*, 1994). These sprouting retinal vessels express EphrinB2, which is phosphorylated (Salvucci *et al.*, 2009), indicative of activity (Fig. 9B). Once the sprouting subsides in the retinal vessels, EphrinB2 phosphorylation also subsides, without change in the levels of EphrinB2 expression (Salvucci *et al.*, 2009). These results show that hyperoxia-induced retinal neovascularization is associated with EphrinB2 activation and thus is similar in this respect to the occurrence of EphrinB2 phosphorylation during physiological postnatal angiogenesis of the retina.

B. Tumor Angiogenesis

Angiogenesis, a hallmark of a growing malignancy, often starts when the tumors are quite small in size (1–2 mm) and is principally driven by VEGF-A produced by the tumor cells and by cells in the tumor microenvironment. Tumor hypoxia is believed to represent an important inducer of tumor-associated VEGF (Dayan *et al.*, 2008; Ferrara and Kerbel, 2005; Mantovani *et al.*, 2002; Murdoch *et al.*, 2008), which is an indispensable driver of tumor angiogenesis in preclinical mouse models (Ferrara and Kerbel, 2005; Kim *et al.*, 1993). Tumor angiogenesis resembles physiological angiogenesis but presents notable differences (Chung *et al.*, 2011). A principal difference is that tumor angiogenesis is not self-limited but appears to be a self-perpetuating and persistent process that is not turned off once vessels are formed. Tumor vessels are disorganized compared to normal vessels and display a number of abnormalities, including fragility, reduced number of pericytes, and leakiness with a propensity to produce exudates and to bleed. These abnormalities have been linked to excess VEGF-A, and thus VEGF neutralization has been reported to somewhat "normalize" the tumor vasculature (Jain, 2005). The tumor cells, particularly melanomas and glioblastomas, may undergo a process of transformation, named "vasculogenic mimicry," such that the tumor cells or the differentiated tumor stem cells acquire phenotypic characteristics of endothelial cells and become components of the vessel wall (Hendrix *et al.*, 2003a,b; Ricci-Vitiani *et al.*, 2010; Wang *et al.*, 2010a).

A number of preclinical and clinical studies using inhibitors of VEGF have identified VEGF-independent pathways of tumor angiogenesis, which may account for the poor responses and/or acquired resistance to VEGF neutralization (Shojaei *et al.*, 2007). One such pathway involves the neutrophil-derived proangiogenic factor Bv8 (also known as prokineticin 2), which drives endothelial cell proliferation and migration in the absence of VEGF-A (Shojaei *et al.*, 2007, 2009). Other pathways markedly modulate VEGF expression and/or VEGF-A-induced responses, both physiologically and in cancer, including the Dll4/Notch (Hellstrom *et al.*, 2007; Noguera-Troise *et al.*, 2006; Ridgway *et al.*, 2006), Angiopoietin-1/Tie2 and Angiopoietin-2/Tie2 (Augustin *et al.*, 2009), platelet-derived growth factors (PDGFs)/PDGF-receptor β (Bergers *et al.*, 2003), and TGFβ1/TGFβ receptor II (Chung *et al.*, 2011; Fridlender *et al.*, 2009). The Eph/Ephrin signaling pathways can also regulate VEGF/VEGFR signaling in the context of cancer, but the system is complex (Bruckner *et al.*, 1997; Pasquale, 2010; Sawamiphak *et al.*, 2010a,b).

The functions of the Eph/Ephrin system in cancer are complex due to the fact that many tumor cells express Eph receptors and ligands, albeit to varying degrees (Davalos *et al.*, 2006; Huang *et al.*, 2007; Noren and Pasquale, 2007; Zhuang *et al.*, 2010), which can stimulate or be stimulated by Eph/Ephrin molecules in the tumor endothelium and in other cells of the tumor microenvironment. Thus, the outcome of Eph receptor and ligand signaling in the tumor cells differs in different tumors (no effect, as opposed to inhibition or stimulation of tumor growth) (Batlle *et al.*, 2002; Dohn *et al.*, 2001; Dopeso *et al.*, 2009; Jin *et al.*, 2006; Kumar *et al.*, 2009; Miao *et al.*, 2009; Noren and Pasquale, 2007; Noren *et al.*, 2006; Ruhe *et al.*, 2007; Smith *et al.*, 2004; Yu *et al.*, 1999).

There are a few situations that best illustrate the complex roles of Eph/Ephrin in tumor growth. For example, a number of studies have provided evidence that EphB2/EphB4 signaling in certain tumor cells suppresses tumor growth or promotes tumor dormancy. In particular, it was reported that soluble EphrinB-Fc reduced colon and breast cancer cell growth *in vitro* and tumor progression in mice by pharmacologically activating tumor cell-associated EphB4 (Batlle *et al.*, 2002; Noren and Pasquale, 2007; Noren *et al.*, 2006). Consistent with a tumor-suppressor role of EphB4 signaling, a dominant-negative form of EphB4 promoted colorectal tumor growth (Dopeso *et al.*, 2009). Also, consistent with the notion that EphB2 and EphB4 act as inhibitors of tumor growth, EphB2 and EphB4 are mostly inactive in tumors as judged by low levels of phosphorylation in many malignant cells and are expressed at higher levels in benign tumors of the same lineage (Kumar *et al.*, 2009; Miao *et al.*, 2005; Noren and Pasquale, 2007; Noren *et al.*, 2006). Thus, it was proposed that EphB signaling in the tumor cells may inhibit oncogenic signaling pathways operating in the tumor cells thereby reducing tumor growth, may promote repulsive interactions between the EphB-expressing tumor cells and the EphrinB1-expressing normal tissue that would restrict tumor cell invasion, or may promote tumor cell adherence to each other preventing local tissue infiltration (Cortina *et al.*, 2007; Kemp *et al.*, 2009; Pasquale, 2010).

In contrast with these results, other studies have indicated that EphB4 signaling promotes tumor growth, perhaps due to a loss of tumor-inhibitory function or to acquisition of ligand-independent activities (Pasquale, 2010). In some studies, activation of tumor-associated EphB4 enhanced tumor growth, motility, and metastasis of prostate, bladder, colorectal, and melanoma cancer cells (Xia *et al.*, 2005, 2006; Yang *et al.*, 2006). Also, knockdown of EphB4 in colorectal and ovarian cancer cells inhibited tumor growth and metastasis (Kumar *et al.*, 2007, 2009). Tumor-associated expression of EphB4 stimulated tumor angiogenesis and tumor growth by activating EphrinB2 on the tumor endothelium (Noren *et al.*, 2006), and overexpression of EphrinB2 in ovarian and melanoma tumor cells correlated with increased tumor invasion and poor prognosis, perhaps by promoting EphB4 activation in the tumor cells (Castellvi *et al.*, 2006; Meyer *et al.*, 2005). Deregulated expression of EphB4 and EphrinB2 in KSHV (Kaposi's sarcoma-associated herpes virus)-infected KS cells is also believed to contribute to tumor growth (Masood *et al.*, 2005).

The complexities of Eph/Ephrin interactions in growing tumors and the difficulties in generalizing their function in cancer suggest that careful experimentation with controlled systems may be useful. In line with this, a simplified tumor model unequivocally showed that EphrinB2 signaling in tumor-associated vessels promotes tumor angiogenesis, much alike the situation in physiological angiogenesis (Sawamiphak *et al.*, 2010b). Using mouse tumor models of intracranial gliomas and subcutaneous astrocytoma, the authors showed that tumor growth and tumor angiogenesis were significantly reduced in EphrinB2-deficient tumor-bearing mice in comparison to wild-type control mice (Sawamiphak *et al.*, 2010b). The interpretation of these results is that sprouting angiogenesis in tumors resembles physiological angiogenesis and as such is similarly regulated by EphrinB2 signaling via modulation of VEGF-A/VEGFR2 function (Sawamiphak *et al.*, 2010b; Wang *et al.*, 2010b). Consistent with EphrinB signaling playing an important role in tumor angiogenesis, the angiogenic tumor vessels in experimental MOPC315 tumors displayed an intensely phosphorylated EphrinB, particularly at the tumor margins where angiogenesis was most prominent (Salvucci *et al.*, 2009; Fig. 9C). In addition, tumors that express EphB4 promote angiogenesis through interaction with EphrinB2 expressed by the tumor endothelium (Noren *et al.*, 2006).

IX. THERAPEUTIC IMPLICATIONS

Pathological angiogenesis in the adult is mostly associated with neovascular diseases of the eye, including wet macular degeneration and diabetic retinopathy, and with tumor angiogenesis, which contributes to tumor growth in many types of cancer. It is now clear that EphrinB2 signaling plays a critical role in promoting VEGF-induced endothelial cell sprouting and orchestrating endothelial-to-endothelial and endothelial-to-pericyte assembly. For these reasons, EphrinB2 and its ligands are desirable therapeutic targets for anti-angiogenic therapy. However, the complexities of cancer, where Eph/Ephrin interactions involve not only the tumor vasculature but also the tumor cells and other cells resulting in strikingly different tumor outcomes, are a challenge for the development of Eph/Ephrin-based cancer drugs. Thus, potential application of Eph/Ephrin targeting to cancer treatment will need additional study and validation for individual cancer types, stage, and other variables.

A number of approaches have demonstrated to effectively inhibit Eph/Ephrin interactions. Monomeric soluble EphB4 ectodomain, which inhibits EphB4 forward signaling and EphrinB2 reverse signaling, has been shown to reduce angiogenesis *in vivo* and experimental tumor growth in mice (Kertesz *et al.*, 2006; Scehnet *et al.*, 2009). EphB4 agonist antibodies could be useful in the context of selected EphB4-expressing tumors as they could promote EphB4 tumor-suppressor activity and at the same time inhibit EphrinB2-induced tumor angiogenesis (Krasnoperov *et al.*, 2010; Noren and Pasquale, 2004; Noren *et al.*, 2006). Specific antibodies to EphB2 have been generated that can effectively block interaction with Ephrins and kill EphB2-expressing tumor cells when conjugated with the cytotoxic drug auristatin

(Mao *et al.*, 2004; Xu *et al.*, 2009). Phage display allowed the identification of specific peptides that bind EphB2 and EphB4 at nanomolar concentrations and can readily inhibit endothelial-to-endothelial cell assembly *in vitro* and experimental angiogenesis *in vivo* (Chrencik *et al.*, 2006, 2007; Koolpe *et al.*, 2005; Salvucci *et al.*, 2006, 2009). A neutralizing antibody to EphrinB2 was reported to reduce vessel number, but not size, in an experimental model of human glioblastoma (Li *et al.*, 2011).

X. CONCLUSIONS

In this review, we have discussed evidence for a role of the B family of Ephs and Ephrins in angiogenesis. We have described how B Ephs and Ephrins signaling play critical roles in developmental and postnatal angiogenesis in physiology and disease. Thus, B Ephs and Ephrins are promising targets to modulate angiogenesis. Several approaches to block EphB/EphrinB function seem to be very effective at reducing angiogenesis in experimental models. However, many complexities of EphB and EphrinB signaling are not understood, particularly how they integrate with other signaling pathways. The context-dependent functions of B Ephs and Ephrins in cancer are poorly understood and may require a better understanding of the role of Eph/Ephrin in cell interactions between tumor cells and the tumor microenvironment as well as understanding how EphB/EphrinB signaling integrates with oncogenic signaling pathways. Research in the coming years will likely decipher the diverse functions of Ephs and Ephrins in the context of cancer. This will be an important advance that may open a wide range of therapeutic opportunities.

ACKNOWLEDGMENTS

This work was supported by the Intramural Program of the Center for Cancer Research, National Cancer Institute, National Institutes of Health, Bethesda, MD 20892. The authors wish to thank Dr. Douglas Lowy for his support and Inn Inn Chen for reviewing the manuscript.

REFERENCES

Adams, R. H., Wilkinson, G. A., Weiss, C., Diella, F., Gale, N. W., Deutsch, U., Risau, W., and Klein, R. (1999). Roles of ephrinB ligands and EphB receptors in cardiovascular development: demarcation of arterial/venous domains, vascular morphogenesis, and sprouting angiogenesis. *Genes Dev.* **13**, 295–306.

Adams, R. H., Diella, F., Hennig, S., Helmbacher, F., Deutsch, U., and Klein, R. (2001). The cytoplasmic domain of the ligand ephrinB2 is required for vascular morphogenesis but not cranial neural crest migration. *Cell* **104**, 57–69.

Alon, T., Hemo, I., Itin, A., Pe'er, J., Stone, J., and Keshet, E. (1995). Vascular endothelial growth factor acts as a survival factor for newly formed retinal vessels and has implications for retinopathy of prematurity. *Nat. Med.* **1**, 1024–1028.

Armulik, A., Abramsson, A., and Betsholtz, C. (2005). Endothelial/pericyte interactions. *Circ. Res.* **97**, 512–523.

Armulik, A., Genove, G., Mae, M., Nisancioglu, M. H., Wallgard, E., Niaudet, C., He, L., Norlin, J., Lindblom, P., Strittmatter, K., Johansson, B. R., and Betsholtz, C. (2010). Pericytes regulate the blood–brain barrier. *Nature* **468**, 557–561.

Augustin, H. G., Koh, G. Y., Thurston, G., and Alitalo, K. (2009). Control of vascular morphogenesis and homeostasis through the angiopoietin–tie system. *Nat. Rev. Mol. Cell Biol.* **10**, 165–177.

Banks, R. E., Forbes, M. A., Kinsey, S. E., Stanley, A., Ingham, E., Walters, C., and Selby, P. J. (1998). Release of the angiogenic cytokine vascular endothelial growth factor (VEGF) from platelets: significance for VEGF measurements and cancer biology. *Br. J. Cancer* **77**, 956–964.

Bao, P., Kodra, A., Tomic-Canic, M., Golinko, M. S., Ehrlich, H. P., and Brem, H. (2009). The role of vascular endothelial growth factor in wound healing. *J. Surg. Res.* **153**, 347–358.

Batlle, E., Henderson, J. T., Beghtel, H., van den Born, M. M., Sancho, E., Huls, G., Meeldijk, J., Robertson, J., van de Wetering, M., Pawson, T., and Clevers, H. (2002). Beta-catenin and TCF mediate cell positioning in the intestinal epithelium by controlling the expression of EphB/ephrinB. *Cell* **111**, 251–263.

Benedito, R., Roca, C., Sorensen, I., Adams, S., Gossler, A., Fruttiger, M., and Adams, R. H. (2009). The Notch ligands Dll4 and Jagged1 have opposing effects on angiogenesis. *Cell* **137**, 1124–1135.

Bentley, K., Mariggi, G., Gerhardt, H., and Bates, P. A. (2009). Tipping the balance: robustness of tip cell selection, migration and fusion in angiogenesis. *PLoS Comput. Biol.* **5**, e1000549.

Bergers, G., Song, S., Meyer-Morse, N., Bergsland, E., and Hanahan, D. (2003). Benefits of targeting both pericytes and endothelial cells in the tumor vasculature with kinase inhibitors. *J. Clin. Invest.* **111**, 1287–1295.

Bong, Y. S., Lee, H. S., Carim-Todd, L., Mood, K., Nishanian, T. G., Tessarollo, L., and Daar, I. O. (2007). ephrinB1 signals from the cell surface to the nucleus by recruitment of STAT3. *Proc. Natl. Acad. Sci. U.S.A.* **104**, 17305–17310.

Boutros, M., and Mlodzik, M. (1999). Dishevelled: at the crossroads of divergent intracellular signaling pathways. *Mech. Dev.* **83**, 27–37.

Brantley-Sieders, D., Schmidt, S., Parker, M., and Chen, J. (2004). Eph receptor tyrosine kinases in tumor and tumor microenvironment. *Curr. Pharm. Des.* **10**, 3431–3442.

Bruckner, K., Pasquale, E. B., and Klein, R. (1997). Tyrosine phosphorylation of transmembrane ligands for Eph receptors. *Science* **275**, 1640–1643.

Castellvi, J., Garcia, A., de la Torre, J., Hernandez, J., Gil, A., Xercavins, J., and Ramon y Cajal, S. (2006). Ephrin B expression in epithelial ovarian neoplasms correlates with tumor differentiation and angiogenesis. *Hum. Pathol.* **37**, 883–889.

Chappell, J. C., Taylor, S. M., Ferrara, N., and Bautch, V. L. (2009). Local guidance of emerging vessel sprouts requires soluble Flt-1. *Dev. Cell* **17**, 377–386.

Chong, L. D., Park, E. K., Latimer, E., Friesel, R., and Daar, I. O. (2000). Fibroblast growth factor receptor-mediated rescue of x-ephrin B1-induced cell dissociation in Xenopus embryos. *Mol. Cell. Biol.* **20**, 724–734.

Chrencik, J. E., Brooun, A., Recht, M. I., Kraus, M. L., Koolpe, M., Kolatkar, A. R., Bruce, R. H., Martiny-Baron, G., Widmer, H., Pasquale, E. B., and Kuhn, P. (2006). Structure and thermodynamic characterization of the EphB4/Ephrin-B2 antagonist peptide complex reveals the determinants for receptor specificity. *Structure* **14**, 321–330.

Chrencik, J. E., Brooun, A., Recht, M. I., Nicola, G., Davis, L. K., Abagyan, R., Widmer, H., Pasquale, E. B., and Kuhn, P. (2007). Three-dimensional structure of the EphB2 receptor in complex with an antagonistic peptide reveals a novel mode of inhibition. *J. Biol. Chem.* **282,** 36505–36513.

Chung, A. S., Lee, J., and Ferrara, N. (2011). Targeting the tumour vasculature: insights from physiological angiogenesis. *Nat. Rev. Cancer* **10,** 505–514.

Cortina, C., Palomo-Ponce, S., Iglesias, M., Fernandez-Masip, J. L., Vivancos, A., Whissell, G., Huma, M., Peiro, N., Gallego, L., Jonkheer, S., Davy, A., Lloreta, J., *et al.* (2007). EphB–ephrin-B interactions suppress colorectal cancer progression by compartmentalizing tumor cells. *Nat. Genet.* **39,** 1376–1383.

Cowan, C. A., and Henkemeyer, M. (2001). The SH2/SH3 adaptor Grb4 transduces B-ephrin reverse signals. *Nature* **413,** 174–179.

Cowan, C. A., Yokoyama, N., Saxena, A., Chumley, M. J., Silvany, R. E., Baker, L. A., Srivastava, D., and Henkemeyer, M. (2004). Ephrin-B2 reverse signaling is required for axon pathfinding and cardiac valve formation but not early vascular development. *Dev. Biol.* **271,** 263–271.

Cowan, C. W., Shao, Y. R., Sahin, M., Shamah, S. M., Lin, M. Z., Greer, P. L., Gao, S., Griffith, E. C., Brugge, J. S., and Greenberg, M. E. (2005). Vav family GEFs link activated Ephs to endocytosis and axon guidance. *Neuron* **46,** 205–217.

Crisan, M., Yap, S., Casteilla, L., Chen, C. W., Corselli, M., Park, T. S., Andriolo, G., Sun, B., Zheng, B., Zhang, L., Norotte, C., Teng, P. N., *et al.* (2008). A perivascular origin for mesenchymal stem cells in multiple human organs. *Cell Stem Cell* **3,** 301–313.

Cuevas, P., Gutierrez-Diaz, J. A., Reimers, D., Dujovny, M., Diaz, F. G., and Ausman, J. I. (1984). Pericyte endothelial gap junctions in human cerebral capillaries. *Anat. Embryol. (Berl)* **170,** 155–159.

Damsky, C. H., Fitzgerald, M. L., and Fisher, S. J. (1992). Distribution patterns of extracellular matrix components and adhesion receptors are intricately modulated during first trimester cytotrophoblast differentiation along the invasive pathway, in vivo. *J. Clin. Invest.* **89,** 210–222.

Davalos, V., Dopeso, H., Castano, J., Wilson, A. J., Vilardell, F., Romero-Gimenez, J., Espin, E., Armengol, M., Capella, G., Mariadason, J. M., Aaltonen, L. A., Schwartz, S., Jr., *et al.* (2006). EPHB4 and survival of colorectal cancer patients. *Cancer Res.* **66,** 8943–8948.

Dayan, F., Mazure, N. M., Brahimi-Horn, M. C., and Pouyssegur, J. (2008). A dialogue between the hypoxia-inducible factor and the tumor microenvironment. *Cancer Microenviron.* **1,** 53–68.

Deroanne, C., Vouret-Craviari, V., Wang, B., and Pouyssegur, J. (2003). EphrinA1 inactivates integrin-mediated vascular smooth muscle cell spreading via the Rac/PAK pathway. *J. Cell Sci.* **116,** 1367–1376.

Dohn, M., Jiang, J., and Chen, X. (2001). Receptor tyrosine kinase EphA2 is regulated by p53-family proteins and induces apoptosis. *Oncogene* **20,** 6503–6515.

Dopeso, H., Mateo-Lozano, S., Mazzolini, R., Rodrigues, P., Lagares-Tena, L., Ceron, J., Romero, J., Esteves, M., Landolfi, S., Hernandez-Losa, J., Castano, J., Wilson, A. J., *et al.* (2009). The receptor tyrosine kinase EPHB4 has tumor suppressor activities in intestinal tumorigenesis. *Cancer Res.* **69,** 7430–7438.

Dravis, C., and Henkemeyer, M. (2011). Ephrin-B reverse signaling controls septation events at the embryonic midline through separate tyrosine phosphorylation-independent signaling avenues. *Dev. Biol.* **355,** 138–151.

Dumont, D. J., Gradwohl, G., Fong, G. H., Puri, M. C., Gertsenstein, M., Auerbach, A., and Breitman, M. L. (1994). Dominant-negative and targeted null mutations in the endothelial receptor tyrosine kinase, tek, reveal a critical role in vasculogenesis of the embryo. *Genes Dev.* **8,** 1897–1909.

Egawa, M., Yoshioka, S., Higuchi, T., Sato, Y., Tatsumi, K., Fujiwara, H., and Fujii, S. (2003). Ephrin B1 is expressed on human luteinizing granulosa cells in corpora lutea of the early luteal phase: the possible involvement of the B class Eph–ephrin system during corpus luteum formation. *J. Clin. Endocrinol. Metab.* **88,** 4384–4392.

Egea, J., and Klein, R. (2007). Bidirectional Eph–ephrin signaling during axon guidance. *Trends Cell Biol.* **17,** 230–238.

Erber, R., Eichelsbacher, U., Powajbo, V., Korn, T., Djonov, V., Lin, J., Hammes, H. P., Grobholz, R., Ullrich, A., and Vajkoczy, P. (2006). EphB4 controls blood vascular morphogenesis during postnatal angiogenesis. *EMBO J.* **25,** 628–641.

Ferrara, N., and Kerbel, R. S. (2005). Angiogenesis as a therapeutic target. *Nature* **438,** 967–974.

Ferrara, N., Gerber, H. P., and LeCouter, J. (2003). The biology of VEGF and its receptors. *Nat. Med.* **9,** 669–676.

Fisher, S. J., Cui, T. Y., Zhang, L., Hartman, L., Grahl, K., Zhang, G. Y., Tarpey, J., and Damsky, C. H. (1989). Adhesive and degradative properties of human placental cytotrophoblast cells in vitro. *J. Cell Biol.* **109,** 891–902.

Foo, S. S., Turner, C. J., Adams, S., Compagni, A., Aubyn, D., Kogata, N., Lindblom, P., Shani, M., Zicha, D., and Adams, R. H. (2006). Ephrin-B2 controls cell motility and adhesion during blood-vessel-wall assembly. *Cell* **124,** 161–173.

Fraisl, P., Mazzone, M., Schmidt, T., and Carmeliet, P. (2009). Regulation of angiogenesis by oxygen and metabolism. *Dev. Cell* **16,** 167–179.

Fridlender, Z. G., Sun, J., Kim, S., Kapoor, V., Cheng, G., Ling, L., Worthen, G. S., and Albelda, S. M. (2009). Polarization of tumor-associated neutrophil phenotype by TGF-beta: "N1" versus "N2" TAN. *Cancer Cell* **16,** 183–194.

Fukai, J., Yokote, H., Yamanaka, R., Arao, T., Nishio, K., and Itakura, T. (2008). EphA4 promotes cell proliferation and migration through a novel EphA4-FGFR1 signaling pathway in the human glioma U251 cell line. *Mol. Cancer Ther.* **7,** 2768–2778.

Fuller, T., Korff, T., Kilian, A., Dandekar, G., and Augustin, H. G. (2003). Forward EphB4 signaling in endothelial cells controls cellular repulsion and segregation from ephrinB2 positive cells. *J. Cell Sci.* **116,** 2461–2470.

Gale, N. W., Baluk, P., Pan, L., Kwan, M., Holash, J., DeChiara, T. M., McDonald, D. M., and Yancopoulos, G. D. (2001). Ephrin-B2 selectively marks arterial vessels and neovascularization sites in the adult, with expression in both endothelial and smooth-muscle cells. *Dev. Biol.* **230,** 151–160.

Gariano, R. F., and Gardner, T. W. (2005). Retinal angiogenesis in development and disease. *Nature* **438,** 960–966.

Georgakopoulos, A., Litterst, C., Ghersi, E., Baki, L., Xu, C., Serban, G., and Robakis, N. K. (2006). Metalloproteinase/Presenilin1 processing of ephrinB regulates EphB-induced Src phosphorylation and signaling. *EMBO J.* **25,** 1242–1252.

Gerety, S. S., and Anderson, D. J. (2002). Cardiovascular ephrinB2 function is essential for embryonic angiogenesis. *Development* **129,** 1397–1410.

Gerety, S. S., Wang, H. U., Chen, Z. F., and Anderson, D. J. (1999). Symmetrical mutant phenotypes of the receptor EphB4 and its specific transmembrane ligand ephrin-B2 in cardiovascular development. *Mol. Cell* **4,** 403–414.

Gerhardt, H., Golding, M., Fruttiger, M., Ruhrberg, C., Lundkvist, A., Abramsson, A., Jeltsch, M., Mitchell, C., Alitalo, K., Shima, D., and Betsholtz, C. (2003). VEGF guides angiogenic sprouting utilizing endothelial tip cell filopodia. *J. Cell Biol.* **161,** 1163–1177.

Hellstrom, M., Gerhardt, H., Kalen, M., Li, X., Eriksson, U., Wolburg, H., and Betsholtz, C. (2001). Lack of pericytes leads to endothelial hyperplasia and abnormal vascular morphogenesis. *J. Cell Biol.* **153,** 543–553.

Hellstrom, M., Phng, L. K., Hofmann, J. J., Wallgard, E., Coultas, L., Lindblom, P., Alva, J., Nilsson, A. K., Karlsson, L., Gaiano, N., Yoon, K., Rossant, J., *et al.* (2007). Dll4 signalling through Notch1 regulates formation of tip cells during angiogenesis. *Nature* **445**, 776–780.

Hendrix, M. J., Seftor, E. A., Hess, A. R., and Seftor, R. E. (2003a). Molecular plasticity of human melanoma cells. *Oncogene* **22**, 3070–3075.

Hendrix, M. J., Seftor, E. A., Hess, A. R., and Seftor, R. E. (2003b). Vasculogenic mimicry and tumour-cell plasticity: lessons from melanoma. *Nat. Rev. Cancer* **3**, 411–421.

Herbert, S. P., Huisken, J., Kim, T. N., Feldman, M. E., Houseman, B. T., Wang, R. A., Shokat, K. M., and Stainier, D. Y. (2009). Arterial-venous segregation by selective cell sprouting: an alternative mode of blood vessel formation. *Science* **326**, 294–298.

Himanen, J. P., and Nikolov, D. B. (2003). Eph signaling: a structural view. *Trends Neurosci.* **26**, 46–51.

Himanen, J. P., Rajashankar, K. R., Lackmann, M., Cowan, C. A., Henkemeyer, M., and Nikolov, D. B. (2001). Crystal structure of an Eph receptor–ephrin complex. *Nature* **414**, 933–938.

Holland, S. J., Gale, N. W., Gish, G. D., Roth, R. A., Songyang, Z., Cantley, L. C., Henkemeyer, M., Yancopoulos, G. D., and Pawson, T. (1997). Juxtamembrane tyrosine residues couple the Eph family receptor EphB2/Nuk to specific SH2 domain proteins in neuronal cells. *EMBO J.* **16**, 3877–3888.

Huang, F., Reeves, K., Han, X., Fairchild, C., Platero, S., Wong, T. W., Lee, F., Shaw, P., and Clark, E. (2007). Identification of candidate molecular markers predicting sensitivity in solid tumors to dasatinib: rationale for patient selection. *Cancer Res.* **67**, 2226–2238.

Hunkapiller, N. M., Gasperowicz, M., Kapidzic, M., Plaks, V., Maltepe, E., Kitajewski, J., Cross, J. C., and Fisher, S. J. (2011). A role for Notch signaling in trophoblast endovascular invasion and in the pathogenesis of pre-eclampsia. *Development* **138**, 2987–2998.

Huynh-Do, U., Vindis, C., Liu, H., Cerretti, D. P., McGrew, J. T., Enriquez, M., Chen, J., and Daniel, T. O. (2002). Ephrin-B1 transduces signals to activate integrin-mediated migration, attachment and angiogenesis. *J. Cell Sci.* **115**, 3073–3081.

Irie, F., Okuno, M., Pasquale, E. B., and Yamaguchi, Y. (2005). EphrinB-EphB signalling regulates clathrin-mediated endocytosis through tyrosine phosphorylation of synaptojanin 1. *Nat. Cell Biol.* **7**, 501–509.

Jager, R. D., Mieler, W. F., and Miller, J. W. (2008). Age-related macular degeneration. *N. Engl. J. Med.* **358**, 2606–2617.

Jain, R. K. (2005). Normalization of tumor vasculature: an emerging concept in antiangiogenic therapy. *Science* **307**, 58–62.

Jin, Y. J., Wang, J., Qiao, C., Hei, T. K., Brandt-Rauf, P. W., and Yin, Y. (2006). A novel mechanism for p53 to regulate its target gene ECK in signaling apoptosis. *Mol. Cancer Res.* **4**, 769–778.

Kalo, M. S., and Pasquale, E. B. (1999). Signal transfer by Eph receptors. *Cell Tissue Res.* **298**, 1–9.

Kemp, H. A., Cooke, J. E., and Moens, C. B. (2009). EphA4 and EfnB2a maintain rhombomere coherence by independently regulating intercalation of progenitor cells in the zebrafish neural keel. *Dev. Biol.* **327**, 313–326.

Kertesz, N., Krasnoperov, V., Reddy, R., Leshanski, L., Kumar, S. R., Zozulya, S., and Gill, P. S. (2006). The soluble extracellular domain of EphB4 (sEphB4) antagonizes EphB4-EphrinB2 interaction, modulates angiogenesis, and inhibits tumor growth. *Blood* **107**, 2330–2338.

Kim, K. J., Li, B., Winer, J., Armanini, M., Gillett, N., Phillips, H. S., and Ferrara, N. (1993). Inhibition of vascular endothelial growth factor-induced angiogenesis suppresses tumour growth in vivo. *Nature* **362**, 841–844.

Kim, I., Ryu, Y. S., Kwak, H. J., Ahn, S. Y., Oh, J. L., Yancopoulos, G. D., Gale, N. W., and Koh, G. Y. (2002). EphB ligand, ephrinB2, suppresses the VEGF- and angiopoietin 1-induced Ras/mitogen-activated protein kinase pathway in venous endothelial cells. *FASEB J.* **16,** 1126–1128.

Klein, R. (2009). Bidirectional modulation of synaptic functions by Eph/ephrin signaling. *Nat. Neurosci.* **12,** 15–20.

Knoll, B., and Drescher, U. (2004). Src family kinases are involved in EphA receptor-mediated retinal axon guidance. *J. Neurosci.* **24,** 6248–6257.

Koolpe, M., Burgess, R., Dail, M., and Pasquale, E. B. (2005). EphB receptor-binding peptides identified by phage display enable design of an antagonist with ephrin-like affinity. *J. Biol. Chem.* **280,** 17301–17311.

Korff, T., Braun, J., Pfaff, D., Augustin, H. G., and Hecker, M. (2008). Role of ephrinB2 expression in endothelial cells during arteriogenesis: impact on smooth muscle cell migration and monocyte recruitment. *Blood* **112,** 73–81.

Krasnoperov, V., Kumar, S. R., Ley, E., Li, X., Scehnet, J., Liu, R., Zozulya, S., and Gill, P. S. (2010). Novel EphB4 monoclonal antibodies modulate angiogenesis and inhibit tumor growth. *Am. J. Pathol.* **176,** 2029–2038.

Kullander, K., and Klein, R. (2002). Mechanisms and functions of Eph and ephrin signalling. *Nat. Rev. Mol. Cell Biol.* **3,** 475–486.

Kumar, S. R., Masood, R., Spannuth, W. A., Singh, J., Scehnet, J., Kleiber, G., Jennings, N., Deavers, M., Krasnoperov, V., Dubeau, L., Weaver, F. A., Sood, A. K., *et al.* (2007). The receptor tyrosine kinase EphB4 is overexpressed in ovarian cancer, provides survival signals and predicts poor outcome. *Br. J. Cancer* **96,** 1083–1091.

Kumar, S. R., Scehnet, J. S., Ley, E. J., Singh, J., Krasnoperov, V., Liu, R., Manchanda, P. K., Ladner, R. D., Hawes, D., Weaver, F. A., Beart, R. W., Singh, G., *et al.* (2009). Preferential induction of EphB4 over EphB2 and its implication in colorectal cancer progression. *Cancer Res.* **69,** 3736–3745.

Kvanta, A., Algvere, P. V., Berglin, L., and Seregard, S. (1996). Subfoveal fibrovascular membranes in age-related macular degeneration express vascular endothelial growth factor. *Invest. Ophthalmol. Vis. Sci.* **37,** 1929–1934.

Lampugnani, M. G., Orsenigo, F., Gagliani, M. C., Tacchetti, C., and Dejana, E. (2006). Vascular endothelial cadherin controls VEGFR-2 internalization and signaling from intracellular compartments. *J. Cell Biol.* **174,** 593–604.

Lanahan, A. A., Hermans, K., Claes, F., Kerley-Hamilton, J. S., Zhuang, Z. W., Giordano, F. J., Carmeliet, P., and Simons, M. (2010). VEGF receptor 2 endocytic trafficking regulates arterial morphogenesis. *Dev. Cell* **18,** 713–724.

Lauterbach, J., and Klein, R. (2006). Release of full-length EphB2 receptors from hippocampal neurons to cocultured glial cells. *J. Neurosci.* **26,** 11575–11581.

Lee, H. S., Bong, Y. S., Moore, K. B., Soria, K., Moody, S. A., and Daar, I. O. (2006). Dishevelled mediates ephrinB1 signalling in the eye field through the planar cell polarity pathway. *Nat. Cell Biol.* **8,** 55–63.

Lee, S., Chen, T. T., Barber, C. L., Jordan, M. C., Murdock, J., Desai, S., Ferrara, N., Nagy, A., Roos, K. P., and Iruela-Arispe, M. L. (2007). Autocrine VEGF signaling is required for vascular homeostasis. *Cell* **130,** 691–703.

Lee, H. S., Mood, K., Battu, G., Ji, Y. J., Singh, A., and Daar, I. O. (2009). Fibroblast growth factor receptor-induced phosphorylation of ephrinB1 modulates its interaction with Dishevelled. *Mol. Biol. Cell* **20,** 124–133.

Li, J. L., Sainson, R. C., Oon, C. E., Turley, H., Leek, R., Sheldon, H., Bridges, E., Shi, W., Snell, C., Bowden, E. T., Wu, H., Chowdhury, P. S., *et al.* (2011). DLL4-Notch signaling mediates tumor resistance to anti-VEGF therapy in vivo. *Cancer Res.* **71,** 6073–6083.

Librach, C. L., Werb, Z., Fitzgerald, M. L., Chiu, K., Corwin, N. M., Esteves, R. A., Grobelny, D., Galardy, R., Damsky, C. H., and Fisher, S. J. (1991). 92-kD type IV collagenase mediates invasion of human cytotrophoblasts. *J. Cell Biol.* **113**, 437–449.

Lin, D., Gish, G. D., Songyang, Z., and Pawson, T. (1999). The carboxyl terminus of B class ephrins constitutes a PDZ domain binding motif. *J. Biol. Chem.* **274**, 3726–3733.

Liu, Y., Wada, R., Yamashita, T., Mi, Y., Deng, C. X., Hobson, J. P., Rosenfeldt, H. M., Nava, V. E., Chae, S. S., Lee, M. J., Liu, C. H., Hla, T., *et al.* (2000). Edg-1, the G protein-coupled receptor for sphingosine-1-phosphate, is essential for vascular maturation. *J. Clin. Invest.* **106**, 951–961.

Lobov, I. B., Renard, R. A., Papadopoulos, N., Gale, N. W., Thurston, G., Yancopoulos, G. D., and Wiegand, S. J. (2007). Delta-like ligand 4 (Dll4) is induced by VEGF as a negative regulator of angiogenic sprouting. *Proc. Natl. Acad. Sci. U.S.A.* **104**, 3219–3224.

Lopez, P. F., Sippy, B. D., Lambert, H. M., Thach, A. B., and Hinton, D. R. (1996). Transdifferentiated retinal pigment epithelial cells are immunoreactive for vascular endothelial growth factor in surgically excised age-related macular degeneration-related choroidal neovascular membranes. *Invest. Ophthalmol. Vis. Sci.* **37**, 855–868.

Lu, Q., Sun, E. E., Klein, R. S., and Flanagan, J. G. (2001). Ephrin-B reverse signaling is mediated by a novel PDZ-RGS protein and selectively inhibits G protein-coupled chemoattraction. *Cell* **105**, 69–79.

Maekawa, H., Oike, Y., Kanda, S., Ito, Y., Yamada, Y., Kurihara, H., Nagai, R., and Suda, T. (2003). Ephrin-B2 induces migration of endothelial cells through the phosphatidylinositol-3 kinase pathway and promotes angiogenesis in adult vasculature. *Arterioscler. Thromb. Vasc. Biol.* **23**, 2008–2014.

Makinen, T., Adams, R. H., Bailey, J., Lu, Q., Ziemiecki, A., Alitalo, K., Klein, R., and Wilkinson, G. A. (2005). PDZ interaction site in ephrinB2 is required for the remodeling of lymphatic vasculature. *Genes Dev.* **19**, 397–410.

Mann, F., Miranda, E., Weinl, C., Harmer, E., and Holt, C. E. (2003). B-type Eph receptors and ephrins induce growth cone collapse through distinct intracellular pathways. *J. Neurobiol.* **57**, 323–336.

Mantovani, A., Sozzani, S., Locati, M., Allavena, P., and Sica, A. (2002). Macrophage polarization: tumor-associated macrophages as a paradigm for polarized M2 mononuclear phagocytes. *Trends Immunol.* **23**, 549–555.

Mao, W., Luis, E., Ross, S., Silva, J., Tan, C., Crowley, C., Chui, C., Franz, G., Senter, P., Koeppen, H., and Polakis, P. (2004). EphB2 as a therapeutic antibody drug target for the treatment of colorectal cancer. *Cancer Res.* **64**, 781–788.

Margolis, S. S., Salogiannis, J., Lipton, D. M., Mandel-Brehm, C., Wills, Z. P., Mardinly, A. R., Hu, L., Greer, P. L., Bikoff, J. B., Ho, H. Y., Soskis, M. J., Sahin, M., *et al.* (2010). EphB-mediated degradation of the RhoA GEF Ephexin5 relieves a developmental brake on excitatory synapse formation. *Cell* **143**, 442–455.

Marston, D. J., Dickinson, S., and Nobes, C. D. (2003). Rac-dependent trans-endocytosis of ephrinBs regulates Eph–ephrin contact repulsion. *Nat. Cell Biol.* **5**, 879–888.

Masood, R., Xia, G., Smith, D. L., Scalia, P., Still, J. G., Tulpule, A., and Gill, P. S. (2005). Ephrin B2 expression in Kaposi sarcoma is induced by human herpesvirus type 8: phenotype switch from venous to arterial endothelium. *Blood* **105**, 1310–1318.

Meyer, S., Hafner, C., Guba, M., Flegel, S., Geissler, E. K., Becker, B., Koehl, G. E., Orso, E., Landthaler, M., and Vogt, T. (2005). Ephrin-B2 overexpression enhances integrin-mediated ECM-attachment and migration of B16 melanoma cells. *Int. J. Oncol.* **27**, 1197–1206.

Miao, H., Strebhardt, K., Pasquale, E. B., Shen, T. L., Guan, J. L., and Wang, B. (2005). Inhibition of integrin-mediated cell adhesion but not directional cell migration requires catalytic activity of EphB3 receptor tyrosine kinase. Role of Rho family small GTPases. *J. Biol. Chem.* **280**, 923–932.

Miao, H., Li, D. Q., Mukherjee, A., Guo, H., Petty, A., Cutter, J., Basilion, J. P., Sedor, J., Wu, J., Danielpour, D., Sloan, A. E., Cohen, M. L., *et al.* (2009). EphA2 mediates ligand-dependent inhibition and ligand-independent promotion of cell migration and invasion via a reciprocal regulatory loop with Akt. *Cancer Cell* **16**, 9–20.

Miller, J. W., Adamis, A. P., Shima, D. T., D'Amore, P. A., Moulton, R. S., O'Reilly, M. S., Folkman, J., Dvorak, H. F., Brown, L. F., Berse, B., *et al.* (1994). Vascular endothelial growth factor/vascular permeability factor is temporally and spatially correlated with ocular angiogenesis in a primate model. *Am. J. Pathol.* **145**, 574–584.

Mohle, R., Green, D., Moore, M. A., Nachman, R. L., and Rafii, S. (1997). Constitutive production and thrombin-induced release of vascular endothelial growth factor by human megakaryocytes and platelets. *Proc. Natl. Acad. Sci. U.S.A.* **94**, 663–668.

Murai, K. K., and Pasquale, E. B. (2003). 'Eph'ective signaling: forward, reverse and crosstalk. *J. Cell Sci.* **116**, 2823–2832.

Murai, K. K., and Pasquale, E. B. (2005). New exchanges in eph-dependent growth cone dynamics. *Neuron* **46**, 161–163.

Murdoch, C., Muthana, M., Coffelt, S. B., and Lewis, C. E. (2008). The role of myeloid cells in the promotion of tumour angiogenesis. *Nat. Rev. Cancer* **8**, 618–631.

Murohara, T., Horowitz, J. R., Silver, M., Tsurumi, Y., Chen, D., Sullivan, A., and Isner, J. M. (1998). Vascular endothelial growth factor/vascular permeability factor enhances vascular permeability via nitric oxide and prostacyclin. *Circulation* **97**, 99–107.

Nishimura, T., Yamaguchi, T., Kato, K., Yoshizawa, M., Nabeshima, Y., Ohno, S., Hoshino, M., and Kaibuchi, K. (2005). PAR-6-PAR-3 mediates Cdc42-induced Rac activation through the Rac GEFs STEF/Tiam1. *Nat. Cell Biol.* **7**, 270–277.

Noguera-Troise, I., Daly, C., Papadopoulos, N. J., Coetzee, S., Boland, P., Gale, N. W., Lin, H. C., Yancopoulos, G. D., and Thurston, G. (2006). Blockade of Dll4 inhibits tumour growth by promoting non-productive angiogenesis. *Nature* **444**, 1032–1037.

Noren, N. K., and Pasquale, E. B. (2004). Eph receptor–ephrin bidirectional signals that target Ras and Rho proteins. *Cell. Signal.* **16**, 655–666.

Noren, N. K., and Pasquale, E. B. (2007). Paradoxes of the EphB4 receptor in cancer. *Cancer Res.* **67**, 3994–3997.

Noren, N. K., Foos, G., Hauser, C. A., and Pasquale, E. B. (2006). The EphB4 receptor suppresses breast cancer cell tumorigenicity through an Abl–Crk pathway. *Nat. Cell Biol.* **8**, 815–825.

Ogita, H., Kunimoto, S., Kamioka, Y., Sawa, H., Masuda, M., and Mochizuki, N. (2003). EphA4-mediated Rho activation via Vsm-RhoGEF expressed specifically in vascular smooth muscle cells. *Circ. Res.* **93**, 23–31.

Palmer, A., Zimmer, M., Erdmann, K. S., Eulenburg, V., Porthin, A., Heumann, R., Deutsch, U., and Klein, R. (2002). EphrinB phosphorylation and reverse signaling: regulation by Src kinases and PTP-BL phosphatase. *Mol. Cell* **9**, 725–737.

Pasquale, E. B. (2010). Eph receptors and ephrins in cancer: bidirectional signalling and beyond. *Nat. Rev. Cancer* **10**, 165–180.

Pitulescu, M. E., and Adams, R. H. (2010). Eph/ephrin molecules—a hub for signaling and endocytosis. *Genes Dev.* **24**, 2480–2492.

Pugh, C. W., and Ratcliffe, P. J. (2003). The von Hippel-Lindau tumor suppressor, hypoxia-inducible factor-1 (HIF-1) degradation, and cancer pathogenesis. *Semin. Cancer Biol.* **13**, 83–89.

Red-Horse, K., Kapidzic, M., Zhou, Y., Feng, K. T., Singh, H., and Fisher, S. J. (2005). EPHB4 regulates chemokine-evoked trophoblast responses: a mechanism for incorporating the human placenta into the maternal circulation. *Development* **132**, 4097–4106.

Ricci-Vitiani, L., Pallini, R., Biffoni, M., Todaro, M., Invernici, G., Cenci, T., Maira, G., Parati, E. A., Stassi, G., Larocca, L. M., and De Maria, R. (2010). Tumour vascularization via endothelial differentiation of glioblastoma stem-like cells. *Nature* **468**, 824–828.

Ridgway, J., Zhang, G., Wu, Y., Stawicki, S., Liang, W. C., Chanthery, Y., Kowalski, J., Watts, R. J., Callahan, C., Kasman, I., Singh, M., Chien, M., *et al.* (2006). Inhibition of Dll4 signalling inhibits tumour growth by deregulating angiogenesis. *Nature* **444,** 1083–1087.

Ruhe, J. E., Streit, S., Hart, S., Wong, C. H., Specht, K., Knyazev, P., Knyazeva, T., Tay, L. S., Loo, H. L., Foo, P., Wong, W., Pok, S., *et al.* (2007). Genetic alterations in the tyrosine kinase transcriptome of human cancer cell lines. *Cancer Res.* **67,** 11368–11376.

Ruhrberg, C., Gerhardt, H., Golding, M., Watson, R., Ioannidou, S., Fujisawa, H., Betsholtz, C., and Shima, D. T. (2002). Spatially restricted patterning cues provided by heparin-binding VEGF-A control blood vessel branching morphogenesis. *Genes Dev.* **16,** 2684–2698.

Sahin, M., Greer, P. L., Lin, M. Z., Poucher, H., Eberhart, J., Schmidt, S., Wright, T. M., Shamah, S. M., O'Connell, S., Cowan, C. W., Hu, L., Goldberg, J. L., *et al.* (2005). Eph-dependent tyrosine phosphorylation of ephexin1 modulates growth cone collapse. *Neuron* **46,** 191–204.

Salvucci, O., de la Luz Sierra, M., Martina, J. A., McCormick, P. J., and Tosato, G. (2006). EphB2 and EphB4 receptors forward signaling promotes SDF-1-induced endothelial cell chemotaxis and branching remodeling. *Blood* **108,** 2914–2922.

Salvucci, O., Maric, D., Economopoulou, M., Sakakibara, S., Merlin, S., Follenzi, A., and Tosato, G. (2009). EphrinB reverse signaling contributes to endothelial and mural cell assembly into vascular structures. *Blood* **114,** 1707–1716.

Sato, T. N., Tozawa, Y., Deutsch, U., Wolburg-Buchholz, K., Fujiwara, Y., Gendron-Maguire, M., Gridley, T., Wolburg, H., Risau, W., and Qin, Y. (1995). Distinct roles of the receptor tyrosine kinases Tie-1 and Tie-2 in blood vessel formation. *Nature* **376,** 70–74.

Sawamiphak, S., Ritter, M., and Acker-Palmer, A. (2010a). Preparation of retinal explant cultures to study ex vivo tip endothelial cell responses. *Nat. Protoc.* **5,** 1659–1665.

Sawamiphak, S., Seidel, S., Essmann, C. L., Wilkinson, G. A., Pitulescu, M. E., Acker, T., and Acker-Palmer, A. (2010b). Ephrin-B2 regulates VEGFR2 function in developmental and tumour angiogenesis. *Nature* **465,** 487–491.

Scehnet, J. S., Ley, E. J., Krasnoperov, V., Liu, R., Manchanda, P. K., Sjoberg, E., Kostecke, A. P., Gupta, S., Kumar, S. R., and Gill, P. S. (2009). The role of Ephs, Ephrins, and growth factors in Kaposi sarcoma and implications of EphrinB2 blockade. *Blood* **113,** 254–263.

Schruefer, R., Sulyok, S., Schymeinsky, J., Peters, T., Scharffetter-Kochanek, K., and Walzog, B. (2006). The proangiogenic capacity of polymorphonuclear neutrophils delineated by microarray technique and by measurement of neovascularization in wounded skin of CD18-deficient mice. *J. Vasc. Res.* **43,** 1–11.

Schultz, G. S., and Wysocki, A. (2009). Interactions between extracellular matrix and growth factors in wound healing. *Wound Repair Regen.* **17,** 153–162.

Segura, I., Essmann, C. L., Weinges, S., and Acker-Palmer, A. (2007). Grb4 and GIT1 transduce ephrinB reverse signals modulating spine morphogenesis and synapse formation. *Nat. Neurosci.* **10,** 301–310.

Semenza, G. L. (2003). Targeting HIF-1 for cancer therapy. *Nat. Rev. Cancer* **3,** 721–732.

Semenza, G. L. (2011). Regulation of metabolism by hypoxia-inducible factor 1. *Cold Spring Harb. Symp. Quant. Biol.* Jul 22. [Epub ahead of print].

Shamah, S. M., Lin, M. Z., Goldberg, J. L., Estrach, S., Sahin, M., Hu, L., Bazalakova, M., Neve, R. L., Corfas, G., Debant, A., and Greenberg, M. E. (2001). EphA receptors regulate growth cone dynamics through the novel guanine nucleotide exchange factor ephexin. *Cell* **105,** 233–244.

Shin, D., Garcia-Cardena, G., Hayashi, S., Gerety, S., Asahara, T., Stavrakis, G., Isner, J., Folkman, J., Gimbrone, M. A., Jr., and Anderson, D. J. (2001). Expression of ephrinB2 identifies a stable genetic difference between arterial and venous vascular smooth muscle as well as endothelial cells, and marks subsets of microvessels at sites of adult neovascularization. *Dev. Biol.* **230,** 139–150.

Shojaei, F., Wu, X., Malik, A. K., Zhong, C., Baldwin, M. E., Schanz, S., Fuh, G., Gerber, H. P., and Ferrara, N. (2007). Tumor refractoriness to anti-VEGF treatment is mediated by CD11b +Gr1+ myeloid cells. *Nat. Biotechnol.* **25,** 911–920.

Shojaei, F., Wu, X., Qu, X., Kowanetz, M., Yu, L., Tan, M., Meng, Y. G., and Ferrara, N. (2009). G-CSF-initiated myeloid cell mobilization and angiogenesis mediate tumor refractoriness to anti-VEGF therapy in mouse models. *Proc. Natl. Acad. Sci. U.S.A.* **106,** 6742–6747.

Smith, L. E., Wesolowski, E., McLellan, A., Kostyk, S. K., D'Amato, R., Sullivan, R., and D'Amore, P. A. (1994). Oxygen-induced retinopathy in the mouse. *Invest. Ophthalmol. Vis. Sci.* **35,** 101–111.

Smith, F. M., Vearing, C., Lackmann, M., Treutlein, H., Himanen, J., Chen, K., Saul, A., Nikolov, D., and Boyd, A. W. (2004). Dissecting the EphA3/Ephrin-A5 interactions using a novel functional mutagenesis screen. *J. Biol. Chem.* **279,** 9522–9531.

Stein, E., Schoecklmann, H., and Daniel, T. O. (1997). Eph family receptors and ligands in vascular cell targeting and assembly. *Trends Cardiovasc. Med.* **7,** 329–334.

Steinle, J. J., Meininger, C. J., Forough, R., Wu, G., Wu, M. H., and Granger, H. J. (2002). Eph B4 receptor signaling mediates endothelial cell migration and proliferation via the phosphatidylinositol 3-kinase pathway. *J. Biol. Chem.* **277,** 43830–43835.

Steinle, J. J., Meininger, C. J., Chowdhury, U., Wu, G., and Granger, H. J. (2003). Role of ephrin B2 in human retinal endothelial cell proliferation and migration. *Cell. Signal.* **15,** 1011–1017.

Stenzel, D., Lundkvist, A., Sauvaget, D., Busse, M., Graupera, M., van der Flier, A., Wijelath, E. S., Murray, J., Sobel, M., Costell, M., Takahashi, S., Fassler, R., *et al.* (2011). Integrin-dependent and -independent functions of astrocytic fibronectin in retinal angiogenesis. *Development* **138,** 4451–4463.

Suchting, S., Freitas, C., le Noble, F., Benedito, R., Breant, C., Duarte, A., and Eichmann, A. (2007). The Notch ligand Delta-like 4 negatively regulates endothelial tip cell formation and vessel branching. *Proc. Natl. Acad. Sci. U.S.A.* **104,** 3225–3230.

Suri, C., Jones, P. F., Patan, S., Bartunkova, S., Maisonpierre, P. C., Davis, S., Sato, T. N., and Yancopoulos, G. D. (1996). Requisite role of angiopoietin-1, a ligand for the TIE2 receptor, during embryonic angiogenesis. *Cell* **87,** 1171–1180.

Suri, C., McClain, J., Thurston, G., McDonald, D. M., Zhou, H., Oldmixon, E. H., Sato, T. N., and Yancopoulos, G. D. (1998). Increased vascularization in mice overexpressing angiopoietin-1. *Science* **282,** 468–471.

Tanaka, M., Kamo, T., Ota, S., and Sugimura, H. (2003). Association of Dishevelled with Eph tyrosine kinase receptor and ephrin mediates cell repulsion. *EMBO J.* **22,** 847–858.

Tanaka, M., Ohashi, R., Nakamura, R., Shinmura, K., Kamo, T., Sakai, R., and Sugimura, H. (2004). Tiam1 mediates neurite outgrowth induced by ephrin-B1 and EphA2. *EMBO J.* **23,** 1075–1088.

Tang, M., Chen, B., Lin, T., Li, Z., Pardo, C., Pampo, C., Chen, J., Lien, C. L., Wu, L., Ai, L., Wang, H., Yao, K., *et al.* (2011). Restraint of angiogenesis by zinc finger transcription factor CTCF-dependent chromatin insulation. *Proc. Natl. Acad. Sci. U.S.A.* **108,** 15231–15236.

Thurston, G., Suri, C., Smith, K., McClain, J., Sato, T. N., Yancopoulos, G. D., and McDonald, D. M. (1999). Leakage-resistant blood vessels in mice transgenically overexpressing angiopoietin-1. *Science* **286,** 2511–2514.

Toth, J., Cutforth, T., Gelinas, A. D., Bethoney, K. A., Bard, J., and Harrison, C. J. (2001). Crystal structure of an ephrin ectodomain. *Dev. Cell* **1,** 83–92.

Vihanto, M. M., Vindis, C., Djonov, V., Cerretti, D. P., and Huynh-Do, U. (2006). Caveolin-1 is required for signaling and membrane targeting of EphB1 receptor tyrosine kinase. *J. Cell Sci.* **119,** 2299–2309.

Wang, H. U., Chen, Z. F., and Anderson, D. J. (1998). Molecular distinction and angiogenic interaction between embryonic arteries and veins revealed by ephrin-B2 and its receptor Eph-B4. *Cell* **93,** 741–753.

Wang, R., Chadalavada, K., Wilshire, J., Kowalik, U., Hovinga, K. E., Geber, A., Fligelman, B., Leversha, M., Brennan, C., and Tabar, V. (2010a). Glioblastoma stem-like cells give rise to tumour endothelium. *Nature* **468,** 829–833.

Wang, Y., Nakayama, M., Pitulescu, M. E., Schmidt, T. S., Bochenek, M. L., Sakakibara, A., Adams, S., Davy, A., Deutsch, U., Luthi, U., Barberis, A., Benjamin, L. E., *et al.* (2010b). Ephrin-B2 controls VEGF-induced angiogenesis and lymphangiogenesis. *Nature* **465,** 483–486.

Wybenga-Groot, L. E., Baskin, B., Ong, S. H., Tong, J., Pawson, T., and Sicheri, F. (2001). Structural basis for autoinhibition of the Ephb2 receptor tyrosine kinase by the unphosphorylated juxtamembrane region. *Cell* **106,** 745–757.

Xia, G., Kumar, S. R., Masood, R., Zhu, S., Reddy, R., Krasnoperov, V., Quinn, D. I., Henshall, S. M., Sutherland, R. L., Pinski, J. K., Daneshmand, S., Buscarini, M., *et al.* (2005). EphB4 expression and biological significance in prostate cancer. *Cancer Res.* **65,** 4623–4632.

Xia, G., Kumar, S. R., Stein, J. P., Singh, J., Krasnoperov, V., Zhu, S., Hassanieh, L., Smith, D. L., Buscarini, M., Broek, D., Quinn, D. I., Weaver, F. A., *et al.* (2006). EphB4 receptor tyrosine kinase is expressed in bladder cancer and provides signals for cell survival. *Oncogene* **25,** 769–780.

Xu, N. J., and Henkemeyer, M. (2009). Ephrin-B3 reverse signaling through Grb4 and cytoskeletal regulators mediates axon pruning. *Nat. Neurosci.* **12,** 268–276.

Xu, Y., Zagoura, D., Keck, C., and Pietrowski, D. (2006). Expression of Eph receptor tyrosine kinases and their ligands in human Granulosa lutein cells and human umbilical vein endothelial cells. *Exp. Clin. Endocrinol. Diabetes* **114,** 590–595.

Xu, Z., Jin, H., and Qian, Q. (2009). Humanized anti-EphB4 antibodies for the treatment of carcinomas and vasculogenesis-related diseases. *Expert Opin. Ther. Pat.* **19,** 1035–1037.

Yang, N. Y., Pasquale, E. B., Owen, L. B., and Ethell, I. M. (2006). The EphB4 receptor-tyrosine kinase promotes the migration of melanoma cells through Rho-mediated actin cytoskeleton reorganization. *J. Biol. Chem.* **281,** 32574–32586.

Yu, J., Zhang, L., Hwang, P. M., Rago, C., Kinzler, K. W., and Vogelstein, B. (1999). Identification and classification of p53-regulated genes. *Proc. Natl. Acad. Sci. U.S.A.* **96,** 14517–14522.

Yuan, K., Jin, Y. T., and Lin, M. T. (2000). Expression of Tie-2, angiopoietin-1, angiopoietin-2, ephrinB2 and EphB4 in pyogenic granuloma of human gingiva implicates their roles in inflammatory angiogenesis. *J. Periodontal Res.* **35,** 165–171.

Yuan, K., Hong, T. M., Chen, J. J., Tsai, W. H., and Lin, M. T. (2004). Syndecan-1 up-regulated by ephrinB2/EphB4 plays dual roles in inflammatory angiogenesis. *Blood* **104,** 1025–1033.

Zhong, T. P., Childs, S., Leu, J. P., and Fishman, M. C. (2001). Gridlock signalling pathway fashions the first embryonic artery. *Nature* **414,** 216–220.

Zhou, Y., Fisher, S. J., Janatpour, M., Genbacev, O., Dejana, E., Wheelock, M., and Damsky, C. H. (1997). Human cytotrophoblasts adopt a vascular phenotype as they differentiate. A strategy for successful endovascular invasion? *J. Clin. Invest.* **99,** 2139–2151.

Zhou, Y., Bellingard, V., Feng, K. T., McMaster, M., and Fisher, S. J. (2003). Human cytotrophoblasts promote endothelial survival and vascular remodeling through secretion of Ang2, PlGF, and VEGF-C. *Dev. Biol.* **263,** 114–125.

Zhuang, G., Brantley-Sieders, D. M., Vaught, D., Yu, J., Xie, L., Wells, S., Jackson, D., Muraoka-Cook, R., Arteaga, C., and Chen, J. (2010). Elevation of receptor tyrosine kinase EphA2 mediates resistance to trastuzumab therapy. *Cancer Res.* **70,** 299–308.

Zimmer, M., Palmer, A., Kohler, J., and Klein, R. (2003). EphB–ephrinB bi-directional endocytosis terminates adhesion allowing contact mediated repulsion. *Nat. Cell Biol.* **5,** 869–878.

Zisch, A. H., Pazzagli, C., Freeman, A. L., Schneller, M., Hadman, M., Smith, J. W., Ruoslahti, E., and Pasquale, E. B. (2000). Replacing two conserved tyrosines of the EphB2 receptor with glutamic acid prevents binding of SH2 domains without abrogating kinase activity and biological responses. *Oncogene* **19,** 177–187.

Semaphorin Signals Tweaking the Tumor Microenvironment

Claudia Muratori and Luca Tamagnone

University of Torino Medical School, Institute for Cancer Research (IRCC), Candiolo, Turin, Italy

Solid tumors not only comprise malignant cells but also other nonmalignant cell types, forming a unique microenvironment that can strongly influence the behavior of tumor cells. Recent advances in the understanding of cancer biology have highlighted the functional role of semaphorins. In fact, semaphorins form a family of molecular signals known to guide and control cell migration during embryo development and in adults. Tumor cells express semaphorins as well as their receptors, plexins and neuropilins. It has been shown that semaphorin signaling can regulate tumor cell behavior. Moreover, semaphorins are important regulators of tumor angiogenesis. Conversely, very little is known about the functional relevance of semaphorin signals for tumor-infiltrating stromal cells, such as leukocytes. In this chapter, we review the current knowledge on the functional role of semaphorins in cancer progression, and we focus on the emerging role of semaphorins in mediating the cross talk between tumor cells and different tumor stromal cells.

I. SEMAPHORINS, THEIR RECEPTORS AND SEMAPHORIN-DEPENDENT SIGNALING PATHWAYS

Semaphorins are a large family of secreted and membrane-bound glycoproteins that were initially implicated in axon guidance and nervous system development (Kolodkin *et al.*, 1993; Luo *et al.*, 1993). However, it is now clear that they are widely expressed outside of the nervous system and are

Advances in CANCER RESEARCH, Volume 114

0065-230X/12 $35.00
DOI: 10.1016/B978-0-12-386503-8.00003-X

involved in several processes including tissue morphogenesis, immune response, and cancer progression (Tamagnone and Comoglio, 2004; Zhou *et al.*, 2008).

Semaphorins are characterized by containing a large sema domain, bearing seven-blade beta-propeller structure, which contains critical residues for receptor binding and specificity (Gherardi *et al.*, 2004; Janssen *et al.*, 2010; Koppel *et al.*, 1997; Nogi *et al.*, 2010). This is followed by a cysteine rich PSI (plexin, semaphorin, and integrin) domain (Bork *et al.*, 1999). Semaphorins have been grouped into eight classes. Classes 1 and 2 comprise invertebrate orthologues, whereas classes 3–7 comprise vertebrate family genes, and class 5 is composed of viral semaphorins. Among vertebrate semaphorins, class 3 comprises secreted members that are distinguished by a conserved basic charged domain at the C-terminus. Classes 4–7 semaphorins are membrane bound and are distinguished by specific structural elements, such as the thrombospondin repeats (class 5 semaphorins) or a glycophosphatidylinositol (GPI) anchor (class 7 semaphorins) (Kruger *et al.*, 2005). Moreover, membrane-attached semaphorins can be further processed and released in soluble form as a result of proteolytic cleavage, as seen for Sema4D (Basile *et al.*, 2007b; Elhabazi *et al.*, 2001).

Plexins and Neuropilins (NPs) are high-affinity receptors for semaphorins (He and Tessier-Lavigne, 1997; Kolodkin *et al.*, 1997; Takahashi *et al.*, 1999; Tamagnone *et al.*, 1999). In vertebrates, nine plexins have been identified, which are further divided in four subfamilies on the basis of overall similarity: PlexinA (1–4), PlexinB (1–3), PlexinC1, and PlexinD1 (Maestrini *et al.*, 1996; Tamagnone and Comoglio, 2000). Like semaphorins, plexins are characterized by the presence of a sema domain. Further, the extracellular domain of plexins contains three PSI domains, three IPT domains, and furin-specific cleavage sites only in members of PlexinB subfamily (Artigiani *et al.*, 2003). The cytoplasmic domain is highly conserved among plexin family members, and it has been shown to associate with several intracellular signal transducers, activating multiple signaling pathways in response to semaphorins (Franco and Tamagnone, 2008; Kruger *et al.*, 2005). Notably, the intracellular domain of plexins contains two amino acid stretches similar to GTPase-activating proteins (GAPs), which catalyze the inactivation of R-Ras monomeric GTPase (Oinuma *et al.*, 2004). Different groups have studied the structure of the cytoplasmic domain of plexins, but the putative mechanism responsible for its functional activation is still controversial (Bell *et al.*, 2011; He *et al.*, 2009; Tong *et al.*, 2009).

The neuropilin family consists of two proteins sharing a similar domain structure: NP1 and NP2. Neuropilins are single-span transmembrane glycoproteins, which were initially characterized as coreceptors for class 3 semaphorins and for members of the vascular endothelial growth factor (VEGF) family (He and Tessier-Lavigne, 1997; Kolodkin and Ginty, 1997; Soker *et al.*, 1998). Membrane-bound semaphorins bind directly to plexins. Conversely, secreted class 3 semaphorins, except for semaphorin 3E (Gu *et al.*, 2005), require NPs as obligate coreceptors.

Increasing evidences indicate that the semaphorin-receptor complex can contain additional transmembrane components associated to plexins, such as the cell-adhesion molecule L1 (Castellani and Rougon, 2002) and multiple receptor tyrosine kinases (RTK), such as VEGFR-2 (Bellon *et al.*, 2010; Toyofuku *et al.*, 2004a), erytroblastic leukemia viral oncogene homolog 2 (ErbB2) (Casazza *et al.*, 2010; Swiercz *et al.*, 2004), offtrack kinase (OKT) (Winberg *et al.*, 2001), and Met (Giacobini *et al.*, 2008; Giordano *et al.*, 2002; Swiercz *et al.*, 2008). Moreover, in dendritic cells (DCs) and osteoclast precursors, PlexinA1 associates with the triggering receptor expressed on myeloid cells 2 (TREM2) (Takegahara *et al.*, 2006). Interestingly, certain semaphorins can also signal through receptors distinct from plexins. For instance, in the immune system, Sema4D and Sema4A interact with CD72 (Kumanogoh *et al.*, 2000) and with T-cell immunoglobulin and mucin domain containing protein 2 (TIM-2) (Kumanogoh *et al.*, 2002a), respectively. Finally, Sema7A uses alpha1–beta1 integrin to exert its functions in both the immune and the nervous system (Pasterkamp *et al.*, 2003; Suzuki *et al.*, 2007).

Semaphorin receptors can trigger multiple signaling cascades (Zhou *et al.*, 2008). Plexins have key roles in these events, whereas the function of NPs is still controversial. Although plexins do not carry an intrinsic kinase activity, they can trigger the activation of associated RTK and nonreceptor-type tyrosine kinases (for a review, see Franco and Tamagnone, 2008). By activating plexin-associated kinases and the intrinsic GAP activity of plexins, semaphorin binding can regulate integrin functions, cytoskeletal dynamics, cell adhesion, and migration. A role in this process may also be played by other intracellular effectors of the plexins called MICALs belonging to the family of flavoprotein oxidoreductases (Terman *et al.*, 2002), which are implicated in the regulation of cytoskeletal dynamics and vesicular trafficking. Notably, the functional effects elicited by one semaphorin may vary according to the activation of distinctive signaling pathways in cell-specific manner.

Several reports have shown that semaphorins can regulate cell migration by impinging on cell-substrate adhesion and integrin function (Casazza *et al.*, 2007). Integrin signaling is central to sustain cell growth, motility, survival, and apoptosis (Hanahan and Weinberg, 2000). The molecular mechanisms by which semaphorin regulates integrins have been investigated and involve both the inactivation of small GTPases and the regulation of tyrosine kinases. As mentioned above, the intrinsic GAP activity of plexins acts on monomeric GTPase R-Ras, which is known to sustain integrin activation (Oinuma *et al.*, 2004) Following semaphorin binding, the GAP activity of plexins leads to R-Ras inactivation, thereby inhibiting integrin-mediated adhesion and other downstream events. A recent report demonstrated that Sema3E-activated PlexinD1 is further activating Arf6 GTPase, which is implicated in vesicular trafficking and integrin turnover (Sakurai *et al.*, 2010). In addition, plexins can regulate the activity of the small GTPase RhoA via signal transducers associated with the cytoplasmic domain (Barberis *et al.*, 2005; Shimizu *et al.*, 2008; Swiercz *et al.*, 2004). Finally, many plexin-associated RTK and non-receptor-type tyrosine kinases can control integrin functions, including the SRC-family kinases FAK and PYK2 and the RTK Met and ErbB2 (Guo *et al.*, 2006; Trusolino *et al.*, 2001).

Tyrosine phosphorylation is a major regulatory mechanism implicated in cell adhesion and migration. Upon semaphorin binding, the above-mentioned RTK become activated, and in turn they phosphorylate plexins on tyrosine residues. Tyrosine phosphorylation of plexin is thought to have a role in semaphorin signaling. For instance, it could mediate the recruitment of transducer molecules through specific phosphotyrosine docking sites (Swiercz *et al.*, 2009) or induce conformational changes in the plexin cytoplasmic domain, thereby allowing for a switch between multiple intracellular signaling routes (reviewed by Franco and Tamagnone, 2008).

Certain semaphorins have been found to trigger multiple and sometimes opposing functional effects, depending on the receptor complex involved. For example, PlexinB1 can associate with the RTKs Met and ErbB2, and it has been shown that the expression in breast carcinoma cells of the two different RTKs caused opposite effects on RhoA activity in response to Sema4D (Swiercz *et al.*, 2008). As another example, Sema6D-mediated signals in different developing cardiac cells can induce opposite migratory effects, due to the involvement of either PlexinA1-VEGFR-2 or PlexinA1-OKT receptor complexes (Toyofuku *et al.*, 2004a). Sema3E can mediate repelling signals in endothelial cells via PlexinD1, while instead it promotes cancer cell migration via the activation of ErbB2 kinase in complex with PlexinD1 (Casazza *et al.*, 2010). Notably, not only plexin-associated RTKs are implicated to explain this signaling "multiplicity" of semaphorins. For instance, it was shown that the interaction of Neuropilin-1 with PlexinD1 can switch axonal repulsion to attraction during brain development (Chauvet *et al.*, 2007). Moreover, Sema7A was found to promote cell spreading and dendricity in melanocytes by interacting with beta1 integrin (Scott *et al.*, 2008), whereas it can inhibit cell-substrate adhesion in DCs through signaling via PlexinC1 (Walzer *et al.*, 2005). Finally, it was shown that, upon association of the adhesive molecule L1-CAM to Neuropilin-1, the repelling activity of Sema3A is turned into attraction (Castellani *et al.*, 2000).

Cytoplasmic tyrosine kinases such as Fyn, FAK, PYK2, and Src have also been implicated in semaphorin signaling. In particular, in endothelial cells, Sema4D stimulation leads to the recruitment and activation of the integrin-associated kinase PYK2, thereby eliciting the PI(3)K/AKT pathway, which is implicated in cell motility and angiogenesis (Basile *et al.*, 2005, 2007a). Cytoplasmic tyrosine kinases have also been implicated in the "reverse" signaling pathways mediated by the intracellular domain of transmembrane semaphorins. In particular, the Abl kinase was reported to be required for Sema6D/PlexinA1-induced migration of endocardiac cells during heart development (Toyofuku *et al.*, 2004b).

A. Semaphorins as Regulators of Tumor Cell Behavior

Semaphorin receptors (plexins and NPs) are widely expressed in cancer cells (for a review, see Capparuccia and Tamagnone, 2009). For example, PlexinA1 expression has been found in ovarian, breast, gastric and colon carcinomas, and in melanomas, whereas PlexinB1 is expressed by ovarian and prostate cancers. In general, plexin gene mutations in human tumors are rarely seen (Balakrishnan *et al.*, 2009). However, somatic missense mutations in *PLXNB1* as well as upregulation of its expression have been associated with poor prognosis in prostate tumors (Wong *et al.*, 2007), as well as breast and ovarian cancers (Valente *et al.*, 2009). Conversely, loss of PlexinB1 expression was associated with poor prognosis in estrogen-receptor-positive breast cancer (Rody *et al.*, 2007) and BRAF-positive melanoma (Argast *et al.*, 2009). PlexinC1 expression is markedly reduced in melanoma cells and melanoma metastatic samples (Scott *et al.*, 2008). Finally, PlexinD1 is expressed in several tumor cell types, including melanoma and breast cancer cell lines (Casazza *et al.*, 2010; Kigel *et al.*, 2008; Roodink *et al.*, 2005, 2008).

NPs are also expressed widely in tumors and their expression usually correlates with tumor progression (for a review Rizzolio and Tamagnone, 2011). NP1 is prevalently expressed in carcinomas, whereas NP2 is frequently expressed in melanomas, leukemia, and neuroblastomas (Pellet-Many *et al.*, 2008). Moreover, the two receptors are often coexpressed (Marcus *et al.*, 2005; Rieger *et al.*, 2003). Overexpression of NP1 in prostate carcinoma cells induced rapid tumor growth and progression *in vivo* (Miao *et al.*, 2000), but there are also reports that show opposite outcome (Gray *et al.*, 2005; Kamiya *et al.*, 2006). High levels of NP2 expression correlate with poor prognosis in osteosarcomas (Handa *et al.*, 2000). Moreover, it has been shown that the inhibition of NP2 expression in colorectal carcinoma cells inhibited tumor development (Gray *et al.*, 2008).

It is known that cancer cells release semaphorins capable of regulating their own behavior in autocrine manner. Most available information about semaphorin-dependent regulation of tumor cells comes from studies on class 3 semaphorins. Class 3 semaphorins can regulate tumor cell functions by multiple mechanisms either by activating plexin-dependent signaling cascades or by binding to NPs coreceptors and competing with VEGF signaling.

Notably, recent literature supports the idea that semaphorins and VEGF bind to NPs independently and thereby trigger antagonistic intracellular signaling pathways (Appleton *et al.*, 2007; Liang *et al.*, 2007; Pan *et al.*, 2007).

Among class 3 semaphorins, Sema3B was initially characterized as a tumor suppressor. In fact, its expression is reduced in small cell lung carcinoma by loss of heterozygosity and promoter hypermethylation (reviewed by Potiron *et al.*, 2009). Consistently with this observation, Sema3B overexpression in tumor cell lines induces apoptosis, inhibits cell proliferation, and anchorage-independent cell growth, confirming its direct inhibitory effects on tumor cells (Tomizawa *et al.*, 2001; Tse *et al.*, 2002). However, Sema3B expression has also been correlated to poor prognosis in glioblastoma (Rich *et al.*, 2005), and tumor subsets that have an increased level of Sema3B expression have been associated with metastatic progression (Rolny *et al.*, 2008). In the latter study, Sema3B was actually found to inhibit tumor cell growth and concomitantly to sustain metastatic dissemination in different tumorigenic models, by an indirect effect on the tumor-associated macrophages (TAMs) in the tumor microenvironment (Rolny *et al.*, 2008).

Sema3F is considered a putative tumor suppressor (Roche *et al.*, 1996; Xiang *et al.*, 1996). Increased expression of Sema3F in tumor cells inhibits proliferation, anchorage-independent cell growth, and tumor growth *in vivo* (reviewed by Potiron *et al.*, 2009). Moreover, Sema3F antitumor activity in lung carcinoma cells was associated with integrin dysregulation and increased substrate adhesion (Brambilla *et al.*, 2000; Kusy *et al.*, 2005; Potiron *et al.*, 2007). Finally, Sema3F inhibits the migration of both breast (Nasarre *et al.*, 2003) and melanoma cells (Bielenberg *et al.*, 2004).

Sema3A is widely expressed by tumor cells and has major functions in the regulation of the tumor microenvironment. In addition, it was shown that Sema3A exerts autocrine functions by inhibiting the proliferation of breast tumor cells overexpressing NP1 (Kigel *et al.*, 2008). Moreover, Sema3A was found to inhibit the migration and metastatic behavior of multiple tumor cells (Bachelder *et al.*, 2003; Casazza *et al.*, 2011).

Contrary to the semaphorins described above, Sema3C is mainly thought to promote tumor progression. Sema3C aberrant expression in ovarian cancer has been correlated with progression (Galani *et al.*, 2002). Sema3C is upregulated in metastatic cells of lung adenocarcinoma (Martin-Satue and Blanco, 1999). Finally, its expression increases the motility and invasion of prostate cancer cells (Herman and Meadows, 2007).

Recent literature highlights the functional role of Sema3E in tumor progression. The expression of Sema3E and its receptor PlexinD1 correlates with the metastatic progression in human tumors (Casazza *et al.*, 2010). Overexpression of Sema3E in cancer cells increases their invasiveness, transendothelial migration, and metastatic spreading. The proinvasive and metastatic activity of Sema3E in tumor cells is dependent on its proteolytic conversion into the p61 active fragment (Christensen *et al.*, 2005), which elicits the transactivation of PlexinD1-associated ErbB2 oncogenic kinase (Casazza *et al.*, 2010).

Among transmembrane semaphorins, Sema4D has been found to inhibit or promote migration depending on the receptor complexes that are expressed in tumor cells. In fact, Sema4D can inhibit cell migration (Barberis *et al.*, 2004; Swiercz *et al.*, 2008) but can also sustain invasive growth in cells where PlexinB1 can associate with Met and ErbB2 (Conrotto *et al.*, 2004; Giordano *et al.*, 2002; Swiercz *et al.*, 2008).

Finally, overexpression of Sema5A has been correlated with tumor growth, invasion, and metastasis in both pancreatic and gastric cancer (Pan *et al.*, 2010; Sadanandam *et al.*, 2010b).

II. MULTIFACETED ROLE OF SEMAPHORINS IN TUMOR ANGIOGENESIS

Blood vessels transport through the organism oxygen, nutrients, and hematopoietic cells to guarantee immune surveillance. The establishment of a functional integrated vascular system is of primary importance for tissue growth and homeostasis.

During the development of mammalian embryo, the process through which endothelial cells differentiate and assemble into a vascular network is known as vasculogenesis. Subsequently, sprouting angiogenesis ensures the expansion of the vascular network by regulating the growth of new capillary vessels out of preexisting ones. Finally, during arteriogenesis, the new vessels are covered with pericytes and vascular smooth muscle cells to ensure vessel stability and adequate tissue perfusion. In adults the vasculature is usually quiescent; neoangiogenesis occurs only in organs that are undergoing physiological growth and in injured tissue.

Semaphorins play an important role in physiological vasculature development. Several semaphorin receptors are expressed by endothelial cells, including NPs, PlexinA family members, and PlexinD1. Moreover, endothelial cells express class 3 semaphorins, such as Sema3A and Sema3F, suggesting that the latter might act in an autocrine manner. Many class 3 semaphorins have been shown to exert inhibitory or antiangiogenic effects by controlling endothelial cells. Some studies have shown that Sema3A and Sema3F inhibit VEGF-induced angiogenesis (Guttmann-Raviv *et al.*, 2007; Kessler *et al.*, 2004). Moreover, VEGF-A and Sema3A have opposing effects on the growth and survival of endothelial cells *in vitro*, indicating that they may act antagonistically *in vivo* to regulate angiogenic sprouting (Miao *et al.*, 1999). As already mentioned above, it is a controversial issue whether there is competition between Sema3A and VEGF-A for binding NP1 (Appleton *et al.*, 2007). Moreover, it has been shown that Sema3A can interfere with VEGF signaling without inhibiting VEGF-induced phosphorylation of VEGFR-2 (Guttmann-Raviv *et al.*, 2007), which supports the idea that this is not the result of competition for a shared binding site on neuropilins. In fact, other studies have suggested that class 3 semaphorins directly inhibit the migration of endothelial cells by blocking integrin activation downstream to plexin signaling (Sakurai *et al.*, 2010; Serini *et al.*, 2003).

Unlike most semaphorins regulating angiogenesis, Sema3E has been shown to bind PlexinD1 independently from NP1 (Gu *et al.*, 2005). PlexinD1 knockout mice die shortly after birth because of cardiovascular abnormalities. Genetic evidences indicate that Sema3E–PlexinD1 signaling is required for developmental angiogenesis (Gitler *et al.*, 2004; Gu *et al.*, 2005; Kim *et al.*, 2011; Zygmunt *et al.*, 2011). Moreover, it has been shown that Sema3E is implicated in regulating postischemic angiogenesis (Fukushima *et al.*, 2011; Moriya *et al.*, 2010) and have a role in regulating tip cells (reviewed by Tamagnone and Mazzone, 2011).

Angiogenesis is characterized by locally controlled balance of pro- and antiangiogenic signals. Conversely, in cancer, this highly coordinated cascade of events is often deregulated. Initially most tumor masses grow avascular until the equilibrium between cell proliferation and apoptosis is reached. Then the initiation of neoangiogenesis (often indicated as "angiogenic switch") ensues, in order to provide adequate oxygen and nutrient supply and allow exponential tumor growth. In tumors, oncogenic signaling, inflammation, and hypoxia (Carmeliet and Jain, 2000) help to shift the balance toward a proangiogenic phenotype. As a consequence, the architecture of tumor vasculature is usually heterogeneous and disorganized. These structural abnormalities cause an impairment of oxygen delivery, which further supports hypoxia and stress in the tumor microenvironment. Further, these conditions contribute to the induction of the metastatic program allowing for the escape of tumor cells through the leaky vasculature.

In addition to their role in physiological vasculogenesis and angiogenesis, many semaphorins (including Sema3A, Sema3B, Sema3D, Sema3E, Sema3F, and Sema4A) were found to inhibit tumor vasculature, whereas Sema3C and Sema4D were found to promote its development. In the following paragraphs, we review the current knowledge on the functional role of antiangiogenic and proangiogenic semaphorins.

A. Semaphorins Inhibiting Tumor Angiogenesis

In accordance with the inhibitory function exerted by several class 3 semaphorins during embryo vasculature development, these factors were further reported to act as inhibitory signals for tumor angiogenesis.

Class 3 semaphorins are expressed by endothelial cells (Serini *et al.*, 2003) and can act in an autocrine manner. Sema3A inhibits the adhesion and migration of endothelial cells (Miao *et al.*, 1999; Serini *et al.*, 2003), exerts a repelling effect, and promotes cell apoptosis (Guttmann-Raviv *et al.*, 2007). Further, Sema3A can inhibit angiogenesis *in vivo* (Acevedo *et al.*, 2008; Guttmann-Raviv *et al.*, 2007) and induce vascular permeability (Acevedo *et al.*, 2008). Similarly, Sema3F repels endothelial cells and inhibits their adhesion, migration, survival, and proliferation (Bielenberg *et al.*, 2004; Guttmann-Raviv *et al.*, 2007). Moreover, Sema3F also inhibits tumor angiogenesis *in vivo* (Bielenberg *et al.*, 2004; Kessler *et al.*, 2004).

Antagonistic autocrine loops of VEGF-A and Sema3A and Sema3F act in endothelial cells both *in vitro* (Serini *et al.*, 2003; Vacca *et al.*, 2006) and during angiogenesis *in vivo* (Ito *et al.*, 2000; Serini *et al.*, 2003). It has been suggested that an imbalance between autocrine loops of proangiogenic VEGF-A and antiangiogenic semaphorins could occur during tumor progression and account at least in part for the disorganized tumor vasculature. This hypothesis is supported by the observation that in patients with multiple myeloma, BM endothelial cells have lost autocrine loops of Sema3A in favor of VEGF-A signals (Vacca *et al.*, 2006).

Interestingly, it has recently been shown that the ectopic reexpression of Sema3A in a transgenic oncogene-driven model of pancreatic islet cell carcinogenesis results in the inhibition and subsequent normalization of tumor vasculature, associated with reduced tumor burden (Maione *et al.*, 2009). Further, in another study it has been demonstrated that the systemic or local expression of Sema3A in multiple preclinical models severely disrupted tumor vessels, impaired tumor growth, and prevented metastatic dissemination (Casazza *et al.*, 2011). These studies provide proof of concept that the antiangiogenic activity of Sema3A may be exploited to effectively inhibit tumor progression.

Similarly to Sema3A and Sema3F, three other class 3 semaphorins, that is, Sema3B, Sema3D, and Sema3G, were reported to repel endothelial cells and/or inhibit tumor angiogenesis (Kigel *et al.*, 2008). Interestingly, it has been proposed that the antiangiogenic effect exerted by Sema3B in tumor may be partly overcome by a rebound effect inducing the production of the angiogenic cytokine IL-8 (Rolny *et al.*, 2008).

Sema3E, too, acting through its receptor PlexinD1, has been found to repel endothelial cells. Moreover, it has recently been published that Sema3E is able to independently regulate endothelial and tumor cells through the involvement of distinctive signaling pathways (Casazza *et al.*, 2010, 2012). For instance, Sema3E dramatically increases invasiveness and metastatic dissemination of tumor cells through the activation of PlexinD1-associated ErbB2 tyrosine kinase. Conversely, the repelling and inhibitory effect of Sema3E on endothelial cells was shown to be dependent on the R-Ras GAP activity of PlexinD1 and on the expression of the small GTPase Rnd2

(Casazza *et al.*, 2010). In addition, other studies have implicated the regulation of additional small GTPases, such as Arf6 and Rhoj, in endothelial cells downstream to Sema3E/PlexinD1 signaling (Fukushima *et al.*, 2011; Sakurai *et al.*, 2010).

Membrane-bound semaphorins, such as Sema4A and Sema6A, were also found to inhibit angiogenesis. Sema4A can inhibit developmental angiogenesis (Toyofuku *et al.*, 2007), but it has not been tested whether it can inhibit tumor angiogenesis. Regarding Sema6A, there are evidences indicating that its extracellular domain functions as an inhibitor of endothelial cell migration and tumor angiogenesis, but there is no information about the role of the full-length molecule in tumor progression (Dhanabal *et al.*, 2005).

B. Semaphorins Promoting Tumor Angiogenesis

Several semaphorins have been found to promote tumor angiogenesis. For example, Sema3C induces endothelial cell proliferation, adhesion, and directional migration *in vitro* (Banu *et al.*, 2006). Among membrane-bound semaphorins, Sema4D has been shown to exert potent proangiogenic activity both *in vitro* and *in vivo* (Basile *et al.*, 2004, 2005; Conrotto *et al.*, 2005). Many human cancers express Sema4D, which acts by binding to its high-affinity receptor PlexinB1 or the intermediate affinity receptor PlexinB2, both expressed by endothelial cells (Fazzari *et al.*, 2007). A soluble form of Sema4D, released by the activity of membrane type-1 matrix metalloproteinase, is capable to induce endothelial cell chemotaxis and blood vessel growth *in vivo* (Basile *et al.*, 2006). Notably, the so-called TAMs may become a major source of Sema4D in the tumor microenvironment (Sierra *et al.*, 2008). As a consequence, tumors grown in Sema4D knockout mice have been found to contain smaller and aberrant vessel, reduced tumor growth, and metastases (Sierra *et al.*, 2008). An additional source of secreted Sema4D in the tumor microenvironment could be activated platelets (Zhu *et al.*, 2007), which are commonly found in this context and are considered a major source of cytokines and paracrine regulatory factors. Finally, Sema5A has been found to induce endothelial cell proliferation, migration, and survival (Sadanandam *et al.*, 2010a). However, whether Sema5A may play a role in tumor angiogenesis is not known.

III. SEMAPHORINS AND TUMOR IMMUNE MICROENVIRONMENT

Interactions between tumor cells and their tissue microenvironment have been shown to play a major role in cancer progression. In addition to resident fibroblasts, blood vessels, and lymphatic vessels, the tumor microenvironment can be infiltrated by bone-marrow (BM)-derived cells, such as monocytes and lymphocytes, recruited from the circulation. Notably, it is believed that tumor-derived paracrine signals activate the BM compartment, resulting in the mobilization and recruitment of discrete subsets of BM-derived cells to the tumor. In the following paragraphs, we review the reported functions of semaphorin signals in the regulation of immune cells both in physiology and in the cancer context.

A. Regulatory Role of Semaphorins in Physiological Immune Responses

Several semaphorins have been found to play important roles in the immune system. Our understanding of the immunoregulatory functions of semaphorins has notably advanced over the past years. For instance, Semaphorins have been shown to exert costimulatory functions and to regulate immune cell trafficking.

Sema4D was the first semaphorin found to have immunoregulatory functions. In the immune system, Sema4D is highly expressed in resting T cells (Suzuki *et al.*, 2008) and weakly expressed in B cells and DCs. However, its expression is significantly upregulated following cellular activation (Kumanogoh *et al.*, 2000). The functional receptor for Sema4D in the immune system is the C-type lectin CD72 (Kumanogoh *et al.*, 2000). CD72 is a negative regulator of B cells activation, by recruiting tyrosine phosphatase SHP-1 to B-cell receptor complex (Parnes and Pan, 2000). On Sema4D-CD72 binding, SHP-1 dissociates from CD72 allowing B-cell activation (Kumanogoh *et al.*, 2000). Notably, Sema4D-deficient mice have defects in B-cell response and altered antibody production (Shi *et al.*, 2000). Sema4D-deficient mice also show defects in T-cell activation. In fact, Sema4D expressed on T cells favors the activation and maturation of DCs that in turn regulate T-cell function (Kumanogoh *et al.*, 2002b).

Another class 4 semaphorin, Sema4A, is involved in T-cell regulation. Sema4A is abundantly and constitutively expressed in DCs (Kumanogoh *et al.*, 2002a), and it is moreover expressed by activated T cells and T-helper-1 (Th1)-polarized cells (Kumanogoh *et al.*, 2005). It has been shown that Sema4A expressed on DC is important for T-cell priming, whereas Sema4A expressed by T cell promotes Th1 differentiation through cell-to-cell interactions and/or an autocrine signaling (Kumanogoh *et al.*, 2005). In the immune system, TIM2 acts as Sema4A receptor (Kumanogoh *et al.*, 2002a). However, as knockout Sema4A and TIM2 mice show some phenotypic differences, it has been proposed that Sema4A could bind also other receptors such as PlexinD1 and PlexinB family members both expressed on T cells.

Class 3 semaphorin, Sema3A, also regulates T-cell functions. Sema3A has an immunosuppressive function on T-cell proliferation. Sema3A is produced by several tumor cell types (Catalano *et al.*, 2006) and by DCs (Lepelletier *et al.*, 2006). It has been published that the treatment of T cell with recombinant or tumor-derived Sema3A inhibits T-cell proliferation and cytokine production (Catalano *et al.*, 2006; Lepelletier *et al.*, 2006). Moreover, Sema3A can trigger a proapoptotic signal in T cells by sensitizing them to Fas-induced apoptosis (Moretti *et al.*, 2008). NP1 and PlexinA1 have been frequently implicated to mediate Sema3A signaling in immune cells. However, PlexinA4 was recently found to be required to mediate the signaling of class 3 semaphorins (including Sema3A) in T cells, DCs, and macrophages (Yamamoto *et al.*, 2008).

Among class 6 semaphorins, Sema6D is expressed in B and T lymphocytes and in natural killer cells. Sema6D signals through its receptor PlexinA1, which in DCs and osteoclasts forms a complex with TREM2 and the adaptor protein DAP12 (Takegahara *et al.*, 2006). PlexinA1- and DAP12-deficient mice have impaired T-cell responses and develop osteopetrosis (Kaifu *et al.*, 2003; Takegahara *et al.*, 2006). Moreover, PlexinA1-deficient DCs poorly stimulate antigen-specific T cells (Takegahara *et al.*, 2006). These observations indirectly indicate that Sema6D is required for the optimal induction of T-cell responses by DCs. Further, the role of Sema6D in T-cell activation has been directly investigated. By targeting Sema6D with anti-Sema6D antibody, it has been shown that Sema6D is an important regulator of the late phases of T-cell activation (O'Connor *et al.*, 2008).

The GPI-anchored Sema7A, too, is expressed by activated T cells (Suzuki *et al.*, 2007). In analogy to that demonstrated in the nervous system (Pasterkamp *et al.*, 2003), Sema7A signals through the alpha1–beta1 integrin in immune cells (Suzuki *et al.*, 2007). Sema7A expressed by activated T cells stimulates macrophages to produce proinflammatory cytokines (Suzuki *et al.*, 2007) suggesting that Sema7A play an important role during inflammatory responses.

Among neuropilins, NP1 is expressed in both DCs and T cells (Tordjman *et al.*, 2002), and it has been proposed to stabilize the interaction between DCs and T cells through homophilic interactions, thereby favoring the initiation of primary immune responses (so-called immune synapse). Recently, NP1 has been identified as a specific marker for CD4+CD25+ regulatory T (Treg) cells (Bruder *et al.*, 2004). Moreover, it has been published that NP1 contributes to the long contact between Treg cells and DCs, thus favoring the inhibition of T-cell activation (Sarris *et al.*, 2008). These data further support the idea that NP1 might stabilize DC-T-cell interactions. Whether secreted semaphorin interacting with NP1 may have a regulatory function in this context is not currently understood.

As described above, the immunological studies on semaphorins have been mainly focused on their costimulatory functions and their involvement in cell-to-cell interactions. However, very recent literature is highlighting the functional role of semaphorins in immune cell trafficking. For instance, it has recently been published that Sema3E–PlexinD1 signaling is necessary for intrathymic cell migration, a prerequisite for proper thymocyte maturation (Choi *et al.*, 2008). The expression of PlexinD1 was found to be highly regulated on thymocytes. PlexinD1 is highly expressed on CD4+CD8+ double positive (DP) thymocytes but decreased in single positive cells. The activation of PlexinD1 by its ligand Sema3E (predominantly expressed in the thymus medulla) regulates CCR9-directed migration of maturing thymocytes from the cortex toward the medulla (Choi *et al.*, 2008).

Several semaphorins have been found to regulate monocyte and DC migration. Among them, Sema3A and Sema4D have been shown to inhibit monocytes and T cells responsiveness to chemokine gradients (Delaire *et al.*, 2001; Ji *et al.*, 2009). Conversely, Sema7A has been found to attract monocytes (Holmes *et al.*, 2002). Finally, it has recently been published that PlexinA1 is crucially involved in the entry of DCs into the lymphatics (Takamatsu *et al.*, 2010). Sema3A produced by the lymphatics induces actomyosin contraction at the rear side of DCs where PlexinA1 is localized, thus allowing DCs transmigration (Takamatsu *et al.*, 2010).

B. Role of Semaphorins in the Regulation of T Lymphocytes in Tumors

Only recently we started to understand the functional plasticity of lymphocyte populations in tumor. Tumor-infiltrating lymphocytes have both pro- and antitumor properties. Indeed, CD8+ cytotoxic T lymphocytes play a critical role in restraining tumor development, whereas CD4+ T and B lymphocytes have been shown to have more paradoxical effects. For example, in colon and lung carcinoma, CD4+ T-cell infiltration correlates with favorable prognosis, whereas in breast and renal cancers, CD4+ T-cell infiltration correlates with poor survival (Ruffell *et al.*, 2010; Siddiqui *et al.*, 2007; Wakabayashi *et al.*, 2003). These observations suggest that the activation of a pro- or antitumor immune program is dependent on the specific organ and on the tumor microenvironment. In fact, lymphocyte effector functions are dependent on the cytokines and chemokines released in the tumor microenvironment as well as on their interaction with stromal cells, other leukocyte populations, and tumor cells.

Upon exposure to polarizing cytokine signals during activation, CD4+ T cells can differentiate into a variety of effector subsets, including classical Th1 and Th2 cells, the more recently defined Th17 cells and inducible regulatory T (iTreg) cells (Zhou *et al.*, 2009). In general, Th1 polarization is thought to promote antitumor effects, whereas Th2 polarization drives protumor effects. Part of the Th1-/Th2 response is the capacity of Th1 and Th2 lymphocytes to shape monocyte-macrophage functions. In fact, Th1 and Th2 cells drive M1 and M2 polarization of macrophages, respectively (Biswas and Mantovani, 2010). The presence of Th17 and Treg cells has been reported in several tumors (Curiel *et al.*, 2004; Wang *et al.*, 2009; Zhang *et al.*, 2009). The IL-17 pathway can have both pro- and antitumor effects (Ji and Zhang, 2010). Treg cells are also thought to play an important role in regulating tumor immunity. In follicular lymphoma, increased infiltration of Treg cells correlates with increased survival (Carreras *et al.*, 2009), whereas Treg infiltration correlates with poor prognosis in non-small cell lung carcinoma (Petersen *et al.*, 2006) and breast carcinoma (Merlo *et al.*, 2009).

Accumulating evidences indicate that semaphorin signals can regulate lymphocyte activities. By inhibiting T lymphocyte-mediated responses, semaphorin signals may contribute to the immunosuppressive environment found in tumors. In this respect, it has been reported that Sema3A produced by several tumor cell types (Catalano *et al.*, 2006) can inhibit T-cell proliferation and cytokine production (Catalano *et al.*, 2006; Lepelletier *et al.*, 2006). In addition, Sema3A was shown to promote apoptosis in T cells (Moretti *et al.*, 2008).

Neuropilin-1 has been proposed to play a critical role in the interaction of Treg cells with DCs. NP1 is actually expressed by most Treg and it promotes their prolonged interaction with immature DCs. In turn, this may provide an advantage to Treg over naïve T cells and, in limiting concentration of antigen, it could favor an immune suppressive response. In support of this hypothesis, blocking of NP1 completely abrogates suppression of T-cell proliferation mediated by Treg (Sarris *et al.*, 2008).

Semaphorin signals have also been found to promote T- and B-cell activation and to induce the release of proinflammatory cytokines. By sustaining inflammatory responses, semaphorins might favor immune dysfunction in the tumor stroma, thus supporting cancer progression. In this respect, it has been found that Sema4D promotes B and T-cell activation (Kumanogoh *et al.*, 2000) (Kumanogoh *et al.*, 2002b).

Moreover, Sema4A has been found to promote Th1 differentiation (Kumanogoh *et al.*, 2005). Among class 6 semaphorins, Sema6D is an important regulator of the late phases of T-cell activation (O'Connor *et al.*, 2008). Finally, Sema7A expressed by activated T cells stimulates macrophages to produce proinflammatory cytokines (Suzuki *et al.*, 2007).

C. Semaphorin-Dependent Regulation of Tumor-Associated Macrophages

Recruited BM-derived cells such as macrophage subpopulations contribute significantly to neovasculature formation and tumor growth (Carmeliet and Jain, 2000). Notably, TAMs have been shown to regulate nearly all steps of tumor progression (Qian and Pollard, 2010). During tumor initiation, TAMs create an inflammatory environment that promotes tumor growth. Further, macrophages stimulate angiogenesis, enhance tumor cell migration and invasion, and suppress antitumor immunity. Finally, macrophages promote tumor cell extravasation, survival, and colonization of metastatic sites. In one meta-analysis, it was reported that over 80% of studies show a correlation between macrophage density and poor patient prognosis (Bingle *et al.*, 2002). For example, there is a strong association between poor survival and increased macrophage density in thyroid, lung, and hepatocellular cancers (Chen *et al.*, 2005; Ryder *et al.*, 2008; Zhu *et al.*, 2008).

Although it is a common theme in cancer biology that the fate of tumor-homing monocytes is modulated directly by the tumor microenvironment, it has recently been suggested that different subsets of circulating monocytes may contribute to distinctive macrophage subpopulations in tumors (Qian and Pollard, 2010). Among them is a particular subset of circulating and tumor-infiltrating monocytes expressing the angiopoietin receptor Tie2, known as TEMs (De Palma *et al.*, 2005). TEMs are proangiogenic when isolated from either tumors or peripheral blood, suggesting their commitment to a proangiogenic function already in the circulation. Moreover, although TEMs are a minor proportion of the bulk of tumor-infiltrating macrophages, they are capable of enhancing angiogenesis to a much greater extent than TAMs (De Palma *et al.*, 2005; Venneri *et al.*, 2007).

Interestingly, the regulation of TAM activity in the tumor microenvironment has recently been associated with semaphorin signals. Among class 4 semaphorins, Sema4D is highly expressed by TAMs (Sierra *et al.*, 2008). Interestingly, TAMs-derived Sema4D has been found to be crucial for tumor angiogenesis (Sierra *et al.*, 2008). Further, Sema4D has been shown to induce the release of proinflammatory cytokines by monocytes (Kikutani and Kumanogoh, 2003). Also Sema7A induces monocytes to release proinflammatory cytokines such as IL-1beta, tumor necrosis factor, IL-8, and IL-6 (Holmes *et al.*, 2002).

It has been proposed that macrophages in tumors are shifted from the activated proinflammatory M1 state to an alternatively activated phenotype named M2 (Mantovani and Sica, 2010). In general, M2 macrophages have immunoregulatory functions such as participating in polarized Th2 responses and promoting tissue remodeling and angiogenesis. Interestingly,

M2 polarization has been associated with the upregulation of NP1 and PlexinA1 (Ji *et al.*, 2009), whereas the expression of the two receptors is decreased in M1 polarized macrophages. These results suggest that semaphorins may differentially regulate macrophages depending on the receptors expressed by the subpopulation found in the tumor.

In addition to regulate their behavior, semaphorin signals have been recently associated with TAM recruitment into the tumor. For instance, it has been found that the expression of Sema3B in tumor cells activates a NP-mediated signaling pathway that leads to the secretion of IL-8. The release of IL-8, in turn, induced the recruitment of TAMs into the tumor stroma (Rolny *et al.*, 2008). Therefore, although it has been suggested that it acts as a tumor suppressor, Sema3B seems to trigger an escape program from growth inhibition by inducing TAMs recruitment which in turn promotes cancer progression and metastatic dissemination (Rolny *et al.*, 2008).

Notably, the expression of NP1 has recently been associated with the recruitment of myeloid cells from the BM to the sites of physiological neoangiogenesis (Zacchigna *et al.*, 2008). These recruited myeloid cells contribute to arterial formation by sustaining the activation and proliferation of smooth muscle cells. In fact, both VEGF165 and Sema3A act as chemoattractants for myeloid cells although exerting opposite effects on vessel growth: in particular, arteriogenesis is stimulated by VEGF165 and inhibited by Sema3A (Zacchigna *et al.*, 2008). Whether NP1 receptor on BM-derived populations is also used for their recruitment at the site of tumor angiogenesis is still unknown. Notably, the proangiogenic monocyte subpopulation known as TEMs has recently been found to express, among others, particularly high levels of NP1 and Sema6D (Pucci *et al.*, 2009). NP1 might be involved in the recruitment of TEMs to the site of tumor angiogenesis. Moreover, the enhanced expression of the proangiogenic NP1 and Sema6D by TEM may suggest activation of endothelial cells by cell-to-cell contact. Finally, Sema3A–PlxnA1 signaling was found to mediate DCs trafficking (Takamatsu *et al.*, 2010). Interestingly, as semaphorins are also expressed by endothelial cells, it is plausible that their signals might regulate leukocyte extravasation, thus promoting their infiltration in tumors.

IV. CONCLUDING REMARKS AND OPEN QUESTIONS

Semaphorin family members are emerging pivotal regulators of the tumor microenvironment, and their expression is often altered in tumor samples compared to normal tissues. Actually, by means of these signals, cancer cells can control the recruitment, behavior, and functional differentiation of multiple stromal cells and thereby tweak the microenvironment in favor of their growth and invasive-metastatic progression. Notably, these functions often recapitulate the role played by semaphorins in embryonal tissue development. In addition, tumor-infiltrating cells derived from the bone marrow represent a main source of diffusible semaphorin signals, which in turn can regulate cancer cell behavior. Importantly, the expression of certain inhibitory semaphorins is instead decreased in aggressive tumor cells, supporting functional data which suggest their identification as tumor suppressing genes. Based on these data, semaphorins have further raised interest as potential therapeutic targets to interfere with tumor progression.

While much is known about the role of semaphorins in the regulation of angiogenesis and tumor vasculature, their multifaceted function in immune cells is less well understood. A number of scattered reports highlight important regulatory functions of selected semaphorins in physiological immune responses, for instance impinging on cell migration and lymphocyte activation. Based on these reports, a relevant regulatory role on immune cells recruited in the tumor microenvironment can be envisaged. This could concern, for example, monocyte-derived TAM and TEM, releasing cytokines and supporting tumor angiogenesis and invasive-metastatic behavior of cancer cells, as well as DC and lymphocytes regulating the antitumor immune response. Additional experiments are required to elucidate which are the most important semaphorin signals regulating immune cells in the tumor microenvironment and whether these pathways can be targeted to promote a more effective antitumor immune response. Moreover, appropriate assays should be developed to investigate the activity of membrane-bound semaphorins in cell-contact mediated signaling.

ACKNOWLEDGMENTS

The authors wish to thank friends and colleagues in the lab, particularly Sabrina Rizzolio, for help in editing this chapter. The ongoing research activity of the authors is supported by the Association for International Cancer Research (AICR-UK; grant 11-0274 to L. T.) and by the Italian Association for Cancer Research (AIRC; IG-11598 to L. T.).

REFERENCES

Acevedo, L. M., Barillas, S., Weis, S. M., Gothert, J. R., and Cheresh, D. A. (2008). Semaphorin 3A suppresses VEGF-mediated angiogenesis yet acts as a vascular permeability factor. *Blood* **111**, 2674–2680.

Appleton, B. A., Wu, P., Maloney, J., Yin, J., Liang, W. C., Stawicki, S., Mortara, K., Bowman, K. K., Elliott, J. M., Desmarais, W., Bazan, J. F., Bagri, A., *et al.* (2007). Structural studies of neuropilin/antibody complexes provide insights into semaphorin and VEGF binding. *EMBO J.* **26**, 4902–4912.

Argast, G. M., Croy, C. H., Couts, K. L., Zhang, Z., Litman, E., Chan, D. C., and Ahn, N. G. (2009). Plexin B1 is repressed by oncogenic B-Raf signaling and functions as a tumor suppressor in melanoma cells. *Oncogene* **28**, 2697–2709.

Artigiani, S., Barberis, D., Fazzari, P., Longati, P., Angelini, P., Van De Loo, J. W., Comoglio, P. M., and Tamagnone, L. (2003). Functional regulation of semaphorin receptors by proprotein convertases. *J. Biol. Chem.* **278**, 10094–10101.

Bachelder, R. E., Lipscomb, E. A., Lin, X., Wendt, M. A., Chadborn, N. H., Eickholt, B. J., and Mercurio, A. M. (2003). Competing autocrine pathways involving alternative neuropilin-1 ligands regulate chemotaxis of carcinoma cells. *Cancer Res.* **63**, 5230–5233.

Balakrishnan, A., Penachioni, J. Y., Lamba, S., Bleeker, F. E., Zanon, C., Rodolfo, M., Vallacchi, V., Scarpa, A., Felicioni, L., Buck, M., Marchetti, A., Comoglio, P. M., *et al.* (2009). Molecular profiling of the "plexinome" in melanoma and pancreatic cancer. *Hum. Mutat.* **30**, 1167–1174.

Banu, N., Teichman, J., Dunlap-Brown, M., Villegas, G., and Tufro, A. (2006). Semaphorin 3C regulates endothelial cell function by increasing integrin activity. *FASEB J.* **20**, 2150–2152.

Barberis, D., Artigiani, S., Casazza, A., Corso, S., Giordano, S., Love, C. A., Jones, E. Y., Comoglio, P. M., and Tamagnone, L. (2004). Plexin signaling hampers integrin-based adhesion, leading to Rho-kinase independent cell rounding, and inhibiting lamellipodia extension and cell motility. *FASEB J.* **18**, 592–594.

Barberis, D., Casazza, A., Sordella, R., Corso, S., Artigiani, S., Settleman, J., Comoglio, P. M., and Tamagnone, L. (2005). p190 Rho-GTPase activating protein associates with plexins and it is required for semaphorin signalling. *J. Cell Sci.* **118**, 4689–4700.

Basile, J. R., Barac, A., Zhu, T., Guan, K. L., and Gutkind, J. S. (2004). Class IV semaphorins promote angiogenesis by stimulating Rho-initiated pathways through plexin-B. *Cancer Res.* **64**, 5212–5224.

Basile, J. R., Afkhami, T., and Gutkind, J. S. (2005). Semaphorin 4D/plexin-B1 induces endothelial cell migration through the activation of PYK2, Src, and the phosphatidylinositol 3-kinase-Akt pathway. *Mol. Cell. Biol.* **25**, 6889–6898.

Basile, J. R., Castilho, R. M., Williams, V. P., and Gutkind, J. S. (2006). Semaphorin 4D provides a link between axon guidance processes and tumor-induced angiogenesis. *Proc. Natl. Acad. Sci. USA* **103**, 9017–9022.

Basile, J. R., Gavard, J., and Gutkind, J. S. (2007a). Plexin-B1 utilizes RhoA and Rho kinase to promote the integrin-dependent activation of Akt and ERK and endothelial cell motility. *J. Biol. Chem.* **282**, 34888–34895.

Basile, J. R., Holmbeck, K., Bugge, T. H., and Gutkind, J. S. (2007b). MT1-MMP controls tumor-induced angiogenesis through the release of semaphorin 4D. *J. Biol. Chem.* **282**, 6899–6905.

Bell, C. H., Aricescu, A. R., Jones, E. Y., and Siebold, C. (2011). A dual binding mode for RhoGTPases in plexin signalling. *PLoS Biol.* **9**, e1001134.

Bellon, A., Luchino, J., Haigh, K., Rougon, G., Haigh, J., Chauvet, S., and Mann, F. (2010). VEGFR2 (KDR/Flk1) signaling mediates axon growth in response to semaphorin 3E in the developing brain. *Neuron* **66**, 205–219.

Bielenberg, D. R., Hida, Y., Shimizu, A., Kaipainen, A., Kreuter, M., Kim, C. C., and Klagsbrun, M. (2004). Semaphorin 3F, a chemorepulsant for endothelial cells, induces a poorly vascularized, encapsulated, nonmetastatic tumor phenotype. *J. Clin. Invest.* **114**, 1260–1271.

Bingle, L., Brown, N. J., and Lewis, C. E. (2002). The role of tumour-associated macrophages in tumour progression: implications for new anticancer therapies. *J. Pathol.* **196**, 254–265.

Biswas, S. K., and Mantovani, A. (2010). Macrophage plasticity and interaction with lymphocyte subsets: cancer as a paradigm. *Nat. Immunol.* **11**, 889–896.

Bork, P., Doerks, T., Springer, T. A., and Snel, B. (1999). Domains in plexins: links to integrins and transcription factors. *Trends Biochem. Sci.* **24**, 261–263.

Brambilla, E., Constantin, B., Drabkin, H., and Roche, J. (2000). Semaphorin SEMA3F localization in malignant human lung and cell lines: a suggested role in cell adhesion and cell migration. *Am. J. Pathol.* **156**, 939–950.

Bruder, D., Probst-Kepper, M., Westendorf, A. M., Geffers, R., Beissert, S., Loser, K., von Boehmer, H., Buer, J., and Hansen, W. (2004). Neuropilin-1: a surface marker of regulatory T cells. *Eur. J. Immunol.* **34**, 623–630.

Capparuccia, L., and Tamagnone, L. (2009). Semaphorin signaling in cancer cells and in cells of the tumor microenvironment—two sides of a coin. *J. Cell Sci.* **122**, 1723–1736.

Carmeliet, P., and Jain, R. K. (2000). Angiogenesis in cancer and other diseases. *Nature* **407**, 249–257.

Carreras, J., Lopez-Guillermo, A., Roncador, G., Villamor, N., Colomo, L., Martinez, A., Hamoudi, R., Howat, W. J., Montserrat, E., and Campo, E. (2009). High numbers of tumor-infiltrating programmed cell death 1-positive regulatory lymphocytes are associated with improved overall survival in follicular lymphoma. *J. Clin. Oncol.* **27**, 1470–1476.

Casazza, A., Fazzari, P., and Tamagnone, L. (2007). Semaphorin signals in cell adhesion and cell migration: functional role and molecular mechanisms. *Adv. Exp. Med. Biol.* **600**, 90–108.

Casazza, A., Finisguerra, V., Capparuccia, L., Camperi, A., Swiercz, J. M., Rizzolio, S., Rolny, C., Christensen, C., Bertotti, A., Sarotto, I., Risio, M., Trusolino, L., *et al.* (2010). Sema3E-Plexin D1 signaling drives human cancer cell invasiveness and metastatic spreading in mice. *J. Clin. Invest.* **120**, 2684–2698.

Casazza, A., Fu, X., Johansson, I., Capparuccia, L., Andersson, F., Giustacchini, A., Squadrito, M. L., Venneri, M. A., Mazzone, M., Larsson, E., Carmeliet, P., De, P. M., *et al.* (2011). Systemic and targeted delivery of semaphorin 3A inhibits tumor angiogenesis and progression in mouse tumor models. *Arterioscler. Thromb. Vasc. Biol.* **31**, 741–749.

Casazza, A., Kigel, B., Maione, F., Capparuccia, L., Kessler, O., Giraudo, E., Mazzone, M., Neufeld, G., and Tamagnone, L. (2012). Tumour growth inhibition and anti-metastatic activity of a mutated furin-resistant Semaphorin 3E isoform. *EMBO Mol. Med.* **4**, 234–250.

Castellani, V., and Rougon, G. (2002). Control of semaphorin signaling. *Curr. Opin. Neurobiol.* **12**, 532.

Castellani, V., Chedotal, A., Schachner, M., Faivre-Sarrailh, C., and Rougon, G. (2000). Analysis of the L1-deficient mouse phenotype reveals cross-talk between Sema3A and L1 signaling pathways in axonal guidance. *Neuron* **27**, 237–249.

Catalano, A., Caprari, P., Moretti, S., Faronato, M., Tamagnone, L., and Procopio, A. (2006). Semaphorin-3A is expressed by tumor cells and alters T-cell signal transduction and function. *Blood* **107**, 3321–3329.

Chauvet, S., Cohen, S., Yoshida, Y., Fekrane, L., Livet, J., Gayet, O., Segu, L., Buhot, M. C., Jessell, T. M., Henderson, C. E., and Mann, F. (2007). Gating of Sema3E/PlexinD1 signaling by neuropilin-1 switches axonal repulsion to attraction during brain development. *Neuron* **56**, 807–822.

Chen, J. J., Lin, Y. C., Yao, P. L., Yuan, A., Chen, H. Y., Shun, C. T., Tsai, M. F., Chen, C. H., and Yang, P. C. (2005). Tumor-associated macrophages: the double-edged sword in cancer progression. *J. Clin. Oncol.* **23**, 953–964.

Choi, Y. I., Duke-Cohan, J. S., Ahmed, W. B., Handley, M. A., Mann, F., Epstein, J. A., Clayton, L. K., and Reinherz, E. L. (2008). PlexinD1 glycoprotein controls migration of positively selected thymocytes into the medulla. *Immunity* **29**, 888–898.

Christensen, C., Ambartsumian, N., Gilestro, G., Thomsen, B., Comoglio, P., Tamagnone, L., Guldberg, P., and Lukanidin, E. (2005). Proteolytic processing converts the repelling signal Sema3E into an inducer of invasive growth and lung metastasis. *Cancer Res.* **65**, 6167–6177.

Conrotto, P., Corso, S., Gamberini, S., Comoglio, P. M., and Giordano, S. (2004). Interplay between scatter factor receptors and B plexins controls invasive growth. *Oncogene* **23,** 5131–5137.

Conrotto, P., Valdembri, D., Corso, S., Serini, G., Tamagnone, L., Comoglio, P. M., Bussolino, F., and Giordano, S. (2005). Sema4D induces angiogenesis through Met recruitment by Plexin B1. *Blood* **105,** 4321–4329.

Curiel, T. J., Coukos, G., Zou, L., Alvarez, X., Cheng, P., Mottram, P., Evdemon-Hogan, M., Conejo-Garcia, J. R., Zhang, L., Burow, M., Zhu, Y., Wei, S., *et al.* (2004). Specific recruitment of regulatory T cells in ovarian carcinoma fosters immune privilege and predicts reduced survival. *Nat. Med.* **10,** 942–949.

De Palma, M., Venneri, M. A., Galli, R., Sergi, S. L., Politi, L. S., Sampaolesi, M., and Naldini, L. (2005). Tie2 identifies a hematopoietic lineage of proangiogenic monocytes required for tumor vessel formation and a mesenchymal population of pericyte progenitors. *Cancer Cell* **8,** 211–226.

Delaire, S., Billard, C., Tordjman, R., Chedotal, A., Elhabazi, A., Bensussan, A., and Boumsell, L. (2001). Biological activity of soluble CD100. II. Soluble CD100, similarly to H-SemaIII, inhibits immune cell migration. *J. Immunol.* **166,** 4348–4354.

Dhanabal, M., Wu, F., Alvarez, E., McQueeney, K. D., Jeffers, M., Macdougall, J., Boldog, F. L., Hackett, C., Shenoy, S., Khramtsov, N., Weiner, J., Lichenstein, H. S., *et al.* (2005). Recombinant semaphorin 6A-1 ectodomain inhibits in vivo growth factor and tumor cell line-induced angiogenesis. *Cancer Biol. Ther.* **4,** 659–668.

Elhabazi, A., Delaire, S., Bensussan, A., Boumsell, L., and Bismuth, G. (2001). Biological activity of soluble CD100. I. The extracellular region of CD100 is released from the surface of T lymphocytes by regulated proteolysis. *J. Immunol.* **166,** 4341–4347.

Fazzari, P., Penachioni, J., Gianola, S., Rossi, F., Eickholt, B. J., Maina, F., Alexopoulou, L., Sottile, A., Comoglio, P. M., Flavell, R. A., and Tamagnone, L. (2007). Plexin-B1 plays a redundant role during mouse development and in tumour angiogenesis. *BMC Dev. Biol.* **7,** 55.

Franco, M., and Tamagnone, L. (2008). Tyrosine phosphorylation in semaphorin signalling: shifting into overdrive. *EMBO Rep.* **9,** 865–871.

Fukushima, Y., Okada, M., Kataoka, H., Hirashima, M., Yoshida, Y., Mann, F., Gomi, F., Nishida, K., Nishikawa, S., and Uemura, A. (2011). Sema3E-PlexinD1 signaling selectively suppresses disoriented angiogenesis in ischemic retinopathy in mice. *J. Clin. Invest.* **121,** 1974–1985.

Galani, E., Sgouros, J., Petropoulou, C., Janinis, J., Aravantinos, G., Donysiou-Asteriou, D., Skarlos, D., and Gonos, E. (2002). Correlation of MDR-1, nm23-H1 and H Sema E gene expression with histopathological findings and clinical outcome in ovarian and breast cancer patients. *Anticancer Res.* **22,** 2275–2280.

Gherardi, E., Love, C. A., Esnouf, R. M., and Jones, E. Y. (2004). The sema domain. *Curr. Opin. Struct. Biol.* **14,** 669–678.

Giacobini, P., Messina, A., Morello, F., Ferraris, N., Corso, S., Penachioni, J., Giordano, S., Tamagnone, L., and Fasolo, A. (2008). Semaphorin 4D regulates gonadotropin hormone-releasing hormone-1 neuronal migration through PlexinB1-Met complex. *J. Cell Biol.* **183,** 555–566.

Giordano, S., Corso, S., Conrotto, P., Artigiani, S., Gilestro, G., Barberis, D., Tamagnone, L., and Comoglio, P. M. (2002). The semaphorin 4D receptor controls invasive growth by coupling with Met. *Nat. Cell Biol.* **4,** 720–724.

Gitler, A. D., Lu, M. M., and Epstein, J. A. (2004). PlexinD1 and semaphorin signaling are required in endothelial cells for cardiovascular development. *Dev. Cell* **7,** 107–116.

Gray, M. J., Wey, J. S., Belcheva, A., McCarty, M. F., Trevino, J. G., Evans, D. B., Ellis, L. M., and Gallick, G. E. (2005). Neuropilin-1 suppresses tumorigenic properties in a human pancreatic adenocarcinoma cell line lacking neuropilin-1 coreceptors. *Cancer Res.* **65,** 3664–3670.

Gray, M. J., Van Buren, G., Dallas, N. A., Xia, L., Wang, X., Yang, A. D., Somcio, R. J., Lin, Y. G., Lim, S., Fan, F., Mangala, L. S., Arumugam, T., *et al.* (2008). Therapeutic targeting of neuropilin-2 on colorectal carcinoma cells implanted in the murine liver. *J. Natl. Cancer Inst.* **100,** 109–120.

Gu, C., Yoshida, Y., Livet, J., Reimert, D. V., Mann, F., Merte, J., Henderson, C. E., Jessell, T. M., Kolodkin, A. L., and Ginty, D. D. (2005). Semaphorin 3E and plexin-D1 control vascular pattern independently of neuropilins. *Science* **307,** 265–268.

Guo, W., Pylayeva, Y., Pepe, A., Yoshioka, T., Muller, W. J., Inghirami, G., and Giancotti, F. G. (2006). Beta 4 integrin amplifies ErbB2 signaling to promote mammary tumorigenesis. *Cell* **126,** 489–502.

Guttmann-Raviv, N., Shraga-Heled, N., Varshavsky, A., Guimaraes-Sternberg, C., Kessler, O., and Neufeld, G. (2007). Semaphorin-3A and semaphorin-3F work together to repel endothelial cells and to inhibit their survival by induction of apoptosis. *J. Biol. Chem.* **282,** 26294–26305.

Hanahan, D., and Weinberg, R. A. (2000). The hallmarks of cancer. *Cell* **100,** 57–70.

Handa, A., Tokunaga, T., Tsuchida, T., Lee, Y. H., Kijima, H., Yamazaki, H., Ueyama, Y., Fukuda, H., and Nakamura, M. (2000). Neuropilin-2 expression affects the increased vascularization and is a prognostic factor in osteosarcoma. *Int. J. Oncol.* **17,** 291–295.

He, Z., and Tessier-Lavigne, M. (1997). Neuropilin is a receptor for the axonal chemorepellent Semaphorin III. *Cell* **90,** 739–751.

He, H., Yang, T., Terman, J. R., and Zhang, X. (2009). Crystal structure of the plexin A3 intracellular region reveals an autoinhibited conformation through active site sequestration. *Proc. Natl. Acad. Sci. USA* **106,** 15610–15615.

Herman, J. G., and Meadows, G. G. (2007). Increased class 3 semaphorin expression modulates the invasive and adhesive properties of prostate cancer cells. *Int. J. Oncol.* **30,** 1231–1238.

Holmes, S., Downs, A. M., Fosberry, A., Hayes, P. D., Michalovich, D., Murdoch, P., Moores, K., Fox, J., Deen, K., Pettman, G., Wattam, T., and Lewis, C. (2002). Sema7A is a potent monocyte stimulator. *Scand. J. Immunol.* **56,** 270–275.

Ito, T., Kagoshima, M., Sasaki, Y., Li, C., Udaka, N., Kitsukawa, T., Fujisawa, H., Taniguchi, M., Yagi, T., Kitamura, H., and Goshima, Y. (2000). Repulsive axon guidance molecule Sema3A inhibits branching morphogenesis of fetal mouse lung. *Mech. Dev.* **97,** 35–45.

Janssen, B. J., Robinson, R. A., Perez-Branguli, F., Bell, C. H., Mitchell, K. J., Siebold, C., and Jones, E. Y. (2010). Structural basis of semaphorin-plexin signalling. *Nature* **467,** 1118–1122.

Ji, Y., and Zhang, W. (2010). Th17 cells: positive or negative role in tumor? *Cancer Immunol. Immunother.* **59,** 979–987.

Ji, J. D., Park-Min, K. H., and Ivashkiv, L. B. (2009). Expression and function of semaphorin 3A and its receptors in human monocyte-derived macrophages. *Hum. Immunol.* **70,** 211–217.

Kaifu, T., Nakahara, J., Inui, M., Mishima, K., Momiyama, T., Kaji, M., Sugahara, A., Koito, H., Ujike-Asai, A., Nakamura, A., Kanazawa, K., Tan-Takeuchi, K., *et al.* (2003). Osteopetrosis and thalamic hypomyelinosis with synaptic degeneration in DAP12-deficient mice. *J. Clin. Invest.* **111,** 323–332.

Kamiya, T., Kawakami, T., Abe, Y., Nishi, M., Onoda, N., Miyazaki, N., Oida, Y., Yamazaki, H., Ueyama, Y., and Nakamura, M. (2006). The preserved expression of neuropilin (NRP) 1 contributes to a better prognosis in colon cancer. *Oncol. Rep.* **15,** 369–373.

Kessler, O., Shraga-Heled, N., Lange, T., Gutmann-Raviv, N., Sabo, E., Baruch, L., Machluf, M., and Neufeld, G. (2004). Semaphorin-3F is an inhibitor of tumor angiogenesis. *Cancer Res.* **64,** 1008–1015.

Kigel, B., Varshavsky, A., Kessler, O., and Neufeld, G. (2008). Successful inhibition of tumor development by specific class-3 semaphorins is associated with expression of appropriate semaphorin receptors by tumor cells. *PLoS One* **3,** e3287.

Kikutani, H., and Kumanogoh, A. (2003). Semaphorins in interactions between T cells and antigen-presenting cells. *Nat. Rev. Immunol.* **3,** 159–167.

Kim, J., Oh, W. J., Gaiano, N., Yoshida, Y., and Gu, C. (2011). Semaphorin 3E-Plexin-D1 signaling regulates VEGF function in developmental angiogenesis via a feedback mechanism. *Genes Dev.* **25,** 1399–1411.

Kolodkin, A. L., and Ginty, D. D. (1997). Steering clear of semaphorins: neuropilins sound the retreat. *Neuron* **19,** 1159–1162.

Kolodkin, A. L., Matthes, D. J., and Goodman, C. S. (1993). The semaphorin genes encode a family of transmembrane and secreted growth cone guidance molecules. *Cell* **75,** 1389–1399.

Kolodkin, A. L., Levengood, D. V., Rowe, E. G., Tai, Y. T., Giger, R. J., and Ginty, D. D. (1997). Neuropilin is a semaphorin III receptor. *Cell* **90,** 753–762.

Koppel, A. M., Feiner, L., Kobayashi, H., and Raper, J. A. (1997). A 70 amino acid region within the semaphorin domain activates specific cellular response of semaphorin family members. *Neuron* **19,** 531–537.

Kruger, R. P., Aurandt, J., and Guan, K. L. (2005). Semaphorins command cells to move. *Nat. Rev. Mol. Cell Biol.* **6,** 789–800.

Kumanogoh, A., Watanabe, C., Lee, I., Wang, X., Shi, W., Araki, H., Hirata, H., Iwahori, K., Uchida, J., Yasui, T., Matsumoto, M., Yoshida, K., *et al.* (2000). Identification of CD72 as a lymphocyte receptor for the class IV semaphorin CD100: a novel mechanism for regulating B cell signaling. *Immunity* **13,** 621–631.

Kumanogoh, A., Marukawa, S., Suzuki, K., Takegahara, N., Watanabe, C., Ch'ng, E., Ishida, I., Fujimura, H., Sakoda, S., Yoshida, K., and Kikutani, H. (2002a). Class IV semaphorin Sema4A enhances T-cell activation and interacts with Tim-2. *Nature* **419,** 629–633.

Kumanogoh, A., Suzuki, K., Ch'ng, E., Watanabe, C., Marukawa, S., Takegahara, N., Ishida, I., Sato, T., Habu, S., Yoshida, K., Shi, W., and Kikutani, H. (2002b). Requirement for the lymphocyte semaphorin, CD100, in the induction of antigen-specific T cells and the maturation of dendritic cells. *J. Immunol.* **169,** 1175–1181.

Kumanogoh, A., Shikina, T., Suzuki, K., Uematsu, S., Yukawa, K., Kashiwamura, S., Tsutsui, H., Yamamoto, M., Takamatsu, H., Ko-Mitamura, E. P., Takegahara, N., Marukawa, S., *et al.* (2005). Nonredundant roles of Sema4A in the immune system: defective T cell priming and Th1/Th2 regulation in Sema4A-deficient mice. *Immunity* **22,** 305–316.

Kusy, S., Nasarre, P., Chan, D., Potiron, V., Meyronet, D., Gemmill, R. M., Constantin, B., Drabkin, H. A., and Roche, J. (2005). Selective suppression of in vivo tumorigenicity by semaphorin SEMA3F in lung cancer cells. *Neoplasia* **7,** 457–465.

Lepelletier, Y., Moura, I. C., Hadj-Slimane, R., Renand, A., Fiorentino, S., Baude, C., Shirvan, A., Barzilai, A., and Hermine, O. (2006). Immunosuppressive role of semaphorin-3A on T cell proliferation is mediated by inhibition of actin cytoskeleton reorganization. *Eur. J. Immunol.* **36,** 1782–1793.

Liang, W. C., Dennis, M. S., Stawicki, S., Chanthery, Y., Pan, Q., Chen, Y., Eigenbrot, C., Yin, J., Koch, A. W., Wu, X., Ferrara, N., Bagri, A., *et al.* (2007). Function blocking antibodies to neuropilin-1 generated from a designed human synthetic antibody phage library. *J. Mol. Biol.* **366,** 815–829.

Luo, Y., Raible, D., and Raper, J. A. (1993). Collapsin: a protein in brain that induces the collapse and paralysis of neuronal growth cones. *Cell* **75,** 217–227.

Maestrini, E., Tamagnone, T., Longati, P., Cremona, O., Gulisano, M., Bione, S., Tamanini, F., Neel, B. G., Toniolo, D., and Comoglio, P. M. (1996). A family of transmembrane proteins with homology to the MET-hepatocyte growth factor receptor. *Proc. Natl. Acad. Sci. USA* **93,** 674–678.

Maione, F., Molla, F., Meda, C., Latini, R., Zentilin, L., Giacca, M., Seano, G., Serini, G., Bussolino, F., and Giraudo, E. (2009). Semaphorin 3A is an endogenous angiogenesis inhibitor that blocks tumor growth and normalizes tumor vasculature in transgenic mouse models. *J. Clin. Invest.* **119**, 3356–3372.

Mantovani, A., and Sica, A. (2010). Macrophages, innate immunity and cancer: balance, tolerance, and diversity. *Curr. Opin. Immunol.* **22**, 231–237.

Marcus, K., Johnson, M., Adam, R. M., O'Reilly, M. S., Donovan, M., Atala, A., Freeman, M. R., and Soker, S. (2005). Tumor cell-associated neuropilin-1 and vascular endothelial growth factor expression as determinants of tumor growth in neuroblastoma. *Neuropathology* **25**, 178–187.

Martin-Satue, M., and Blanco, J. (1999). Identification of semaphorin E gene expression in metastatic human lung adenocarcinoma cells by mRNA differential display. *J. Surg. Oncol.* **72**, 18–23.

Merlo, A., Casalini, P., Carcangiu, M. L., Malventano, C., Triulzi, T., Menard, S., Tagliabue, E., and Balsari, A. (2009). FOXP3 expression and overall survival in breast cancer. *J. Clin. Oncol.* **27**, 1746–1752.

Miao, H. Q., Soker, S., Feiner, L., Alonso, J. L., Raper, J. A., and Klagsbrun, M. (1999). Neuropilin-1 mediates collapsin-1/semaphorin III inhibition of endothelial cell motility: functional competition of collapsin-1 and vascular endothelial growth factor-165. *J. Cell Biol.* **146**, 233–242.

Miao, H. Q., Lee, P., Lin, H., Soker, S., and Klagsbrun, M. (2000). Neuropilin-1 expression by tumor cells promotes tumor angiogenesis and progression. *FASEB J.* **14**, 2532–2539.

Moretti, S., Procopio, A., Lazzarini, R., Rippo, M. R., Testa, R., Marra, M., Tamagnone, L., and Catalano, A. (2008). Semaphorin3A signaling controls Fas (CD95)-mediated apoptosis by promoting Fas translocation into lipid rafts. *Blood* **111**, 2290–2299.

Moriya, J., Minamino, T., Tateno, K., Okada, S., Uemura, A., Shimizu, I., Yokoyama, M., Nojima, A., Okada, M., Koga, H., and Komuro, I. (2010). Inhibition of semaphorin as a novel strategy for therapeutic angiogenesis. *Circ. Res.* **106**, 391–398.

Nasarre, P., Constantin, B., Rouhaud, L., Harnois, T., Raymond, G., Drabkin, H. A., Bourmeyster, N., and Roche, J. (2003). Semaphorin SEMA3F and VEGF have opposing effects on cell attachment and spreading. *Neoplasia* **5**, 83–92.

Nogi, T., Yasui, N., Mihara, E., Matsunaga, Y., Noda, M., Yamashita, N., Toyofuku, T., Uchiyama, S., Goshima, Y., Kumanogoh, A., and Takagi, J. (2010). Structural basis for semaphorin signalling through the plexin receptor. *Nature* **467**, 1123–1127.

O'Connor, B. P., Eun, S. Y., Ye, Z., Zozulya, A. L., Lich, J. D., Moore, C. B., Iocca, H. A., Roney, K. E., Holl, E. K., Wu, Q. P., van Deventer, H. W., Fabry, Z., *et al.* (2008). Semaphorin 6D regulates the late phase of CD4+ T cell primary immune responses. *Proc. Natl. Acad. Sci. USA* **105**, 13015–13020.

Oinuma, I., Ishikawa, Y., Katoh, H., and Negishi, M. (2004). The Semaphorin 4D receptor Plexin-B1 is a GTPase activating protein for R-Ras. *Science* **305**, 862–865.

Pan, Q., Chanthery, Y., Liang, W. C., Stawicki, S., Mak, J., Rathore, N., Tong, R. K., Kowalski, J., Yee, S. F., Pacheco, G., Ross, S., Cheng, Z., *et al.* (2007). Blocking neuropilin-1 function has an additive effect with anti-VEGF to inhibit tumor growth. *Cancer Cell* **11**, 53–67.

Pan, G., Lv, H., Ren, H., Wang, Y., Liu, Y., Jiang, H., and Wen, J. (2010). Elevated expression of semaphorin 5A in human gastric cancer and its implication in carcinogenesis. *Life Sci.* **86**, 139–144.

Parnes, J. R., and Pan, C. (2000). CD72, a negative regulator of B-cell responsiveness. *Immunol. Rev.* **176**, 75–85.

Pasterkamp, R. J., Peschon, J. J., Spriggs, M. K., and Kolodkin, A. L. (2003). Semaphorin 7A promotes axon outgrowth through integrins and MAPKs. *Nature* **424**, 398–405.

Pellet-Many, C., Frankel, P., Jia, H., and Zachary, I. (2008). Neuropilins: structure, function and role in disease. *Biochem. J.* **411**, 211–226.

Petersen, R. P., Campa, M. J., Sperlazza, J., Conlon, D., Joshi, M. B., Harpole, D. H., Jr., and Patz, E. F., Jr. (2006). Tumor infiltrating Foxp3+ regulatory T-cells are associated with recurrence in pathologic stage I NSCLC patients. *Cancer* **107**, 2866–2872.

Potiron, V. A., Sharma, G., Nasarre, P., Clarhaut, J. A., Augustin, H. G., Gemmill, R. M., Roche, J., and Drabkin, H. A. (2007). Semaphorin SEMA3F affects multiple signaling pathways in lung cancer cells. *Cancer Res.* **67**, 8708–8715.

Potiron, V. A., Roche, J., and Drabkin, H. A. (2009). Semaphorins and their receptors in lung cancer. *Cancer Lett.* **273**, 1–14.

Pucci, F., Venneri, M. A., Biziato, D., Nonis, A., Moi, D., Sica, A., Di, S. C., Naldini, L., and De, P. M. (2009). A distinguishing gene signature shared by tumor-infiltrating Tie2-expressing monocytes, blood "resident" monocytes, and embryonic macrophages suggests common functions and developmental relationships. *Blood* **114**, 901–914.

Qian, B. Z., and Pollard, J. W. (2010). Macrophage diversity enhances tumor progression and metastasis. *Cell* **141**, 39–51.

Rich, J. N., Hans, C., Jones, B., Iversen, E. S., McLendon, R. E., Rasheed, B. K., Dobra, A., Dressman, H. K., Bigner, D. D., Nevins, J. R., and West, M. (2005). Gene expression profiling and genetic markers in glioblastoma survival. *Cancer Res.* **65**, 4051–4058.

Rieger, J., Wick, W., and Weller, M. (2003). Human malignant glioma cells express semaphorins and their receptors, neuropilins and plexins. *Glia* **42**, 379–389.

Rizzolio, S., and Tamagnone, L. (2011). Multifaceted role of neuropilins in cancer. *Curr. Med. Chem.* **18**, 3563–3575.

Roche, J., Boldog, F., Robinson, M., Robinson, L., Varella-Garcia, M., Swanton, M., Waggoner, B., Fishel, R., Franklin, W., Gemmill, R., and Drabkin, H. (1996). Distinct 3p21.3 deletions in lung cancer and identification of a new human semaphorin. *Oncogene* **12**, 1289–1297.

Rody, A., Holtrich, U., Gaetje, R., Gehrmann, M., Engels, K., von Minckwitz, G., Loibl, S., Diallo-Danebrock, R., Ruckhaberle, E., Metzler, D., Ahr, A., Solbach, C., *et al.* (2007). Poor outcome in estrogen receptor-positive breast cancers predicted by loss of plexin B1. *Clin. Cancer Res.* **13**, 1115–1122.

Rolny, C., Capparuccia, L., Casazza, A., Mazzone, M., Vallario, A., Cignetti, A., Medico, E., Carmeliet, P., Comoglio, P. M., and Tamagnone, L. (2008). The tumor suppressor semaphorin 3B triggers a prometastatic program mediated by interleukin 8 and the tumor microenvironment. *J. Exp. Med.* **205**, 1155–1171.

Roodink, I., Raats, J., Van Der, Z. B., Verrijp, K., Kusters, B., Van Bokhoven, H., Linkels, M., de Waal, R. M., and Leenders, W. P. (2005). Plexin D1 expression is induced on tumor vasculature and tumor cells: a novel target for diagnosis and therapy? *Cancer Res.* **65**, 8317–8323.

Roodink, I., Kats, G., van Kempen, L., Grunberg, M., Maass, C., Verrijp, K., Raats, J., and Leenders, W. (2008). Semaphorin 3E expression correlates inversely with Plexin D1 during tumor progression. *Am. J. Pathol.* **173**, 1873–1881.

Ruffell, B., DeNardo, D. G., Affara, N. I., and Coussens, L. M. (2010). Lymphocytes in cancer development: polarization towards pro-tumor immunity. *Cytokine Growth Factor Rev.* **21**, 3–10.

Ryder, M., Ghossein, R. A., Ricarte-Filho, J. C., Knauf, J. A., and Fagin, J. A. (2008). Increased density of tumor-associated macrophages is associated with decreased survival in advanced thyroid cancer. *Endocr. Relat. Cancer* **15**, 1069–1074.

Sadanandam, A., Rosenbaugh, E. G., Singh, S., Varney, M., and Singh, R. K. (2010a). Semaphorin 5A promotes angiogenesis by increasing endothelial cell proliferation, migration, and decreasing apoptosis. *Microvasc. Res.* **79**, 1–9.

Sadanandam, A., Varney, M. L., Singh, S., Ashour, A. E., Moniaux, N., Deb, S., Lele, S. M., Batra, S. K., and Singh, R. K. (2010b). High gene expression of semaphorin 5A in pancreatic cancer is associated with tumor growth, invasion and metastasis. *Int. J. Cancer* **127,** 1373–1383.

Sakurai, A., Gavard, J., Annas-Linhares, Y., Basile, J. R., Amornphimoltham, P., Palmby, T. R., Yagi, H., Zhang, F., Randazzo, P. A., Li, X., Weigert, R., and Gutkind, J. S. (2010). Semaphorin 3E initiates antiangiogenic signaling through plexin D1 by regulating Arf6 and R-Ras. *Mol. Cell. Biol.* **30,** 3086–3098.

Sarris, M., Andersen, K. G., Randow, F., Mayr, L., and Betz, A. G. (2008). Neuropilin-1 expression on regulatory T cells enhances their interactions with dendritic cells during antigen recognition. *Immunity* **28,** 402–413.

Scott, G. A., McClelland, L. A., and Fricke, A. F. (2008). Semaphorin 7a promotes spreading and dendricity in human melanocytes through beta1-integrins. *J. Invest. Dermatol.* **128,** 151–161.

Serini, G., Valdembri, D., Zanivan, S., Morterra, G., Burkhardt, C., Caccavari, F., Zammataro, L., Primo, L., Tamagnone, L., Logan, M., Tessier-Lavigne, M., Taniguchi, M., *et al.* (2003). Class 3 semaphorins control vascular morphogenesis by inhibiting integrin function. *Nature* **424,** 391–397.

Shi, W., Kumanogoh, A., Watanabe, C., Uchida, J., Wang, X., Yasui, T., Yukawa, K., Ikawa, M., Okabe, M., Parnes, J. R., Yoshida, K., and Kikutani, H. (2000). The class IV semaphorin CD100 plays nonredundant roles in the immune system: defective B and T cell activation in CD100-deficient mice. *Immunity* **13,** 633–642.

Shimizu, A., Mammoto, A., Italiano, J. E., Jr., Pravda, E., Dudley, A. C., Ingber, D. E., and Klagsbrun, M. (2008). ABL2/ARG tyrosine kinase mediates SEMA3F-induced RhoA inactivation and cytoskeleton collapse in human glioma cells. *J. Biol. Chem.* **283,** 27230–27238.

Siddiqui, S. A., Frigola, X., Bonne-Annee, S., Mercader, M., Kuntz, S. M., Krambeck, A. E., Sengupta, S., Dong, H., Cheville, J. C., Lohse, C. M., Krco, C. J., Webster, W. S., *et al.* (2007). Tumor-infiltrating Foxp3-CD4+CD25+ T cells predict poor survival in renal cell carcinoma. *Clin. Cancer Res.* **13,** 2075–2081.

Sierra, J. R., Corso, S., Caione, L., Cepero, V., Conrotto, P., Cignetti, A., Piacibello, W., Kumanogoh, A., Kikutani, H., Comoglio, P. M., Tamagnone, L., and Giordano, S. (2008). Tumor angiogenesis and progression are enhanced by Sema4D produced by tumor-associated macrophages. *J. Exp. Med.* **205,** 1673–1685.

Soker, S., Takashima, S., Miao, H. Q., Neufeld, G., and Klagsbrun, M. (1998). Neuropilin-1 is expressed by endothelial and tumor cells as an isoform-specific receptor for vascular endothelial growth factor. *Cell* **92,** 735–745.

Suzuki, K., Okuno, T., Yamamoto, M., Pasterkamp, R. J., Takegahara, N., Takamatsu, H., Kitao, T., Takagi, J., Rennert, P. D., Kolodkin, A. L., Kumanogoh, A., and Kikutani, H. (2007). Semaphorin 7A initiates T-cell-mediated inflammatory responses through alpha1-beta1 integrin. *Nature* **446,** 680–684.

Suzuki, K., Kumanogoh, A., and Kikutani, H. (2008). Semaphorins and their receptors in immune cell interactions. *Nat. Immunol.* **9,** 17–23.

Swiercz, J. M., Kuner, R., and Offermanns, S. (2004). Plexin-B1/RhoGEF-mediated RhoA activation involves the receptor tyrosine kinase ErbB-2. *J. Cell Biol.* **165,** 869–880.

Swiercz, J. M., Worzfeld, T., and Offermanns, S. (2008). ErbB-2 and met reciprocally regulate cellular signaling via plexin-B1. *J. Biol. Chem.* **283,** 1893–1901.

Swiercz, J. M., Worzfeld, T., and Offermanns, S. (2009). Semaphorin 4D signaling requires the recruitment of phospholipase C gamma into the plexin-B1 receptor complex. *Mol. Cell. Biol.* **29,** 6321–6334.

Takahashi, T., Fournier, A., Nakamura, F., Wang, L. H., Murakami, Y., Kalb, R. G., Fujisawa, H., and Strittmatter, S. M. (1999). Plexin-neuropilin-1 complexes form functional semaphorin-3A receptors. *Cell* **99,** 59–69.

Takamatsu, H., Takegahara, N., Nakagawa, Y., Tomura, M., Taniguchi, M., Friedel, R. H., Rayburn, H., Tessier-Lavigne, M., Yoshida, Y., Okuno, T., Mizui, M., Kang, S., *et al.* (2010). Semaphorins guide the entry of dendritic cells into the lymphatics by activating myosin II. *Nat. Immunol.* **11**, 594–600.

Takegahara, N., Takamatsu, H., Toyofuku, T., Tsujimura, T., Okuno, T., Yukawa, K., Mizui, M., Yamamoto, M., Prasad, D. V., Suzuki, K., Ishii, M., Terai, K., *et al.* (2006). Plexin-A1 and its interaction with DAP12 in immune responses and bone homeostasis. *Nat. Cell Biol.* **8**, 615–622.

Tamagnone, L., and Comoglio, P. M. (2000). Signalling by semaphorin receptors: cell guidance and beyond. *Trends Cell Biol.* **10**, 377–383.

Tamagnone, L., and Comoglio, P. M. (2004). To move or not to move? Semaphorin signalling in cell migration. *EMBO Rep.* **5**, 356–361.

Tamagnone, L., and Mazzone, M. (2011). Semaphorin signals on the road of endothelial tip cells. *Dev. Cell* **21**, 189–190.

Tamagnone, L., Artigiani, S., Chen, H., He, Z., Ming, G. I., Song, H., Chedotal, A., Winberg, M. L., Goodman, C. S., Poo, M., Tessier-Lavigne, M., and Comoglio, P. M. (1999). Plexins are a large family of receptors for transmembrane, secreted, and GPI-anchored semaphorins in vertebrates. *Cell* **99**, 71–80.

Terman, J. R., Mao, T., Pasterkamp, R. J., Yu, H. H., and Kolodkin, A. L. (2002). MICALs, a family of conserved flavoprotein oxidoreductases, function in plexin-mediated axonal repulsion. *Cell* **109**, 887–900.

Tomizawa, Y., Sekido, Y., Kondo, M., Gao, B., Yokota, J., Roche, J., Drabkin, H., Lerman, M. I., Gazdar, A. F., and Minna, J. D. (2001). Inhibition of lung cancer cell growth and induction of apoptosis after reexpression of 3p21.3 candidate tumor suppressor gene SEMA3B. *Proc. Natl. Acad. Sci. USA* **98**, 13954–13959.

Tong, Y., Hota, P. K., Penachioni, J. Y., Hamaneh, M. B., Kim, S., Alviani, R. S., Shen, L., He, H., Tempel, W., Tamagnone, L., Park, H. W., and Buck, M. (2009). Structure and function of the intracellular region of the plexin-b1 transmembrane receptor. *J. Biol. Chem.* **284**, 35962–35972.

Tordjman, R., Lepelletier, Y., Lemarchandel, V., Cambot, M., Gaulard, P., Hermine, O., and Romeo, P. H. (2002). A neuronal receptor, neuropilin-1, is essential for the initiation of the primary immune response. *Nat. Immunol.* **3**, 477–482.

Toyofuku, T., Zhang, H., Kumanogoh, A., Takegahara, N., Suto, F., Kamei, J., Aoki, K., Yabuki, M., Hori, M., Fujisawa, H., and Kikutani, H. (2004a). Dual roles of Sema6D in cardiac morphogenesis through region-specific association of its receptor, Plexin-A1, with off-track and vascular endothelial growth factor receptor type 2. *Genes Dev.* **18**, 435–447.

Toyofuku, T., Zhang, H., Kumanogoh, A., Takegahara, N., Yabuki, M., Harada, K., Hori, M., and Kikutani, H. (2004b). Guidance of myocardial patterning in cardiac development by Sema6D reverse signalling. *Nat. Cell Biol.* **6**, 1204–1211.

Toyofuku, T., Yabuki, M., Kamei, J., Kamei, M., Makino, N., Kumanogoh, A., and Hori, M. (2007). Semaphorin-4A, an activator for T-cell-mediated immunity, suppresses angiogenesis via Plexin-D1. *EMBO J.* **26**, 1373–1384.

Trusolino, L., Bertotti, A., and Comoglio, P. M. (2001). A signaling adapter function for alpha6beta4 integrin in the control of HGF-dependent invasive growth. *Cell* **107**, 643–654.

Tse, C., Xiang, R. H., Bracht, T., and Naylor, S. L. (2002). Human semaphorin 3B (SEMA3B) located at chromosome 3p21.3 suppresses tumor formation in an adenocarcinoma cell line. *Cancer Res.* **62**, 542–546.

Vacca, A., Scavelli, C., Serini, G., Di, P. G., Cirulli, T., Merchionne, F., Ribatti, D., Bussolino, F., Guidolin, D., Piaggio, G., Bacigalupo, A., and Dammacco, F. (2006). Loss of inhibitory semaphorin 3A (SEMA3A) autocrine loops in bone marrow endothelial cells of patients with multiple myeloma. *Blood* **108**, 1661–1667.

Valente, G., Nicotra, G., Arrondini, M., Castino, R., Capparuccia, L., Prat, M., Kerim, S., Tamagnone, L., and Isidoro, C. (2009). Co-expression of plexin-B1 and Met in human breast and ovary tumours enhances the risk of progression. *Cell. Oncol.* **31**, 423–436.

Venneri, M. A., De, P. M., Ponzoni, M., Pucci, F., Scielzo, C., Zonari, E., Mazzieri, R., Doglioni, C., and Naldini, L. (2007). Identification of proangiogenic TIE2-expressing monocytes (TEMs) in human peripheral blood and cancer. *Blood* **109**, 5276–5285.

Wakabayashi, O., Yamazaki, K., Oizumi, S., Hommura, F., Kinoshita, I., Ogura, S., Dosaka-Akita, H., and Nishimura, M. (2003). CD4+ T cells in cancer stroma, not CD8+ T cells in cancer cell nests, are associated with favorable prognosis in human non-small cell lung cancers. *Cancer Sci.* **94**, 1003–1009.

Walzer, T., Galibert, L., and De Smedt, T. (2005). Poxvirus semaphorin A39R inhibits phagocytosis by dendritic cells and neutrophils. *Eur. J. Immunol.* **35**, 391–398.

Wang, L., Yi, T., Kortylewski, M., Pardoll, D. M., Zeng, D., and Yu, H. (2009). IL-17 can promote tumor growth through an IL-6-Stat3 signaling pathway. *J. Exp. Med.* **206**, 1457–1464.

Winberg, M. L., Tamagnone, L., Bai, J., Comoglio, P. M., Montell, D., and Goodman, C. S. (2001). The transmembrane protein Off-track associates with Plexins and functions downstream of Semaphorin signaling during axon guidance. *Neuron* **32**, 53–62.

Wong, O. G., Nitkunan, T., Oinuma, I., Zhou, C., Blanc, V., Brown, R. S., Bott, S. R., Nariculam, J., Box, G., Munson, P., Constantinou, J., Feneley, M. R., *et al.* (2007). Plexin-B1 mutations in prostate cancer. *Proc. Natl. Acad. Sci. USA* **104**, 19040–19045.

Xiang, R. H., Hensel, C. H., Garcia, D. K., Carlson, H. C., Kok, K., Daly, M. C., Kerbacher, K., van den Berg, A., Veldhuis, P., Buys, C. H., and Naylor, S. L. (1996). Isolation of the human semaphorin III/F gene (SEMA3F) at chromosome 3p21, a region deleted in lung cancer. *Genomics* **32**, 39–48.

Yamamoto, M., Suzuki, K., Okuno, T., Ogata, T., Takegahara, N., Takamatsu, H., Mizui, M., Taniguchi, M., Chedotal, A., Suto, F., Fujisawa, H., Kumanogoh, A., *et al.* (2008). Plexin-A4 negatively regulates T lymphocyte responses. *Int. Immunol.* **20**, 413–420.

Zacchigna, S., Pattarini, L., Zentilin, L., Moimas, S., Carrer, A., Sinigaglia, M., Arsic, N., Tafuro, S., Sinagra, G., and Giacca, M. (2008). Bone marrow cells recruited through the neuropilin-1 receptor promote arterial formation at the sites of adult neoangiogenesis in mice. *J. Clin. Invest.* **118**, 2062–2075.

Zhang, J. P., Yan, J., Xu, J., Pang, X. H., Chen, M. S., Li, L., Wu, C., Li, S. P., and Zheng, L. (2009). Increased intratumoral IL-17-producing cells correlate with poor survival in hepatocellular carcinoma patients. *J. Hepatol.* **50**, 980–989.

Zhou, Y., Gunput, R. A., and Pasterkamp, R. J. (2008). Semaphorin signaling: progress made and promises ahead. *Trends Biochem. Sci.* **33**, 161–170.

Zhou, L., Chong, M. M., and Littman, D. R. (2009). Plasticity of CD4+ T cell lineage differentiation. *Immunity* **30**, 646–655.

Zhu, L., Bergmeier, W., Wu, J., Jiang, H., Stalker, T. J., Cieslak, M., Fan, R., Boumsell, L., Kumanogoh, A., Kikutani, H., Tamagnone, L., Wagner, D. D., *et al.* (2007). Regulated surface expression and shedding support a dual role for semaphorin 4D in platelet responses to vascular injury. *Proc. Natl. Acad. Sci. USA* **104**, 1621–1626.

Zhu, X. D., Zhang, J. B., Zhuang, P. Y., Zhu, H. G., Zhang, W., Xiong, Y. Q., Wu, W. Z., Wang, L., Tang, Z. Y., and Sun, H. C. (2008). High expression of macrophage colony-stimulating factor in peritumoral liver tissue is associated with poor survival after curative resection of hepatocellular carcinoma. *J. Clin. Oncol.* **26**, 2707–2716.

Zygmunt, T., Gay, C. M., Blondelle, J., Singh, M. K., Flaherty, K. M., Means, P. C., Herwig, L., Krudewig, A., Belting, H. G., Affolter, M., Epstein, J. A., and Torres-Vazquez, J. (2011). Semaphorin-PlexinD1 signaling limits angiogenic potential via the VEGF Decoy receptor sFlt1. *Dev. Cell* **21**, 301–314.

Guidance for Life, Cell Death, and Colorectal Neoplasia by Netrin Dependence Receptors

Christian Gespach

INSERM U938, UMR_S938, Molecular and Clinical Oncology, UPMC Université Paris VI, Hôpital Saint-Antoine, Paris Cedex, France

---

This review is focusing on a critical mediator of embryonic and postnatal development with multiple implications in inflammation, neoplasia, and other pathological situations in brain and peripheral tissues. These morphogenetic guidance and dependence processes are involved in several malignancies targeting the epithelial and immune systems including the progression of human colorectal cancers. We consider the most important findings and their impact on basic, translational, and clinical cancer research. Expected information can bring new cues for innovative, efficient, and safe strategies of personalized medicine based on molecular markers, protagonists, signaling networks, and effectors inherent to the Netrin axis in pathophysiological states.

Advances in CANCER RESEARCH, Volume 114

0065-230X/12 $35.00
DOI: 10.1016/B978-0-12-386503-8.00004-1

I. INTRODUCTION AND CONCEPTS

During their intrepid and long-distance journey across the seas, invaders and conquistadors were using magnetic compass with the help of the sextant, sun, moon, and stars to determine their location and direction in the oceans. Discovery of *terra incognita* and precise navigation to new colonies was accomplished by composite systems using celestial navigation, semaphore, and flag signals at the harbor entrance. These maritime explorations, migrations, and discoveries helped to establish fruitful overland connections and reciprocal exchanges between distant worlds.

The formation of specific connections between distant territories and structures during fertilization, embryogenesis, and postnatal development is also based on pathfinding, migration, guidance, and tissue invasion of cellular systems by neural, endothelial, immune, and stromal cells through specific trajectories and locomotion systems. Chemical gradients, electrical signals, and transient and persistent cell–cell contacts between different tissues allow the assemblage, rearrangement, and differentiation of new functional domains and behaviors adapted to integrative morphogenesis and normal development. These mechanisms are the basis of the generation and organization of relatively simple organisms such as worms and of more advanced forms, particularly vertebrates with a highly developed cerebellum controlling sensory and motor functions connected with all parts of the body in mammals. Axonal guidance, cellular migration, and tropism are playing major roles on the growth, development, and functions of mature brain and peripheral organs. Axons are guided by a combination of attractive and repulsive cues. Cellular guidance, migration, and tropism are involved in neural development, spatial arrangement of skeletal elements, tissue remodeling during organogenesis, vasculogenesis, lymphangiogenesis, wound repair, inflammatory situations, neoplasia, tumor angiogenesis, and metastasis. As for the navigators in oceans and seas, local and distant cellular migration in tissues and blood vessels is tightly dependent on cell-surface adhesion and survival signals while crossing frontiers and new territories. Mechanisms for migration and navigation include chemotaxis induced by the secretion and diffusion of chemical signals via autocrine, neuroparacrine, and systemic pathways. Identification of paracrine signaling elements by global analyses using proteomics, biological assays, and gene capture will provide major advances in cancer research (Skarnes *et al.*, 1995). Chemotactic and migration factors such as chemokines in the CXCL12–CXCR4 axis are potent regulators of cell adhesion and conciliate cellular migration and invasion, angiogenesis, cell growth, survival, and cancer cell chemoresistances (Dessein *et al.*, 2010). In addition, gradients of extracellular matrix (ECM) components in basement membranes (BMs) such as laminin, fibronectin, collagen type-I, and vitronectin determine the haptotatic migration, adhesion, and survival of epithelial cells in navigation. Reciprocal invasive migrations and inductive and survival process between epithelial cells and mesenchymal layers are pivotal for the formation of vertebrate organs and tumors during embryonic development and neoplasia. BM also contains soluble or diffusible chemotactic substances, cell adhesion molecules, and growth factors trapped or secreted in the ECM network. These systems determine the haptotatic migration and invasion of normal and transformed epithelial cells in the ECM microenvironment (Debruyne *et al.*, 2002). Conversely, disruption of cell adhesion generates a form of apoptosis termed anoikis or cell death by loss of cell–matrix adhesion (Frisch and Francis, 1994). This property confers strict anchorage dependence. Focal adhesions mediated by the ECM receptor Integrins and their signaling adaptors support both static adhesion and dynamic migration in normal development and cancer biology (Cabodi *et al.*, 2010). For example, clustering of $\alpha6\beta4$ Integrin by laminin can stimulate E-cadherin dependent cell–cell adhesions, a critical mechanism involved in adherens junctions and cohesive layers of epithelial cells in digestive tract (Boissan *et al.*, 2010; Hintermann *et al.*, 2005; Zhou *et al.*, 2011). Survival of normal colonic epithelial cells is dependent on matrix adhesion and $\beta1$ Integrin along the crypt axis (Sträter *et al.*, 1996). Clustering of $\alpha3\beta1$ Integrin by laminin stimulates focal adhesion kinase (FAK) involved in the turnover of focal adhesions, activation of the cytoskeleton, and cell migration. Of note, signaling convergences between epidermal growth factor (EGF) receptors HER-1 and Integrin-associated cell adhesion molecules are described (Fig. 1) in human colon cancer cells (Ayoub *et al.*, 2009). Cancer cells frequently retain and overexpress Integrin subunits in relation with invasive and metastatic phenotypes. Previous studies showed that immortalized intestinal epithelial cells displayed chondroitin sulfate proteoglycans (CSP) with a shorter chain length implicating their role in inflammation and the premalignant phenotype (Lévy *et al.*, 1990; Theocharis *et al.*, 2010). The endogenous lectin Galectin-1 has been identified as a novel ligand that functionally interacts with the fibronectin receptor $\alpha5\beta1$ Integrins via binding to glycans. In HepG2 hepatocellular carcinoma cells, Galectin-1 attenuated cell cycle progression and facilitated anoikis in a caspase-8-dependent manner (Sanchez-Ruderisch *et al.*, 2010). Hypophosphorylation and cleavage of FAK and p130Cas are two signaling events required for anoikis, increased cellular detachment, and ERK1/2 activation in colorectal cancer (CRC) cells (Wendt *et al.*, 2008). ECM cues and Integrins also play important roles in axonal guidance. Tenascin provides neurons with cooperative attractive or repulsion signals and orient neuronal migration, while other molecules such as CSP provide permissive or exclusive signals (Murase and Horwitz, 2002; Porcionatto, 2006). Thus appropriate interactions between ECM components, cell adhesion complexes, and several Integrin subunits

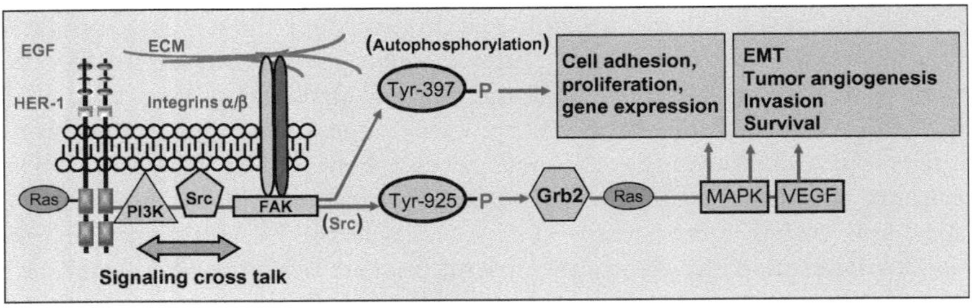

Fig. 1 Signaling convergences of HER-1 and Integrin dependence receptors at the focal adhesion kinase (FAK). Integrin-mediated cellular adhesions at extracellular matrix (ECM) components and HER-1 activation by epidermal growth factor (EGF) initiate several signaling interferences connected with the Ras/PI3K/Src and FAK cascades. As a consequence, tyrosines Y-397 and Y-925 in FAK are phosphorylated to further communicate molecular signals to ras, mitogen-activated protein kinases (MAPK) and vascular endothelial growth factor (VEGF) induction through the Grb2 adaptor. This FAK bifurcation contributes to the organization of a large range of biological responses involved in cancer cell invasion and survival. (See Page 1 in Color Section at the back of the book.)

allow epithelial cells and neurons to survive in their niche and during navigation. This situation provides a state of cellular and receptor dependence to the ligands and survival signals to escape from anoikis and apoptosis (Nagaprashantha *et al.*, 2011). Recent advances in this field were accomplished by the determination of the nanoscale organization of several structural components in focal adhesions and pointed out the importance of talin as a molecular organizer of Integrin-based cell adhesions (Kanchanawong *et al.*, 2010).

Initially discovered as an axon outgrowth and guidance promoter (Kennedy *et al.*, 1994; Serafini *et al.*, 1994), the multifunctional ligand Netrin-1 appears to interact with several dependence receptors: deleted in colon cancer (DCC), UNC-5 family, Neogenin, Integrins, and the Down's syndrome cell adhesion molecule (DSCAM; Fig. 2). The term "Netrin" derives from the Indo-Aryan language "*netr*" which means "one who guides." Netrin-1 also interacts directly and indirectly with α3β1, α6β4, and α6β1 Integrins in epithelial cells, suggesting convergent functions of ECM components and Netrin at dependence Integrin receptors (Stanco *et al.*, 2009; Staquicini *et al.*, 2009; Yebra *et al.*, 2003). The β4 Integrin subunit can form a constitutive complex with the Met scatter factor receptor suggesting possible signaling intersections between the Netrin-1-α6β4 axis and Met signaling during cellular dispersion and migration (Yebra *et al.*, 2003). The mechanisms driving the transitions between cellular guidance, migration, survival, and death through dependence receptors are complex and still poorly understood regarding the diversity and intricacy of the

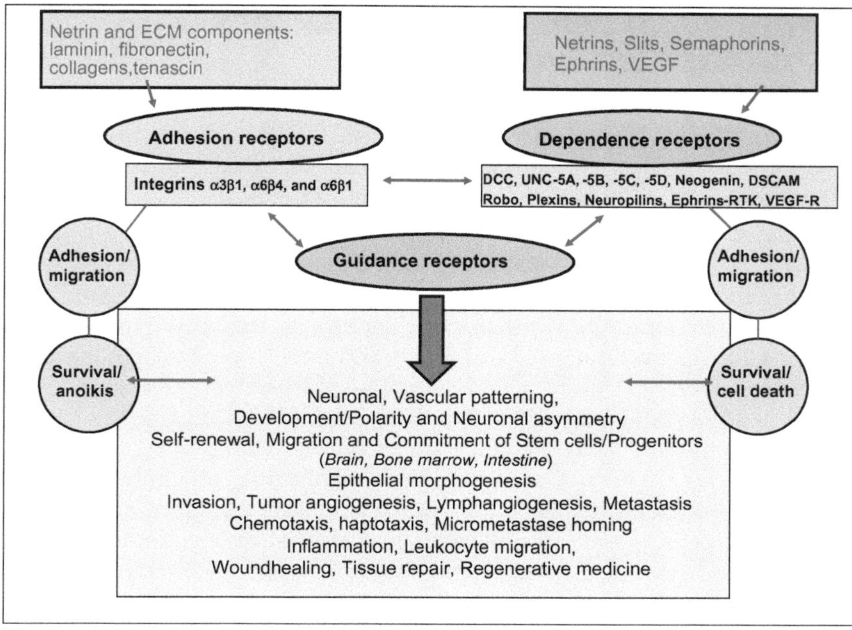

Fig. 2 Interplay between cellular adhesion, guidance, and dependence. (For color version of this figure, the reader is referred to the Web version of this book.)

receptors, coreceptors, and ligands involved in these convergent and divergent mechanisms (Montell, 1999). These pathways support the importance of the guidance and dependence receptors in tissue morphogenesis, differentiation, neural and vascular patterning, and neoplasia (reviewed in Nikolopoulos and Giancotti, 2005). Resident stem cell progenitors, bone marrow stem cells, and neuronal precursors are the main targets of the guidance molecules. In addition, some biological activities of Netrin are not fully explained by the expression of known Netrin receptors. Additional pathways (receptors, ligands, signaling, and cellular cross talk) are not yet elucidated.

The purpose of this review is to summarize and clarify recent advances of our knowledge on the Netrin axis with special emphasis to epithelial tumors and colon cancer progression. To document further this Introduction, a more comprehensive picture is provided on the molecular and cellular mechanisms underlying the guidance and dependence receptor concepts and theory. Finally, new therapeutic approaches are considered for the treatment of developmental defects, neuroendocrine degenerative diseases, and other disorders associated with development and aging.

II. NETRIN AND DEPENDENCE RECEPTORS IN DEVELOPMENT AND DISEASES

A. Neural Development and Netrin

Netrin guidance receptors are now recognized as essential regulators of many developmental processes during the brain development and organogenesis in peripheral tissues (reviewed in Sun *et al.*, 2011). Netrin pathways were originally discovered in the nematode *Caenorhabditis elegans* as a laminin-related molecule through their Uncoordinated (Unc) phenotypes after loss-of-function mutations in *unc-6* and *unc-40* genes encoding their counterparts Netrin-1 and DCC in mammals, respectively (Brenner, 1974; Fearon *et al.*, 1990; Ishii *et al.*, 1992; *Kim et al.*, 1999). Conserved signals that control midline axon guidance are induced by attractive cues such as Netrins and repulsive cues such as Slits, Semaphorins, and Ephrins (Ladle *et al.*, 2007). Netrins display bifunctional opposing guidance activities in both neural cells (repelling and attracting axons) and endothelial cells, acting as angiogenic or antiangiogenic factors. Dependence receptor family includes the canonical human Netrin-1 receptor family UNC-40/DCC/Frazzled (Fra, the *Drosophila* ortholog of DCC), UNC-5A, UNC-5B, UNC-5C, UNC-5D, enhancer of UNC (ENU-3), and DSCAM (reviewed in Cirulli and Yebra, 2007; Fearon *et al.*, 1990; Hiramoto *et al.*, 2000; Keino-Masu *et al.*, 1996; Kim and Ackerman, 2011; Leonardo *et al.*, 1997; Llambi *et al.*, 2001; Moore *et al.*, 2007; Reale *et al.*, 1994; reviewed in Sin *et al.*, 2011; Yang *et al.*, 2009; Yee *et al.*, 2011). Vertebrates express several secreted Netrins (1, 3, and 4) or intracellular Netrins G1 and G2 that are tethered to the plasma membrane. All Netrin members belong to the laminin superfamily (Rajasekharan and Kennedy, 2009; reviewed in Sun *et al.*, 2011; Ziel and Sherwood, 2012). Netrin-4 interacts with laminin in ECM and during BM assembly (Schneiders *et al.*, 2007). Importantly, laminin plays significant role in the persistence of Netrin/DCC signals in the attraction of vagal sensory neurons (Ratcliffe *et al.*, 2008). Neogenin, a receptor for the repulsive guidance molecule (RGM), and Netrins were also reported to interact functionally with the bone morphogenetic proteins (BMPs; De Vries and Cooper, 2008; Hagihara *et al.*, 2011; Vielmetter *et al.*, 1997). RGM-A is confined in areas of inflammation after injury of the central nervous system (CNS), regulates neuronal survival and differentiation, and dampers inflammatory situations by inhibiting leukocyte migration (Hata *et al.*, 2006; Mirakaj *et al.*, 2011b). Exogenous soluble Netrin-1 was also able to induce the differentiation of human embryonic carcinoma cells to neuroectodermal lineage accompanied with reduced levels of Cripto-1 (an EGF family member) and the pluripotency genes Oct4 and Nanog (Mancino *et al.*, 2009). Additional morphogens are the neurotrophic (NT) factors, nerve growth factor (NGF) interacting with NGF receptors (p75NTR) and Trk receptors, the glial cell line-derived NT factor GDNF interacting with Rearranged during Transfection (RET) receptors involved in the development of enteric ganglia derived from vagal neural crest cells, and the Sonic Hedgehog receptor Patched (PTCH) (Mehlen and Bredesen, 2004; Mille *et al.*, 2009b; Schuchardt *et al.*, 1994). Mutations in *RET* gene are associated with multiple endocrine neoplasia *men* type 2 and the megacolon in human, also called Hirschsprung's disease where the normal enteric nerves are absent. The axon guidance cues Netrins and Semaphorins are involved in the development of the nervous system, vascular patterning, and cancer cell invasion (Adams and Eichmann, 2010; Nguyen *et al.*, 2006; Rodrigues *et al.*, 2007; Serafini *et al.*, 1994; Tessier-Lavigne and Goodman, 1996; Trusolino and Comoglio, 2002; Varela-Echavarría and Guthrie, 1997). Mice deficient in the expression of the *RET* protooncogene display renal agenesis and lack enteric neurons throughout the digestive tract (Schuchardt *et al.*, 1994).

Netrin-1 regulates critical guidance functions in the auditory and optic systems, corticospinal tract, cranial nerves, interhemispheric connections, thalamocortical tract, and cerebellum (reviewed in Bradford *et al.*, 2009). In brain and peripheral tissues, the formation of Netrin and Netrin receptor gradients, receptor and coreceptor complexes, homodimers, heterodimers, and receptor clusters promotes additional mechanisms to orchestrate axonal attraction, repulsion, and the neuronal plasticity of the directions followed by the growth cone during the journey. Additional mechanisms are achieved through differential regulation of axon and dendritic growth, as shown in *C. elegans* motor neurons (Teichmann and Shen, 2011). In this model, UNC-6 (Netrin) is required for both axon and dendrite development, but the serine–threonine kinase PAR-4 (LKB1) is selectively required for the activity of UNC-40 (DCC) in dendrite outgrowth. The latter combination is supported by the contribution of the repulsive receptor in the ventral body, whereas the attractive receptor DCC is dendritically enriched. Endogenous ventrally secreted Netrin controls axon guidance, whereas dorsal and membrane-tethered Netrin supports dendrite development to establish appropriate circuits (Teichmann and Shen, 2011). Also in support with this model, recent advances indicate that the Netrin/DCC axis promotes transient increase in cAMP and brief increase in filopodia Ca^{2+} oscillations, and reciprocally, a Ca^{2+}-dependent cAMP increase associated with a sustained increase in Ca^{2+} transient frequency in growth cone center during growth cone tuning (Nicol *et al.*, 2011). So the signaling codes of the guidance and dependence receptors are increasing in number and complexity every day.

Recent findings show that, in brain slices from a mouse model of Alzheimer's disease (AD), Netrin-1 restricts the accumulation of β-amyloid peptides and toxic plaques as the central event triggering neurodegeneration in the pathogenesis of AD (Lourenço *et al.*, 2009). Most importantly, Netrin-treated mice showed improved memory function in AD transgenic animals. In support with this therapeutic perspective, dominantly inherited mutations in the genes encoding Presenilin-1 (PS1) and the amyloid-β precursor protein are the major causes of familial AD. PS1 is an essential component of the γ-secretase complex involved in the cleavage of amyloid-β precursor protein that disrupts neuronal connections. This neural circuit builder is required to control the axonal responses to Netrin by coordinating DCC receptor processing and the interplay between Netrin/DCC and Slit/Robo signaling (Bai *et al.*, 2011). Previous studies showed that some laminin isoforms display abnormal expression and distribution in toxic plaques of AD patients (Palu and Liesi, 2002), but their possible links with the guidance–dependence system have not yet been established. In fetal-derived dopaminergic neurons, Netrin and Slit receptors DCC as well as Robo receptors respectively function as chemoattractant and repellant for neurite outgrowth (Lin and Isacson, 2006). Although embryonic stem cells (ESC)-derived dopaminergic neurons also express DCC and Robo receptors inducing neurite outgrowth, they do not react to the guidance signals from their respective ligands. Additional studies are therefore expected to develop these tools in the restoration of the dopaminergic circuits damaged or lost in Parkinson's disease. Other developmental defects in brain originate from mutations in the *Robo3* gene required for the growing corticospinal and somatosensory axons crossing the midline in the medulla to reach their targets to form the contralateral motor control and sensory input (Jen *et al.*, 2004). In patients with horizontal gaze palsy (HGPPS) with progressive scoliosis, motor and sensory axonal projections appeared uncrossed, leading to hindbrain dysplasia. This HGPPS phenotype is recapitulated in *Robo3*-deficient mice because of the failure of axons to cross the midline of the nervous system in brain (Sabatier *et al.*, 2004; Woods, 2004). Epilepsy is another chronic brain dysfunction induced by abnormal synapses, recurrent excitatory loops, and neuronal discharges in the CNS. In patients with intractable epilepsy, Netrin-G2 expression was found upregulated in the temporal neocortex (Pan *et al.*, 2010). Abnormal expression of this membrane-associated form of Netrin is also observed during memory and learning defects, abnormal acoustic startle response, schizophrenia, and bipolar disorders. Luteinizing hormone-releasing hormone (LHRH) neurons regulating pituitary functions in mammals migrate from the nasal epithelium, cross the cribriform plate medial to the olfactory bulbs, and continue into the developing hypothalamus (Schwarting *et al.*, 2004). Netrin-1 and DCC are also expressed in the olfactory system and may participate to the guidance of LHRH neurons to cross great distances during this long journey. In *DCC*-null mice, trajectories of the caudal vomeronasal nerve and positions of LHRH neurons are abnormal. Genetic deletion of either DCC or Netrin-1 results in the migration of LHRH neurons to inappropriate destinations (Schwarting *et al.*, 2004).

B. Neural Development, Ephrins, Slits and Wnts

Transmembrane Ephrins-B (B1, B2, and B3) interact with the secreted neuronal guidance glycoprotein Reelin, its two lipoprotein receptors (very-low-density lipoprotein receptor and apolipoprotein E receptor 2), and signaling machinery connected with Src activation and phosphorylation of the adaptor protein Disabled1 (Dab1; Sentürk *et al.*, 2011). Reelin signaling is required for proper nervous system functions during development and synaptic plasticity in the adult brain (reviewed in Herz and Chen, 2006). Of note, direct activation of Ephrins-B is sufficient to rescue Reelin signaling in the absence of Reelin, supporting the therapeutic potential of the Ephrins-B axis against neurological disorders resulting in loss of Reelin in humans (Sentürk *et al.*, 2011). Ephrins-B family negates cerebellar granule cell migration by controlling stromal-derived factor-1 (SDF-1)-mediated chemoattraction and activation of CXCR4 (Lu *et al.*, 2001).

The large family guidance molecules include the roundabouts Robo receptors binding the secreted Slit proteins, Plexin receptors interacting with membrane associated and secreted Semaphorins, Neuropilin receptors acting as coreceptors for Plexin signaling, and Ephrin receptor tyrosine kinases (RTKs) binding the Ephrin ligand family and regulating cell adhesion, cardiac stem cell migration, and colon cancer progression (Bogan *et al.*, 2009; Goichberg *et al.*, 2011; reviewed in Mehlen *et al.*, 2011). The Neuropilins also function in molecular complexes at the vascular endothelial growth factor (VEGF) and transforming growth factor β (TGFβ) pathways and their expression is often abnormal in tumors. Membrane NRP-1 promotes the formation of a signaling complex between VEGF-165, VEGF receptor-2 (VEGF-R2), and heparan sulfate (HS), whereas the soluble NRP-1 is thought to act as an antagonist of signaling complex formation. Targeted disruption or overexpression of the repellent cues Semaphorin-3A, Neuropilin-1, and Ephrin-A2 leads to reduplication, distortion, or abnormal bone patterning and growth, cartilage and limb morphogenesis (Schubert *et al.*, 2009). *Semaphorins 3B* and *3F* were identified as tumor suppressor genes in ovarian cancer cells and A9 mouse fibrosarcoma cells, while Sema-3A functions as a proinvasive factor in human colon cancer cells (Nguyen *et al.*, 2006; Tomizawa *et al.*, 2001; Tse *et al.*, 2002; Xiang *et al.*, 2002). Sema-3F also functions to block apoptosis in A9 cells treated by Taxol or Adriamycin (Xiang *et al.*, 2002).

Remodeling of the actin cytoskeleton and dynamic microtubule filaments generates dilatation and suppression of filopodia underlying spatial turns in response to repulsive and attractive signals toward the appropriate connections (reviewed in Dickson, 2002). Repulsive Slit ligands may participate in these trajectories. Interplay between the guidance family receptors and other morphogens supports these patterns (Bai *et al.*, 2011). Spatiotemporal secretion and distribution of the ligands and guidance receptors also contribute to these signaling and functional instructions during development and repair. For example, Wnt signals are involved in asymmetric cell divisions in many organisms (reviewed in Killeen and Sybingco, 2008) according to at least three main signaling axis (Fig. 3): canonical Wnt/β-catenin; PLC/PKC/calcium/NLK and NF-AT; the planar cell polarity (reviewed in Le Floch *et al.*, 2006).

C. Connections with Peripheral Tissues and Epithelial Systems

Sensory neurons innervating the digestive tract integrate secretion, motility, and display excitatory responses by relaying the changing conditions of the external milieu. Intraepithelial vagal sensory nerve terminals are also revealed in pulmonary neuroepithelial bodies as hypoxia sensors (Brouns *et al.*, 2000). The enteric nervous system (ENS) is a component of the autonomic nervous system with the ability to function independently of the CNS. Vagal neural crest-derived precursors of the ENS colonize the bowel by descending within the enteric mesenchyme (reviewed in Goyal and Hirano, 1996; Le Douarin and Teillet, 1973). A secondary migration step toward the mucosa initiates submucosa and pancreatic ganglia through attraction by Netrin of DCC-expressing, crest-derived cells (Jiang *et al.*, 2003). Both the fetal gut mucosal epithelium and pancreas express Netrins, Neogenin, DCC, and adenosine A2b coreceptors (A2b-R). DCC-expressing, crest-derived cells migrate toward the mucosa of cultured rings of fetal bowel (Corset *et al.*, 2000; Jiang *et al.*, 2003). In transgenic mice lacking DCC, submucosal and pancreatic ganglia are absent. Accordingly, Netrin promotes the survival and development of enteric crest-derived cells. The segmental absence of the ENS in Hirschspung's disease causes a pseudo-obstruction that restricts the gastrointestinal (GI) transit. The intrinsic myenteric plexus of ganglia coordinates intestinal motor patterns such as peristalsis, and the submucosal plexus of ganglia regulates the absorption and secretion of fluids, mucins, and electrolytes (Osorio and Delmas, 2010). However, the ENS is also connected with extrinsic innervation composed of vagal and spinal sensory nerves, whereas the majority of vagal axons signal from the gut to brain. Most interestingly, the development of the enteric vagal sensory innervation is dependent upon Netrin-expressing enteric neurons (Ratcliffe, 2011, 2006). Functionally, it is anticipated that these neurons are connected, attracted, or activated by the perception of local visceral sensory pain, mechanodetection, luminal irritants and pathogens, ulceration, neuroinflammation, and inflammatory diseases of the gut, irritable bowel syndrome, impaired mucosal barrier functions, and neoplasia (reviewed in Balkwill and Mantovani, 2010; Buhner *et al.*, 2009; reviewed in Mantovani *et al.*, 2008; Mirakaj *et al.*, 2011a). For example, neurons in the CNS and

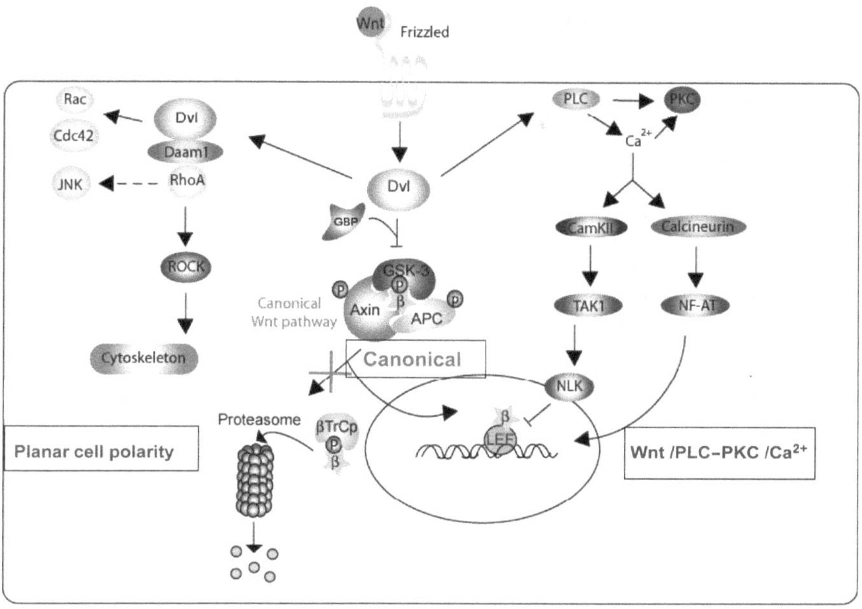

Fig. 3 Wnt signaling highways. From Wnt signals and Frizzled (Fz) G-protein-coupled receptors (GPCRs) to gene expression and activation of the cytoskeleton. The main pathways of the Wnt signalosome (reviewed in Le Floch *et al.*, 2006) include: *Middle*, the canonical Wnt/β-catenin signals via the β-catenin-degradation complex (containing APC, Axin-GSK-3β, β-catenin, and accessory partners), leading to nuclear β-catenin-mediated transcription. When the Ser/Thr protein kinase GSK-3β is inhibited by upstream Wnt–Fz activation and other growth factor receptors, β-catenin escapes from its degradation by the β-transducin repeat-containing protein (βTrCp)-dependent proteasome pathway; *Right*, the Wnt–Ca^{2+} axis operating via phospholipase C (PLC), protein kinase C (PKC), and Ca^{2+} at β-catenin/LEF convergence and gene promoters in the nucleus; Similar to the Fz/PCP pathway, the Wnt/Ca^{2+} axis is involved in the regulation of morphogenetic events during gastrulation. Wnt/Ca^{2+} signals induced by Wnt-5a upstream of calcium/calmodulin-dependent kinase type-II (CamKII)–TAK1 (TGFβ-activated kinase-1) and –NLK (Nemo-like kinase) axis exert an inhibitory action on the canonical Wnt/β-catenin pathways at two levels targeting respectively disheveled (Dvl) and lymphoid-enhancer factor/T-cell factor (LEF/TCF) transcription factors. Bifurcation of the signalome toward a noncanonical Wnt pathway is operating through signaling elements using the cytoplasmic Ca^{2+}/calmodulin-dependent protein phosphatase Calcineurin, dephosphorylation of the nuclear factor of activated T cells (NF-AT) in one hand and transcription of NF-AT-sensitive genes under the control of the Wnt-5a ligand on the other. *WNT5A* is a cancer-associated gene expressed in CRC (Vider *et al.*, 1996) and is implicated in the invasion and metastasis of several human cancers, including colorectal, breast, and pancreas; *Left*, the planar cell polarity (PCP) pathways using Dvl, Dvl-associated activator of morphogenesis 1 (Daam1), Rho-like GTPases (RhoA, Rac1, and Cdc42), c-Jun NH2-terminal kinase (JNK), Rho-kinase (ROCK), and regulation of the actin cytoskeleton via mediation of linear actin assembly. TAK1 can suppress cell death pathways and thus promotes resistance to chemotherapeutic drugs. PCP controls the polarity of epithelial cells within a plane orthogonal to their apical–basal axis. (See Page 2 in Color Section at the back of the book.)

adaptive sensory and motor responses of the autonomic nervous system in GI vagal sensory neurons are glucose sensitive secondary to glucose absorption (Browning, 2010; Grabauskas *et al.*, 2010). Vagal sensory axons are initially attracted by Netrin and dependence receptor agonists and regulators in the developing gut. Thus, Netrin-1 is one of several factors mediating the guidance of vagal axons to the fetal GI tract to establish the brain–gut axis (reviewed in Ratcliffe *et al.*, 2011b). Recent reports indicate that neurogenesis and perineural invasion of clinical tumors play a critical role in the adverse outcomes in patients with CRC (Albo *et al.*, 2011). Netrin- and DCC-positive normal and transformed intestinal epithelial cells as well as components of the ECM may provide additional signals. For example, laminin was shown to convert Netrin-mediated attraction into repulsion (Ratcliffe *et al.*, 2011a). One can postulate that nerve fibers that extend from enteric neurons to the epithelium can have the property to stably or transiently express Netrins and other members of the Netrin dependence ligand and receptor family (Gershon, 2010; Seaman *et al.*, 2001), including ECM components and unknown related mediators. In addition, the general organization of the intestinal mucosa and its anatomical architecture is critically disrupted during the adenoma–carcinoma transition. Illegitimate connections between enteric neurons and related nerve fibers with malignant tumors are likely to provide Netrin messages. Similar situation has been described for several neurotransmitters in submucosal neurons in the ENS and innervations of intestinal epithelial crypt cells (reviewed in Goyal and Hirano, 1996). Extrinsic sensory innervations from the bowel are in convergence with the spinal cord and brain, in close association with subepithelial domains (reviewed in Goyal and Hirano, 1996; Ratcliffe, 2011). Thus, functional and anatomical interactions of Netrin-positive neural cells with Peyer's patches lymphoid nodules and Paneth cells in human ileum and small intestine, as well as epithelial stem cells and progenitors in the small and large intestine are not excluded (Crosnier *et al.*, 2006).

III. NETRIN AND DEPENDENCE RECEPTORS IN CANCER PROGRESSION

A. Epithelial Morphogenesis, Angiogenesis, and Neoplasia

Besides nervous development, guidance receptors, Netrins, laminins, and laminin-γ chain-related molecules are also playing more subtle roles in matrix assembly and BM formation, epithelial cell adhesion to ECM, cell–cell adhesions, as well as migration and differentiation of normal and transformed progenitors such as mammary cap cells (reviewed in Cirulli and Yebra, 2007;

Srinivasan *et al.*, 2003; reviewed in Slorach and Werb, 2003 and in Yurchenco and Wadsworth, 2004). Netrin-1 is expressed at various embryonic stages and in several mesodermally and endodermally derived tissues in adults, including heart, lung, and ovary. Netrin-1, Slit-1, Slit-3, and their respective receptors DCC/UNC-5B and Robo-1/-2 are expressed and modulated spatially in the mesenchyme and epithelium during lung development (Anselmo *et al.*, 2003; Dalvin *et al.*, 2003). Netrin is required for the embryonic development, morphogenesis of vascular smooth-muscle cells (VSMC), vascular patterning, and morphogenesis of several organs, including the mammary gland, intestine, and endocrine and exocrine pancreas (De Breuck *et al.*, 2003; Jiang *et al.*, 2003; Strizzi *et al.*, 2005). Vertebrate Netrin-1 was discovered as a chemotrophic factor (Kennedy *et al.*, 1994). By virtue of the related domains VI of laminin and Netrin, polymerization and matrix assembly in BMs lead to recruitment of Integrins and guidance receptors through chemotaxis, haptotaxis, and cell–matrix adhesion (*vide supra*). ECM components may also include a large diversity of Netrin-like molecules such as Netrins 4, G1, and G2 expressed in both neural and nonneural tissues in the kidney, ovary, heart, and vasculature (Staquicini *et al.*, 2009; Yebra *et al.*, 2011; Yurchenco and Wadsworth, 2004).

Netrin-1 enhanced the viability of isolated human placental microvascular endothelial cells and was found reduced in pregnant women with fetal growth restriction (Qian-Hua *et al.*, 2011). In the *C. elegans* developing uterus, mutations in the guidance system UNC-40 (DCC) and UNC-6 (Netrin) showed important defects in the polarization of invasive protrusions of the anchor cell through enrichment of actin regulators to mediate a uterine–vulval connection (Hagedorn *et al.*, 2009; Ziel *et al.*, 2009). This elegant study provides the basis for the mechanism regulating invasive behavior during normal development and cancer progression. In the human placenta, Netrin-1 is expressed with UNC-5B receptors in villous and extravillous cytotrophoblasts, while DCC receptors are localized to syncytiotrophoblasts and invasive extravillous cytotrophoblasts during the first trimester of gestation (Dakouane-Giudicelli *et al.*, 2010). Thus, the Netrin-1/DCC axis might play a role in the control of extravillous migration and invasion. In contrast, Netrin-1 appears to act locally through Neogenin to mediate cell clustering during mammary gland morphogenesis. Netrin-1 stabilizes the multipotent progenitor cap cell layers, as shown by cell-aggregation assays, and therefore prevents cap cell movements (Srinivasan *et al.*, 2003). Disruption of the Netrin-1–Neogenin axis between adjacent cap cells and terminal end buds may initiate susceptibility to breast cancer.

In the pancreas, Netrin gradients can induce $\alpha3\beta1$ and $\alpha6\beta4$ Integrin-dependent haptotactic migration of embryonic epithelial progenitors for both the endocrine and exocrine cell lineages. Only a discrete subpopulation marked with cytokeratins CK19 and pancreatic homeodomain PDX-1 protein expressed islet hormones insulin or glucagon, suggesting that Netrin-1 is not a critical mediator for the migration of differentiated islet cells (Yebra *et al.*, 2003). Netrin-1 and Neogenin were found coexpressed in fetal pancreas during late gestation, which is a period of active islet morphogenesis, but was not detected in neonates and adults (De Breuck *et al.*, 2003). The two guidance cues Netrin and Slit are working as chemoattractant and repellent for proper migration of *Drosophila* salivary glands (Kolesnikov and Beckendorf, 2005). In submandibular glands of embryonic mice, Netrin-4, a component of BMs integrated with the laminin polymer, retards salivary gland branching morphogenesis (Schneiders *et al.*, 2007). Netrin-induced migratory responses were observed in transformed human pancreatic duct cells CAPAN-2 and in fetal porcine islet cells. Interestingly, reinduction of Netrin-1 was observed during islet neogenesis observed following duct ligation, suggesting that Netrin plays a role in islet development, maintenance, survival, and regeneration. Therapeutic applications for the treatment of diabetes type-I and -II and compensation of β-cell loss are therefore conceivable. Toxic effects of hyperglycemia and high glucose concentrations have been extensively reported in diabetic patients and experimental conditions (Phan *et al.*, 2000). Several Netrin genes and receptors (UNC-5 family, Neogenin, Integrins), as well as multiple-receptor-interacting secreted factors are indentified during the development in pancreatic islets (Cabrera-Vásquez *et al.*, 2009; Yang *et al.*, 2011). Netrins1 and 4 were shown to decrease the activity of the cell death effector caspase-3 in the MIN6 β-cell line incubated in the presence of 25 mM glucose.

Netrin-1 and guidance receptors UNC-5B and Neogenin regulate adhesion, proliferation, and migration of renal proximal tubular epithelial cells (Wang *et al.*, 2009a). Netrin-1 is induced *in vitro* at edges of wounds in the scratch-wound assay and is highly upregulated in kidney epithelial cells in response to ischemia-reperfusion injury (IRI; Wang *et al.*, 2008). High expression of Netrin-1 is detected in kidney, lung, heart, and to a lesser extent, the intestine, liver, and spleen (Ly *et al.*, 2005). Abundant expression of Netrin-1 is detected in the vascular endothelium of large and small blood vessels in lungs. Postcapillary venules were also positive for Netrin-1 in tissue sections of the upper small intestine. Netrin-1 appears to be associated with the luminal surface of endothelial cells. During acute inflammation in a mouse model of *Staphylococcus aureus* infection targeting lungs, Netrin-1 was rapidly downregulated in association with accumulation of leukocytes in the lung. In human umbilical vein epithelial cells, the proinflammatory cytokines TNF-α and IFN-γ reduced Netrin-1 expression. DCC was not detected in human granulocytes, monocytes, or lymphocytes, while UNC-5B was abundantly expressed on the surface of monocytes, granulocytes, and lymphocytes. Coherently, the authors revealed that Netrin-1 inhibits leukocyte migration *in vivo* and *in vitro* using the chemoattractants fMLP and IL-8 (Ly *et al.*, 2005). In this context, Netrin-1 was therefore considered as a potential anti-inflammatory mediator. Supporting evidence for a protective role of Netrin-1 signaling in inflammatory diseases was recently presented in an experimental model of dextran sodium sulfate (DSS)-induced inflammatory bowel disease (IBD) through limitation of neutrophil trafficking, infiltration across the epithelial barrier, and implication of adenosine A2b-R (Aherne *et al.*, 2012). In degenerative osteoarthritis (OA), DCC was shown to be specifically upregulated compared to normal chondrocytes *in vitro* and *in vivo* through transcriptional controls mediated by Sox9 and AP-2 in a synergistic manner (Schubert *et al.*, 2009). Transient transfection of DCC in healthy chondrocytes induced significant migration and chemotaxis to Netrin-1.

Netrins may also act bifunctionally during normal vascularization and tumor angiogenesis. Several families of axon guidance molecules and receptors including Netrins, Slits, and Semaphorins have also been implicated in vessel pathfinding and network formation in the body. Netrin-1 was initially identified as a proangiogenic factor stimulating the proliferation and migration of normal endothelial cells and VSMC (Park *et al.*, 2004). This coherence is in line with the patterning of the vascular system by Netrin-1a to form an integrated network of blood vessels in concerted interactions with axonal motoneurons, muscle pioneers, and endothelial cells to delineate the guidance of lymphendothelial cells, as shown in the parachordal chain in the zebrafish (Lim *et al.*, 2011). Netrin-1-induced vascular sprouting in the chorioallantoic membrane (CAM) of chick yolk sac with a specific activity comparable to the canonical angiogenic factors VEGF, PDGF, and basic FGF. Netrin-1 promoted the adhesion of VSMC but not endothelial cells. VSMC lines expressed Neogenin but were DCC and Robo-4 negative, whereas primary endothelial cells HUVEC were Robo-4 positive and DCC negative. Thus, Neogenin was shown to mediate Netrin signaling in VSMC and proangiogenic functions via an unidentified endothelial receptor (Park *et al.*, 2004). Similar conclusions were presented using human microvascular endothelial cells (HMVEC) where Netrins 1, 2, and 4 all stimulated cell migration and tube morphogenesis dose dependently (Wilson *et al.*, 2006). High levels of endothelial receptors Robo-4 and VEGF-R2 were detected by real-time reverse transcription polymerase chain (qPCR). However, Neogenin, DCC, and UNC-5B receptors were not significantly expressed (Wilson *et al.*, 2006), although UNC-5B was shown to mediate Netrin-1 signaling in the endothelium (Lu *et al.*, 2004). Functional interactions of Netrins1 and 4 with putative A2b adenosine/DCC coactivators were also excluded. In addition, the same authors demonstrated that knockdown of *Netrin1a* in zebrafish revealed the requirement for Netrin signaling in vascular sprouting and formation of the parachordal vessels during vertebrate development (Wilson *et al.*, 2006). Consistently, Netrins can activate blood vessel formation and accelerate revascularization and reperfusion of ischemic tissue in murine disease models. This report provided the proof of concept for treatment of vascular and neural degenerative diseases and to treat associated diabetic complications (Wilson *et al.*, 2006). It is therefore conceivable that Netrins may participate in VSMC contraction and relaxation mechanisms, regulation of blood pressure, pathogenesis, or management of hypertension in cardiovascular and renal diseases and other disorders associated with hypertensive states.

In UNC-5B-negative endothelial cell lines, Netrins1 and 4 initiate promigratory, trophic, and proangiogenic actions via unidentified receptors (Wilson *et al.*, 2006). Conversely, repulsive actions of the Netrin-1/UNC-5B axis and endothelial filopodial retraction are reported in developmental angiogenesis (Lu *et al.*, 2004). This scenario recapitulates the opposing roles of the guidance receptor family in axonal chemoattraction and repulsion. Moreover, UNC-B activation by Netrin-1 inhibits sprouting angiogenesis, supporting the notion that UNC-5B receptor agonists may be considered as potential therapeutic tools in pathological angiogenic states (Larrivée *et al.*, 2007). In HUVEC expressing the Netrin-1 cognate receptors Neogenin, UNC-5B, and UNC-5C, Netrin-4 was shown to inhibit VEGF- or fibroblast growth factor-2 (FGF-2)-dependent angiogenesis through Neogenin and recruitment of UNC-5B receptors (Lejmi *et al.*, 2008). It appears that the determination of positive and negative signals induced by Netrins on angiogenesis is determined from a combination of dependence receptors, their respective interactions and signaling cross talk, and by the cellular and biological context. For example, Netrin-4 promotes the proliferation, migration, adhesion, tube formation, survival, and permeability of human lymphatic endothelial cells (Larrieu-Lahargue *et al.*, 2010). Chemotaxis was observed at low ligand concentration and was maximally induced at 0.6nM Netrin-4. Although UNC-5B, Neogenin, and A2b receptors are expressed in this model, their knockdown or blockade by siRNA or chemical antagonists does not suppress Netrin-4-induced lymphangiogenesis. Recent findings indicate that Integrins $\alpha6\beta1$ are activated by Netrin-4 to induce phosphorylation of Src tyrosine kinase family in lymphatic endothelium to stimulate endothelial cell migration (Larrieu-Lahargue *et al.*, 2011). Integrins $\alpha6\beta1$, $\alpha6\beta4$, $\alpha2\beta1$, and $\alpha3\beta1$ function as cell surface receptors for laminin and Netrins 1 and 4 (Stanco *et al.*, 2009; Staquicini *et al.*, 2009; Yebra *et al.*, 2003, 2011). Netrin-4 also acts as a prodifferentiation cue in the developing human pancreas to support as an ECM component efficient pancreatic epithelial cell adhesion and differentiation and elicits significant induction of insulin and glucagon gene expression (Yebra *et al.*, 2011). In this context, Netrin-4 is produced by pancreatic vascular and ductal cells to interact with $\alpha2\beta1$ and $\alpha3\beta1$ Integrin receptors in adjacent endocrine precursors. It should be stressed that interactions and combinations of Netrins with ECM components and other cell surface molecules may induce heterotypic activation of nonclassical guidance and survival/dependence pathways. Another functional basis supporting the opposing roles of the Netrin ligands during developmental processes and angiogenesis arise from their ability to modulate differentially survival and apoptosis states pending on the expression and engagement of guidance receptors by their respective agonists. Neutralization of the UNC-5B death signals in HUVEC treated with Netrin-1 and siRNA inhibits endothelial cell apoptosis and activity of the death signaling effector DAPK and leads to an increase in neovessel ramifications in the CAM assay (Castets *et al.*, 2009). Moreover, recent advance in the field supports the notion that Robo-4 receptors act as a juxtacrine membrane-bound ligand for UNC-5B receptors to maintain vessel integrity and barrier functions by blocking VEGF signaling through VEGF-R2 (Koch *et al.*, 2011). It is likely that these responses are associated with maintenance of efficient cell–cell contacts in the endothelium. Robo-4 receptor extracellular domain shedding may also provide alternative UNC-5B/VEGF-R2 autocrine and paracrine regulation. UNC-5B and UNC-5D receptors have been recently shown to bind with high affinity to the ectodomain of fibronectin- and leucine-rich transmembrane protein-3 (FLRT-3) that is thought to act autonomously as a coreceptor or as a cell adhesion molecule. The shed FLRT-2 and -3 ectodomains are acting as RGMs for axons and somata of UNC-5-positive neurons (Yamagishi *et al.*, 2011).

Taken together with the antiangiogenic activity of the Netrin-4–Neogenin/UNC-5B axis, this mechanism based on molecular interactions between the Robo-4 receptors and UNC-5B endothelial receptors has potential applications against pathological angiogenesis during cancer progression and other diseases (Eveno *et al.*, 2011; Lejmi *et al.*, 2008). It is anticipated that these antiangiogenic pathways may provide a new paradigm and benefits for the prevention and treatment of macular degeneration in patients who develop new blood vessels under the retina. Limitations of anticancer therapy are also associated by the formation of abnormal and nonfunctional new blood vessels that exclude malignant tumors and cancer cells to the penetration and therapeutic effects of anticancer drugs. It is suspected that cellular, pericellular, and systemic expression of Netrin(s) exerts critical impact on survival and death in normal tissues, cancer cells, and tumor stroma containing, at least in part, fibroblasts, adipocytes, and immune and endothelial cells (reviewed in Hendrix *et al.*, 2011; Slorach and Werb, 2003). This prediction has been confirmed during the early stages of pancreatic adenocarcinoma progression (Dumartin *et al.*, 2010).

B. Epithelial Cancers

Premalignant states are primed and initiated by losses of cell death and proliferation controls, several morphogenetic alterations, and complex interactions of epithelial cells with the growing tumor stroma. Acquisition of the invasive potential by cancer cells is the hallmark of the malignant state (reviewed in Christofori, 2006; Empereur *et al.*, 1997). Subsequent angiogenic responses and the metastatic cascade are directly linked to the fatal issue of epithelial cancers (reviewed in Hanahan and Weinberg, 2011).

The epithelial–mesenchymal transition (EMT) is another critical mechanism implicated in normal embryonic, fetal, and postnatal development as well as multiple pathological states including wound healing, tissue repair, and cancer progression (reviewed in Sabbah *et al.*, 2008; Thiery *et al.*, 2009). EMT contributes to cancer cell plasticity and exacerbates the invasive and metastatic potential of cancer cells as well as anticancer drug resistances connected with survival pathways. These mechanisms also involve vascular endothelial cells that are subjected to endothelial to mesenchymal transition (EndMT) in solid tumors. Several observations support the possible role of dependence and guidance receptors in the restriction of EMT-dependent mechanisms and preservation of intercellular adhesions, namely: (i) the establishment and maintenance of E-cadherin-dependent cell–cell adhesions and cohesive epithelial layers by laminins and $\alpha 6\beta 4$ Integrins (Hintermann *et al.*, 2005), (ii) the maintenance of endothelial cell–cell contacts by Robo-4 and UNC-5B (Koch *et al.*, 2011), and (iii) the adhesive functions of Netrin-1 and Neogenin in cell–cell adhesion and aggregation during mammary gland morphogenesis (Srinivasan *et al.*, 2003). Conversely, Netrin-4 induced the deconstruction and delocalization of endothelial vascular endothelial (VE)-cadherin and downregulation of the tight junction protein *Zona occludens*-1 (ZO-1), suggesting partial EndMT as observed for E-cadherin and EMT in epithelial cancer cells (Boissan *et al.*, 2010; Koch *et al.*, 2011). Accordingly, VEGF-A and VEGF-R2 have been shown to promote vascular permeability through the endocytosis of VE-cadherin (reviewed in Dejana *et al.*, 2008). The relevance of the Netrin pathways in the control of cell plasticity and polarity is further illustrated by the downregulated expression of Netrin-1 and increased expression of its receptor Neogenin in mammary epithelial cells stably transfected by Cripto-1 (Strizzi *et al.*, 2005). Restitution of exogenous soluble Netrin-1 resulted in the reversion of Cripto-induced EMT, as judged by increased expression of E-cadherin, depletion of vimentin, and AKT inhibition. In trying to establish a mechanistic link between Netrin-4 and stimulation of lymphatic permeability, Larrieu-Lahargue *et al.* demonstrated that both ZO-1 and VE-cadherin were downregulated in HMVEC after the addition of the soluble peptide (Larrieu-Lahargue *et al.*, 2010). Downregulation, endocytosis, and neutralization of VE-cadherin have been shown to be associated with vascular permeability, lung metastasis and metastatic breast, and liver tumors.

Normal epithelial crypts isolated from the human colon were positive for the expression of the DCC tumor suppressor gene. DCC-deficient human colon adenoma and adenocarcinoma cells PCAA/C1, HCT8/S11, and HT29 are expressing the UNC-5A, UNC-5B, UNC-5C, and adenosine A2b-R. Addition of soluble Netrin-1 at low concentrations (0.01–1 nM) promoted collagen type-I invasion through signaling cascades using Rac1, Rho–Rho kinase (ROCK), and phosphatidylinositol 3-kinase (PI3K) connected with activation of the actin cytoskeleton (Kotelevets *et al.*, 1998; Rodrigues *et al.*, 2007). Introduction of wt-DCC in DCC-deficient epithelial cells blocked invasion induced by Netrin-1 and metastasis in lymph nodes and lungs, and increased sensitivity to apoptosis in hypoxia, via caspase-8 and PARP cleavage (Rodrigues *et al.*, 2007). Caspase-8 was suggested to function in cell death and survival as genetic ablation and deficiency of this classical cell death caspase by siRNA sensitizes fibroblasts for necrotic cell death in response to TNF (Frisch, 2008; Oberst *et al.*, 2011). Interestingly, forced expression of wt-DCC failed to promote cell death in kidney epithelial cells transformed by Src, suggesting the possibility that this oncogenic tyrosine kinase frequently activated in epithelial cancers interferes with the cell death activity of this dependence receptor (Rodrigues *et al.*, 2007). Taken together with other studies, these data provided evidence that Netrin-1 is a potent invasion promoter in epithelial cancer cells and that DCC is a metastatic suppressor targeting both proinvasive and survival pathways in a cumulative manner (Rodrigues *et al.*, 2007; VanZomeren-Dohm *et al.*, 2011). As described for the angiogenic factor Netrin-1, Netrin-4 is a lymphangiogenic factor contributing to tumor dissemination and metastasis (Larrieu-Lahargue *et al.*, 2010).

Invasive and metastatic cancer cells are fully addicted to their survival conditions during their migrations in primary tumors, systemic blood circulation, lymphatics, as well as homing and invasive growth at secondary sites. It is important to consider the divergences of the survival responses induced by soluble and physiological concentrations of Netrin-1 from those induced by ectopic and forced Netrin overexpression at uncontrolled protein levels in both intracellular sites and secretory pathways. In p53-deficient epithelial cells, Netrin-1 ectopic expression was shown to induce apoptosis via the TAp73α tumor suppressor and to abolish the apoptosis induced by DCC, suggesting that the Netrin-1 and DCC/UNC-5 pathways are mutually exclusive (Llambi *et al.*, 2001; Mehlen *et al.*, 1998; Rodrigues *et al.*, 2007; Roperch *et al.*, 2008; Tanikawa *et al.*, 2003). Netrin-1 ectopic expression and repeated administration of soluble Netrin-1 at high concentrations (100 ng/ml) also potentiated alterations of cell viability induced by the DNA-damaging agent cisplatin (Roperch *et al.*, 2008). It is therefore possible that opposing mechanisms may coexist *in vitro* and *in vivo* for the regulation of survival and cell death signals in cancer cells exposed to transient paracrine actions of soluble and ECM-bound Netrins versus constitutive overexpression of ectopic Netrin-1. Concomitant and sequential actions of the Netrin-1/DCC actors and internalization of DCC/UNC-5 receptors by high Netrin and guidance receptor levels may interfere with the global cell survival and death responses in a given context of neoplasia. According to the dependence receptor theory, high concentrations of Netrin-1 (300 ng/ml) were shown to inhibit DCC proapoptotic activity in commissural neurons (Furne *et al.*, 2008). UNC-5B induction by p53 might be a determinant for the Netrin balance between cell survival and apoptosis (Tanikawa *et al.*, 2003; Thiebault *et al.*, 2003). Indeed, the *UNC-5B* promoter is a direct target of p53 in HCT116 human colon cancer cells treated with the DNA-damaging agent adriamycin. In strong support with the cell survival activity of Netrin-1, disruption of Netrin autocrine loops and signaling by siRNA or decoy-soluble DCC receptor ectodomain in breast cancer cells and neuroblastoma overexpressing Netrin-1 triggers cell death and prevents metastasis formation (Delloye-Bourgeois *et al.*, 2009b; Fitamant *et al.*, 2008).

Netrin-1 expression is frequently observed in epithelial cancers and human cancer cell lines characterized by the accumulation of multiple deficiencies in the guidance and dependence receptor family (Delloye-Bourgeois *et al.*, 2009b; Duman-Scheel, 2009; Fitamant *et al.*, 2008; Hilgers *et al.*, 2000; Meyerhardt *et al.*, 1999; Reale *et al.*, 1994; Rodrigues *et al.*, 2007; Roperch *et al.*, 2008; Shin *et al.*, 2007; Thiebault *et al.*, 2003). However, the mechanisms driving Netrin overexpression in several epithelial cancers remain poorly understood. Similarly, *UNC-5A*, *UNC-5B*, and *UNC-5C* transcripts are often detected in epithelial cancers, while DCC expression is either deficient, barely detectable, or associated with molecular alterations and somatic mutations (Rodrigues *et al.*, 2007; Roperch *et al.*, 2008; Thiebault *et al.*, 2003). Downregulation of UNC-5B expression in cancer cell lines is difficult to interpret from global comparisons between resections of colon mucosa or nontransformed tissues adjacent to tumors versus their transformed counterparts in primary tumors or cultured cancer cell lines of diverse origin. Other studies, however, reported the frequent expression of Netrin-4, UNC-5B, and Neogenin in human cancer cell lines of diverse origin, but showed no or little expression of Netrin-1 or Netrin-3 (Larrivée *et al.*, 2007). Convincing evidence now supports the notion that Netrins and guidance–dependence receptors are critically involved in tumorigenesis (reviewed in Arakawa, 2004; Bernet and Fitamant, 2008; Mehlen *et al.*, 2011). Status of the Netrin and dependence receptors in epithelial cancers as well as future diagnosis and treatment approaches is summarized in Table I. Other alterations in guidance molecules are also described for Slit-2 in gliomas, lymphocytic leukemia, lung, breast, colon, cervix and neck, bladder and prostate cancers (Loss of Heterozygosity, LOH) and Robo receptors in breast, kidney, and lung cancers prostate and cervix (reviewed in Arakawa, 2004; Mehlen *et al.*, 2011; Nasarre *et al*, 2010; Strickland *et al.*, 2006).

Table I Netrins and Guidance–Dependence Systems in Morphogenesis and Tumorigenesis

References	Ligands	Receptors	Diagnosis, prognosis, functions, and therapy	
Brain tumors, neuroblastoma				
Delloye-Bourgeois *et al.* (2009a)	Netrin-1 (over)expression	UNC-5H autocrine activation	Aggressive neuroblastoma, metastasis, and survival	
Meyerhardt *et al.* (1999)	Netrin-1 deficiency, downregulation	Missense mutation DCC coreceptor		
Breast cancer				
Esseghir *et al.* (2007)	Netrin-4 expression		Disease-free patient survival	Invasion
Fitamant *et al.* (2008)	Netrin-1 (over)expression	DCC deficiency	Netrin-1 metastasis marker	Cell survival
Srinivasan *et al.* (2003)	Netrin-1	Neogenin	Gland morphogenesis	Cell aggregation
Strizzi *et al.* (2005)	Netrin-1 downregulation	Neogenin upregulation		EMT
Ovarian cancer				
Papanastasiou *et al.* (2011)	Netrin-1 overexpression	DCC downregulated	Biomarker for Ovarian cancer	
Prostate cancer				
Latil *et al.* (2003)	Netrin-1/4 downregulation	DCC/neogenin downregulation		
	Slit 2 and 3 (over)expression	Robo (over)expression	Hormone-refractory and high-grade tumors	
Huusko *et al.* (2004)	Ephrin-B2		Mutations, LOH, splicing, migration, epithelial architecture	
Colon cancer				
Aschele *et al.* (2004)		DCC expression in primary tumors and metastases	Predictor of survival in patients with 5-FU chemotherapy Not related to tumor stage	
Dallol *et al.* (2003)	Slit-2		LOH, promoter methylation, apoptosis, tumor suppressor	
Rodrigues *et al.* (2007)	Netrin-1 expression	DCC deficiency; UNC-5A, UNC-5B, UNC-5C, adenosine A2b expression	DCC metastasis suppressor	Cell survival

(*continues*)

Table I (*continued*)

References	Ligands	Receptors	Diagnosis, prognosis, functions, and therapy	
Roperch *et al.* (2008)	Netrin-1 ectopic expression	DCC deficiency	Missense mutation, apoptosis, TAp73α	
Thiebault *et al.* (2003)	Netrin-1 (over)expression	UNC-5H downregulation	Missense mutation	Cell survival
Biliary cancers				
Hilgers *et al.* (2000)		DCC inactivation	Homologous deletions	
Pancreatic cancer				
Dumartin *et al.* (2010)	Netrin-1 (over) expression	DCC deficiency, UNC-5A, UNC-5B, neogenin expression		Invasion, Survival
Hilgers *et al.* (2000)		DCC deficiency	Homologous deletions–inactivation	
Link *et al.* (2007)	Netrin-1 expression	DCC, UNC-5C, neogenin	Worse outcome in poorly differentiated carcinomas	
Head and neck squamous cell carcinoma				
Carvalho *et al.* (2006)		DCC deficiency	Promoter methylation	Cell survival
Lung cancer				
Delloye-Bourgeois *et al.* (2009b)	Netrin-1 (over) expression	DCC downregulation, UNC-5B, unknown receptors	Targeting Netrin-1/UNC-5H	Therapeutic approach
Liu *et al.* (2004b)	Netrins 1 and 4		Branching morphogenesis	FGF-10 modulation
Kuroki *et al.* (2003)	Sema-3B		Loss of expression, LOH, promoter hypermethylation	
Tomizawa *et al.* (2001)	Sema-3B	Neuropilin-1 and -2 expression	Loss of Sema-3B expression, promoter hypermethylation apoptosis, tumor suppressor	
Melanoma				
Kaufmann *et al.* (2009)	*Netrin-1* transcriptional activation through ETS-1	UNC-5B upregulated, UNC-5C downregulated		Invasion, migration

HS, Heparan sulfate; GLCE, glucuronyl C5-epimerase; HSPGs, HS proteoglycans.

IV. NETRIN SIGNALING AND CRC

A. Colon Cancer: Genetic, Molecular, and Cellular Aspects

1. COLON STEM CELLS, PROGENITORS, AND STEM CELL NICHE

In order to integrate the Netrin axis and related systems in the progression of CRC, the major molecular defects and mechanisms at the origin of this pathology are now presented. In the colon mucosa, epithelial stem cell progenitors localized in the stem cell niche at the crypt base are characterized by high proliferation rate and potential to differentiate into several lineages through the dynamic migration of mature epithelial cells to the apical side of the colon crypt (Fig. 4). Multilineage differentiation capacity

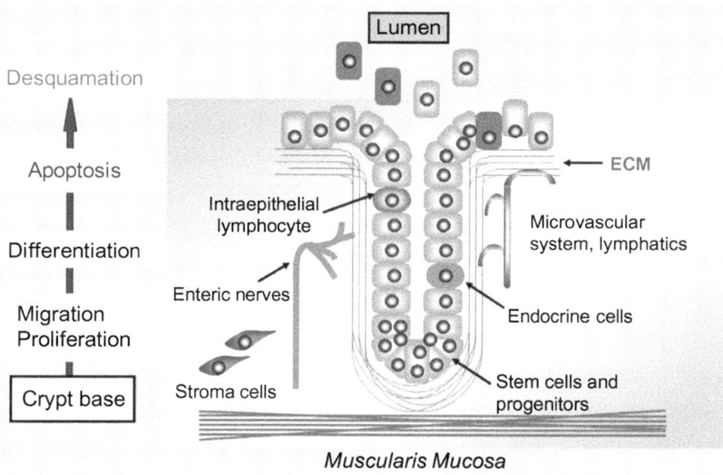

Fig. 4 The human colon epithelial crypt unit. A stem cell progenitor factory producing differentiated epithelial cells is located at the crypt base niche inside the mucosal stroma. The maintenance and protection of the mucosal barrier are mediated by the intense and continuous renewal of the stem cells and epithelial progenitors and migration of their counterparts toward the surface lumen. During this journey, epithelial cells differentiate and establish several paracrine interfaces with intraepithelial lymphocytes and the crypt unit microenvironment (enteric nerves, microvascular system and blood supply, stromal cells, and other resident and bone marrow-derived mesenchymal stem cells). At the mucosal surface, desquamated epithelial cells are scattered in the lumen according to cell death signals and other mechanisms. (See Page 2 in Color Section at the back of the book.)

and EMT are critical features of human ESCs and of normal adult colon stem cells as well (Gespach, 2010). Epithelial progenitors and their differentiated counterparts in colon crypt units are in direct interface with ECM molecules and growth factors produced in a complementary manner by epithelial crypts and stroma cells. Two distinct compartments of proliferating progenitors are operational in colon crypts, the producers of epithelial stem cells with self-renewal properties at the crypt base and a subset of dividing multipotent progenitor cells escaping from this stem cell factory. In the adult human colon, the epithelium contains about 10^{11} cells, representing the production of 6×10^{14} colonocytes over the lifetime of an individual. However, the mediators and signals that regulate colonic cell replacement, stem cell proliferation, migration, survival, and differentiation along the crypt base–apical axis are still poorly understood. Luminal and dietary factors, fasting and refeeding situations are potent physiological stimulants.

Stem- and progenitor-cell lineages in colon crypts are subjected to epigenetic changes (chromatin-structure remodeling, DNA methylation, histone modifications, genomic imprinting) and to chromosomal and microsatellite instability (MSI) leading to morphogenetic aberrations. Colon stem cells in their niche are positive for several biomarkers and targets of the Wnt signaling pathways namely Lgr5, the lineage regulator Achaete-scute like 2 (ASCL2), and the antiapoptotic mediator Survivin. Positivity was also detected for CD133, CD44, CXCR4, the *E-cadherin* gene repressor Slug (a transcription involved in EMT, cancer cell invasion, and anticancer drug resistance), and the detoxifying enzyme aldehyde dehydrogenase 1 (reviewed in Gespach, 2010). These progenitors are the target of the neoplastic transformation. This assumption is based on their continuous active mitosis associated with increased risk of DNA alterations and sequential accumulation of molecular defects inherent to the complex mechanisms of DNA replication and repair and cell division. For this reason, permanent epithelial cell lines have been established from these "immortal" resident intestinal cell precursors by vectorization of viral and cellular oncogenes or insertion of nononcogenic DNA sequences (Emami *et al.*, 1989; Roig *et al.*, 2010). Additionally, complex combinations of extrinsic survival and growth factors are required to maintain and establish permanent cultures of epithelial cells from normal colon mucosa and clinical colon tumors, suggesting that both oncogenic sequences and stromal factors collaborate to determine the "immortal" phenotype. Several lines of evidence support the existence of a subpopulation of cancer cells with stem cell-like (SCL) phenotypes in solid tumors of the digestive system (Gespach, 2010). The cancer stem cell concept is therefore at the edge between a transient and reversible state and identity of self-renewing pluripotent progenitors that are produced during tissue regeneration, remodeling, and carcinogenesis. Stress conditions linked to digestive processes and environmental pressure, infections, nutritional restriction, dietary antigens, inflammation, luminal pathogens, gut microbiome (Arumugam *et al.*, 2011), enteric nematodes, carcinogens, wounding, and adaptation to oxygen and hypoxia strongly influence these integrated mechanisms at the proliferation–differentiation interface and maintenance of colon stem cells and progenitors (Fig. 5). Dietary restriction and reduced calorific intake are the most robust environmental interventions to extend lifespan, to slow the aging process in stem cells and individuals (reviewed in Leanne Jones and Rando, 2011). Thus, delaying stem cell aging by modulation of environmental factors and therapeutic interventions at aging signaling pathways may provide interesting perspectives for the prevention of

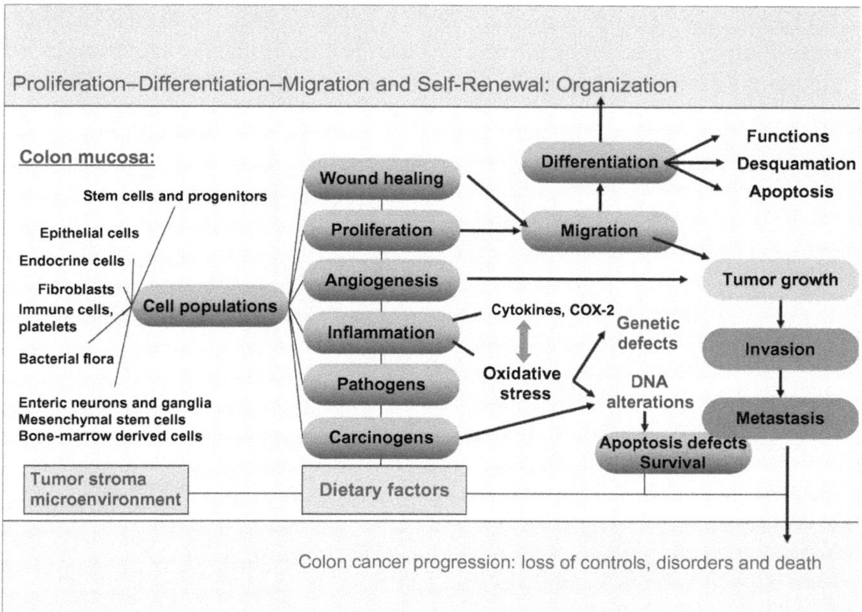

Fig. 5 Environmental factors, intrinsic pressures, and genetic alterations underlying cancer progression from the human colon epithelial crypt unit. Roles of dietary factors, carcinogens, and local pathogens on the proliferation/differentiation balance, survival, and neoplasia in colorectal mucosa: the order and the disorder. Several resident and foreign cell populations living in the colon mucosa establish physical contacts and molecular cross talks with the stem cell niche and the epithelium. Beneficial and deleterious cellular functions are regulated in the colorectal mucosa during normal development, inflammation, and neoplasia. Familial and acquired oncogenic defects cooperate reciprocally with inflammatory mediators, cytokines, and growth factors to induce convergent cellular disorders and fatal issues inherent to the neoplasia: uncontrolled cell proliferation and tumor growth, cell death resistances, uncoordinated epithelial cell migration, tumor growth and angiogenesis, cancer cell invasion, and metastasis. (See Page 3 in Color Section at the back of the book.)

CRC. In addition, the existence of enterotypes in the human gut microbiome may be associated with the health status of individuals, diet and drug intake, CRC, and obesity-related pathologies (Arumugam *et al.*, 2011). Local and systemic interfaces between the stem cell niche and the epithelial crypt microenvironment establish complex neuroendocrine, paracrine, and immune mechanisms via dense mucosal microvasculature networks, enteric nerve fibers, intraepithelial lymphocytes, myofibroblasts, resident mesenchymal stem cells, and interstitial fluids in the stroma. The extrinsic pressure of the tumor stroma microenvironment and the immune system are also playing critical roles in tumor initiation and the emergence of metastatic cancer cells. Cancer cell spreading and dissemination from primary tumors are influenced by the invasive behavior of resident tumor myofibrobasts induced by TGFβ (De Wever *et al.*, 2004a,b; Denys *et al.*, 2008). Myofibroblasts function as accessory partners to enable cancer cells to cross the ECM and tissue barriers during collective invasive processes that ressemble the military strategy. Invasive cohorts of cancer cells and tumor-associated invasive cells escape from growing tumors initiating hypoxia, cell death, quiescence, and senescence signals. Necrosis, oxidative and metabolic stress, and autophagy malfunction are also described in colon cancer. Regarding the genetic and molecular diversity of the cancer pathways that initiate sporadic and hereditary colon cancers (Markowitz and Bertagnolli, 2009), further identification and characterization of the tumor-initiating cells and SCL cancer cells during colon cancer progression will help to define more effective strategies to target selectively or simultaneously SCL and non-SCL cancer cells in solid neoplasms (Inoda *et al.*, 2011).

2. GENETICS AND ANIMAL MODELS OF COLON CANCER PROGRESSION

Colon cancer is the second leading cause of cancer in North America and Western Europe. Cancer pathogenesis is a degenerative disease strongly associated with aging. Approximately 5–6% of individuals will develop a CRC within their lifetime. In 2011, it is estimated that about 140,000 people will be diagnosed with CRC in the United States and approximately 50,000 will die from this disease (reviewed in Gellad and Provenzale, 2010). While stage I of the Tumor-Node-Metastasis (TNM) classification is usually cured by surgery alone, adjuvant and neoadjuvant chemotherapy is currently recommended for stage III and high-risk stage II cancers. The prognosis for 5-year survival is over 95% for stage T1 but only 5% for stage M1.

The emergence of colorectal adenocarcinomas is a complex multistep process linked to genomic and chromosomal instability (CIN), DNA aneuploidy, the progressive and generalized deregulation of gene expression, and signal transduction pathways (Fig. 6). The implication of the adenomatous polyposis coli (APC) protein in the origin of CIN and aneuploidy has been also suggested (Fodde *et al.*, 2001). Familial adenomatous polyposis (FAP) is induced by mutations of the *APC* gene, a defect that contributes to Wnt pathways and CIN in sporadic cancers and appearance of more than 100 colorectal adenomas in FAP. As observed for the maintenance of the stem cell niche (Gespach, 2010), Wnt/APC, TGFβ, Notch and Hedgehog signaling pathways appear to play key roles in colon neoplasia. Several transgenic animal models indicate that molecular alterations targeting the Wnt pathways and other modifiers genes are linked to stem cells/progenitors and Paneth-cell differentiation programs, as well as priming of adenocarcinoma transitions in the mucosa of the small intestine (Boivin *et al.*, 2003; Fodde *et al.*, 1994; Gupta *et al.*, 2004; Philipp-Staheli *et al.*, 2002; Sonoshita *et al.*, 2001; reviewed in Taketo and Edelmann, 2009; van Es *et al.*, 2005; Yamada *et al.*, 2003). However, genetically engineered mice models of colon cancer are mainly associated to the emergence of small adenomas with wide multiplicity (3–300 per animal) in the small intestine, bleeding benign tumors, and animal death linked to anemia. No APC genetic mouse models are available to reproduce the neoplasia cascade in the colon, including the adenoma–carcinoma sequence and fatal colon metastases in target organs such as liver and lungs (Taketo and Edelmann, 2009). This restriction is currently linked to the failure to identify a colon-specific promoter targeting selectively APC transgenesis in the large bowel. The influence of genetic factors, the life style, and diets applied to laboratory animals are also plausible. Of note, colon cancer metastasis has been recently established through depletion of the metastasis suppressor and Notch signaling inhibitor Aes (amino-terminal enhancer of Split) in a syngeneic and orthotopic transplantation model and APC-mutant mice (Sonoshita *et al.*, 2011). Acceleration of intestinal polyposis was observed by crossing APC-mutant genotypes with other modifier genetic alterations in transgenic mice, such as p27 cell cycle regulator and prostaglandin receptor EP2-deficient mice (Philipp-Staheli *et al.*, 2002; Sonoshita *et al.*, 2001). Treatment of APC mice with the GW501516 agonist of nuclear hormone receptor peroxisome proliferator-activated receptor (PPAR)-δ produced similar transforming responses (Gupta *et al.*, 2004). Another example of strategy, targeting the TGFβ signaling intermediate Smad3 in mice, was associated with invasive tumorigenesis almost exclusively in the distal colon with lymph node metastases in homozygous mutant Smad3 of $Apc^{Min/+}$ mice (Sodir *et al.*, 2006; Zhu *et al.*, 1998). Finally, the absence of DCC in mice following inactivation of the mouse homologue ($Dcc^{-/-}$ mice) did not affect growth, differentiation, morphogenesis, or tumorigenesis in mouse intestine (Fazeli *et al.*, 1997). These alternative transgenic models further support the notion that cumulative genetic defects are required to prime the full neoplastic cascade.

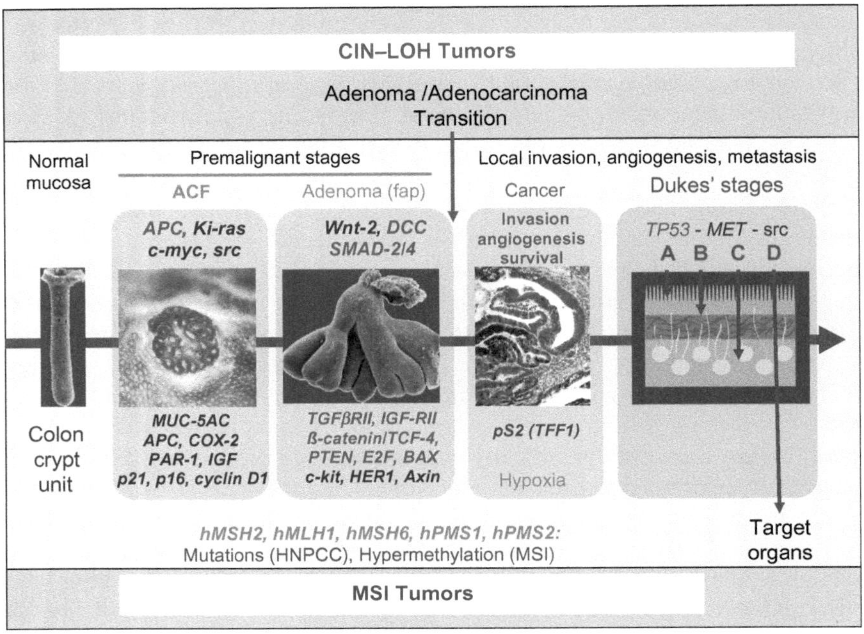

Fig. 6 Genetic, molecular, and cellular alterations linked to the multistep progression of familial and sporadic human colorectal cancers. Molecular and cellular defects operating in CRC are summarized, from the epithelial crypt unit, toward the progression of aberrant crypt foci (ACF), the adenoma formed of several branched and fused crypt units in familial adenomatous polyposis (FAP), the adenoma–adenocarcinoma transition (trefoil factor TFF1 staining is represented in a human CRC specimen), local invasions (intestinal wall and proximal ganglia), and distant spreading to metastases in target organs (e.g., liver, lungs). Chromosomal instability and Loss of Heterozygosity (CIN–LOH) are associated with molecular alterations in *APC*, *DCC*, and *SMAD-2/4*. DNA microsatellite instability (MSI) is observed in both sporadic CRC and familial hereditary nonpopyposis colon cancer (HNPCC). The MSI phenotype is induced by epigenetic silencing and mutations targeting the MSH, MLH, and PMS family of DNA-repair enzymes. As a consequence, mutations in TGFβRII, IGF-RII, β-catenin/TCF-4, PTEN tumor suppressor, E2F, and BAX are observed. Sporadic CRC is also driven by somatic alterations, overexpression of protooncogenes and several receptor and nonreceptor tyrosine kinases, and inactivation or loss of tumor suppressors. These defects include APC, Ki-Ras, c-myc, Src, COX-2, PAR-1 thrombin receptors, IGF signaling, and cell proliferation cycle regulators such as p21, p16, and cyclin D1 (early events). Subsequent alterations are linked to c-Kit, HER-1, axin alterations, overexpression, or silencing of Netrin-1 (secondary events). Additional defects and combinations of early and secondary oncogenic events are supporting further malignant alterations such as trefoil factor induction/overexpression (TFF1, TFF3), tumor angiogenesis, and invasive phenotypes induced by inflammatory situations, hypoxia, and other mechanisms. (See Page 3 in Color Section at the back of the book.)

The most proximal and frequent changes are the activation of Ki-Ras or N-Ras found to occur in about 50% of all CRC and mutations in the PI3K pathway. Wild-type APC may be involved in maintaining proper connections of microtubules with chromosomes. These deregulations target cell proliferation cycle checkpoints and loss of the functional identity of the epithelial cancer cells and tumor stromal cells. Such genetic and molecular changes lead to the formation of aberrant crypt foci (ACF), which precede the appearance of adenomas anchored in the colon mucosal wall. The next stage is the evolution of the adenoma toward more aggressive lesions and irreversible acquisition of a dominant and anarchic invasive potential, chronic inflammation, and oxidative DNA damage. Accordingly, autocrine, paracrine, and systemic regulatory loops are induced by insulin-like growth factor (IGF) receptors IGF-R, HGF/Met receptors, VEGF-R, HER-1 ligands, bile acid nuclear receptors and leptin receptors, G-protein-coupled receptors (GPCRs) PAR-1 (protease-activated receptor family) and neurotensin, trefoil factor pS2 (TFF1), transcription factors AP-1, STAT3, NF-κB, the immediate response gene cyclooxygenase *COX-2*, and the repulsive guidance cue Sema-3A activating Neuropilin NRP-1 receptors (Attoub *et al.*, 2000; Debruyne *et al.*, 2002; reviewed in Degirolamo *et al.*, 2011; Endo *et al.*, 2011; Emami *et al.*, 2001, 2004; Nguyen *et al.*, 2005a, 2006; Rodrigues *et al.*, 2001, 2003a,b, 2004; Souazé *et al.*, 2006; reviewed in Terzi *et al.*, 2010; Zimber and Gespach, 2008).

3. FAMILIAL PREDISPOSITIONS TO COLON CANCER, HNPCC, AND FAP

About 30% of all CRC are linked to familial predispositions and to highly penetrant inherited mutations seen in FAP (described in more details in Paragraph 6 below) and Lynch syndrome resulting of a germine mutation in DNA-repair genes (Jasperson *et al.*, 2010). Familial predispositions are induced by genetic alterations with less penetrance such as promoter or coding sequence polymorphism, promoter silencing and loss of imprinting (LOI), and cumulative molecular defects in multiple susceptibility loci (reviewed in de la Chapelle, 2004). Alterations of the serrated signaling pathways linked to *B-RAF* mutations and CpG island methylator phenotype (CIMP) are involved in the progression of CRC arising from serrated polyps in patients with hyperplastic polyposis syndrome (Hinoue *et al.*, 2009; reviewed in Leggett and Whitehall, 2010). LOI at the *IGF2* gene or activation of the normally silent maternally inherited allele was discovered in embryonal tumors of childhood, such as Wilms' tumor, and occurs in many common cancers, including colorectal tumors. Genomic imprinting is regulated by epigenetic modifications, such as CpG methylations coordinated by imprinting centers containing differentially methylated regions. Epigenetic LOI can be detected in peripheral blood lymphocytes in 10% of the normal population, corresponding to 3.5–5-fold increased risk of neoplastic disease. LOI affecting the *IGF2* locus contributes to overexpression of the autocrine factor IGF2, tumorigenesis, and dedifferentiation in intestinal epithelia of $APC^{Min/+}$ mice and may be a valuable predictive marker of CRC (Cui *et al.*, 2002, 2003; Sakatani *et al.*, 2005). By this model, *IGF2* and *H19* promoters compete at the same chromosome 11p15 region for a shared enhancer located distal of *H19*. Increased or ectopic activity of DNA methyltransferase might lead to aberrant methylation of the maternal differentially methylated region of the *H19* promoter, leading to reciprocal cross activation of the maternal *IGF2* gene (Cui *et al.*, 2002). Access to these enhancers is regulated by binding of the insulator protein CTCF, a CCCTC-binding transcription factor, to the unmethylated maternal imprinting center-1 (IC-1) allele, thus creating a functional chromatin boundary and blocking the interaction of *IGF2* promoters and enhancers. The methylation of IC-1 on the paternal chromosome prevents CTCF binding, thus allowing *IFG2* expression. Human IC-1 contains seven CTCF target sites. Microdeletions at the *H19/IGF2*-imprinting center are necessary for manifestation of Wilms' tumors (Prawitt *et al.*, 2005). In aging-related cancers, altered imprinting patterns of *IGF2* may contribute to the development of replicative senescence in human prostate and urothelial epithelial cells (Fu *et al.*, 2004). Paternal expression of the positive Wnt pathway regulator Frat3 has been identified in the 7C imprinted region of the mouse genome (Kobayashi *et al.*, 2002).

Nonpolyposis forms of the hereditary colon cancer (HNPCC) involved in the Lynch syndrome is more frequent than FAP and is caused by germ cell mutations that invalidate the DNA-repair systems. DNA mismatch repair is deficient in 90% of the HNPCC patients. The mutations concern mostly the hMSH2 and hMLH1 DNA-repair enzymes, less frequently hPMS1 and hPMS2. Genetic factors predisposing to familial CRC are also involved in sporadic cancers in families at risk, defined either by cancers in siblings or by adenomas appearing before the age of 60. Relatively recently, attenuated form of FAP (MHY-associated polyposis, MAP) has been found to result from germ-line mutations in MHY involved in base excision repair (BER) and oxidative DNA damage repair (Al-Tassan *et al.*, 2002). Accordingly, MAP patients show p53 overexpression, and point mutations at guanine nucleotides in APC and Ki-Ras, because of the MYH failure to repair lesions induced by the variant base 8-oxo-guanine (Lipton *et al.*, 2003; Sieber *et al.*, 2004).

4. SPORADIC COLON CANCERS AND LOH

A large majority of CRCs are sporadic (70%) and are strongly associated with environmental causes, nutrition-related complex diseases (e.g. obesity and diabetes mellitus type II), inflammatory situations, ulcerative colitis, and Crohn's disease (Terzi *et al.*, 2010). Sporadic cancers can be divided into two groups according to the initiating mechanisms, generally a loss or inactivation of tumor suppressors and constitutive, dominant activation of proto-oncogenes. The role of recessive tumor suppressor genes (*RB1*, *TP53*, *APC*, *BRCA1*, and *BRCA2*) has been exemplified in a continuum model of tumor progression in dominantly inherited cancer susceptibility syndromes. Haploinsufficiency was seen in *p53*-driven tumorigenicity, and quasi-sufficiency was based to subtle downregulation in (obligate) haploinsufficiency and dosage sensitivity described in cancer initiation controlled by *PTEN* and cell cycle regulator *Rad17* (reviewed in Berger *et al.*, 2011). Similar mechanisms are suspected to occur in dosage/activity of the metastasis suppressor *NM23H1* (Boissan *et al.*, 2010; *vide infra*).

Tumor progression in the digestive tract is controlled by the lifestyle of patients and several parameters such as dietary factors, high caloric intake, obesity, sedentary life, and tobacco smoking inducing epigenetic events such as promoter gene methylation and cyclooxygenase COX-2 induction (Rodrigues *et al.*, 2004). Several hormones, cytokines, bioamines, and growth factors participate in the early and late stages of the neoplastic progression in CRC. The first mechanism, which accounts for 80% of sporadic cases, is connected with CIN and LOH targeting the tumor suppressors *APC* (5q), *TP53* (17p), and *DCC* (18q) and the canonical TGFβ pathway signaling elements *SMAD2*, *SMAD3*, and *SMAD4* (18q) (reviewed in Pino and Chung, 2010; Prunier *et al.*, 1999). LOH at a genetic locus can arise by chromosome duplication, subchromosomal rearrangements and loss, and somatic recombinations and deletions, and represents the loss of a given allele and parent's contribution to part of the genome. Molecular genetic and cytogenetic studies identified losses of only part of chromosomes rather than a whole chromosome in half of the LOH CRCs (Thiagalingam *et al.*, 2001). During adenoma formation, loss and inversion in polarity gradients of proliferating and apoptotic epithelial cells occur along the crypt–villus axis together with delocalization of regulatory cytokines, such as TGFβ, and progressive disorganization of the mucosal architecture. These disorders lead to inappropriate and persistent release of paracrine and autocrine agents acting at illegitimate sites in the nascent tumor. Early dysregulations in the proliferation/differentiation balance occur in sporadic premalignant lesions in the colon, as shown by induction of the gastric *MUC-5AC* mucin gene in ACF.

The formation of premalignant adenoma and progression to the adenocarcinoma sequence in CRC are associated with increased expression of the nucleoside diphosphate kinase (NDPK) and metastasis suppressor NM23H1 (human nonmetastatic clone 23 type 1), as observed in clinical colon tumors (Boissan *et al.*, 2010). Thus, NM23H1 overexpression is already observed at early stages of CRC evolution, in premaligant polyps: NM23H1 was detected at very low levels by Western blotting and immunohistochemistry (IHC) in histologically normal colon epithelial crypts and mucosa adjacent to clinical CRC. In contrast, NM23H1 protein is found strongly overexpressed in adjacent adenomatous tumors and cancer cells within the main body of clinical primary carcinoma, in cirrhotic liver samples, and during liver regeneration in mice. These studies of NM23H1 expression provide an additional link with tissue regeneration, cancer cell proliferation, and intense metabolic activity inherent to growing tumors. However, ablation of this NDPK isoform in *NM23M1−/−* mice was found to be compatible with liver regeneration, suggesting that other growth and survival factors, such as HGF/scatter factor, ligands of the HER family receptors, and many other environmental protagonists. Accordingly, the NM23M2 NDPK isoform can rescue the absence of *NM23M1* in this multitrophic context. Most interestingly, depletion and loss of the NM23H1 protein at the invasion front was observed in clinical colon tumors and hepatocellular carcinomas (Boissan *et al.*, 2010). Thus, transcriptional, posttranscriptional, and -translational regulations are expected to occur through epigenetic controls, protein–protein interactions, NM23H1 sequestration and degradation via proteasome-dependent and -independent mechanisms, microRNA (miRs), and competing endogenous RNAs pathways as well (Berger *et al.*, 2011). Of note, *NM23H1* was originally discovered highly expressed in cells and tumors with relatively low metastatic potential in murine melanoma cells and carcinogen-induced rat mammary carcinoma cells (Steeg *et al.*, 1988). Such a correlation should be extrapolated and interpreted with caution as NM23H1 overexpression is also associated with a mitotic signature. Global NM23H1 overexpression observed in colorectal and hetatocellular primary tumors can mask its depletion or inactivation in a subset of invasive cells and subcellular NM23H1 microdomains implicated in cancer cell scattering and other invasive processes *in vivo*. This prediction is strongly reinforced by the critical role played by NM23H1 expression and silencing in the maintenance and disassembly of E-cadherin/β-catenin-mediated adherens junctions, respectively (Boissan *et al.*, 2010). Thus, discrete subpopulations of NM23H1-depleted epithelial cancer cells in partial EMT states at the invasive fronts may acquire invasive and survival potentials to initiate the metastatic pathway from their primary tumor sites. It is also conceivable that reversible and versatile variations of NM23H1 protein levels and functions in growing epithelial tumors can be observed together with cell cycle-dependent parameter variations and reversible EMT phenotypes as observed for tumor suppressor genes during normal development, cancer progression, and invasion (Berger *et al.*, 2011; Sabbah *et al.*, 2008). Further studies are expected to provide complementary information on the mechanisms upregulating NM23H1 at the proliferation step and downregulating NM23H1 at invasive frontiers in primary tumors. Finally, NM23H1 dosage provides a new paradigm of a metastasis suppressor working at the early steps of the dissemination and metastasis of resident cancer cells from their initial tumor microenvironment. The role of NM23H1 at the transitions between local invasion in the tumor main body (invasive tumor growth in primary lesions) and scattering away from the primary tumor to distant sites (early metastatic seeding spreads in interstitial fluids, blood vessels, lymphatics) remains to be analyzed in more details. Of note, extracellular NM23H1 was shown to promote neurite outgrowth from dorsal root ganglia neurons, independently of its NDPK activity (Wright *et al.*, 2010). In addition, translocation of NM23H1 to the nucleus may cause minor DNA damages to induce subsequently CIN by nicking DNA at early stages of the neoplasia (Conery *et al.*, 2010; Li *et al.*, 2010b). Finally, overexpression and nuclear localization of NM23H1 are a strong predictor of radiation resistance in patients with head and neck squamous cell carcinoma (Park *et al.*, 2011). Conversely, *NM23H1* silencing is associated with resistance to apoptosis in human hepatocellular cancer cells treated *in vitro* with the topoisomerase inhibitors camptothecin and etoposide (Boissan *et al.*, 2010).

5. SPORADIC COLON CANCERS AND MSI

The second mechanism driving CRC progression is characterized by MSI caused by the loss of DNA mismatch repair activity and targeting alterations of short repetitive DNA sequences at the immediate vicinity of critical genes (reviewed in Boland and Goel, 2010). MSI tumors are linked to inactivating mutations or methylation of gene promoters encoding DNA-repair enzymes. These epigenetic changes are observed, for instance in the hMLH1 promoter, hypermethylated in a subset of patients. The hypermethylation of the hMLH1 locus reflects that of numerous genes and the "CpG Island Methylator Phenotype" or CIMP. The hMSH2, hMSH6, hPMS1, and hPMS2 gene promoters are also regulated by hypermethylation. Inactivation of DNA-repair genes favors generalized genetic instability and increases the frequency of subsequently acquired mutations ("Mutator phenotype"). Other forms of DNA repair include the short, single-nucleotide BER system involved in point mutation repair (Ki-Ras activation) and the long patch multinucleotide 2–13 nucleotide repair BER (Wood *et al.*, 2001). Sporadic MSI tumors are frequently mucinous, predominantly localized in the right colon, and generally diploid. In MSI patients, alterations in type-II TGFβ and IGF-Rs, β-catenin, TCF-4 (Duval *et al.*, 1999), and E2F transcription factors, as well as loss and inactivating mutations of the tumor suppressor PTEN (phosphatase and tensin homolog) and the apoptosis regulator BAX, are frequently reported. Loss-of-function events in the TGFβ pathways observed at early steps of the neoplastic progression (inhibition of cell proliferation, induction of apoptosis) are switched to TGFβ transforming functions at later stages: induction of cell proliferation, survival, EMT, angiogenesis, invasion, and metastasis (De Wever *et al.*, 2004a,b; Liu *et al.*, 2011). Activation of Wnt and TGFβ signaling occurs together with other transforming factors related to the transcriptional machinery, cell–cell communication, ECM–BM interactions, and cytoskeleton activation. Recent progress in molecular biology, however, has shown that colorectal carcinogenesis is not necessarily clearly divided into these two pathways, but is in fact more complicated. Other routes, including the serrated and epigenetic pathways have been reported (*vide supra*).

6. Sporadic Colon Cancers and Associated Oncogenic Pathways

Among somatic mutations and oncogenic pathways implicated in sporadic CRC, early events concern—at least in part—(i) the Wnt pathway (including APC, Axin-2) leading to overexpression of Wnt target genes; (ii) *PI3KCA* mutations; (iii) overexpression or activation of the ECM molecules SPARC (matricellular glycoprotein secreted protein acidic and rich in cysteine) and tenascin C (an ECM of stem cell niches), small GTPases pathways Ki-Ras/B-Raf, Rac1b splice variant, c-myc, STAT-3, Src, PKCα (Chastre *et al.*, 1993; De Wever *et al.*, 2004a; Delage *et al.*, 1993; Empereur *et al.*, 1997; Jordan *et al.*, 1999; Porte *et al.*, 1995; Rivat *et al.*, 2003, 2005), and RTKs IGFI-R, HER-1, c-Kit, and c-Met (André *et al.*, 2004a; Attoub *et al.*, 2002; Chastre *et al.*, 1993; Di Renzo *et al.*, 1995); (iv) loss of tumor suppressors and cell cycle regulator genes *DCC*, *p16*, and $p21^{WAF1}$ by promoter hypermethylation in CRC and multiple tumor types (Carvalho *et al.*, 2006; Hibi *et al.*, 2009; Sancho *et al.*, 2004). At the adenoma–adenocarcinoma transition and later stages characterized by metastatic dissemination, other genetic changes become dominant such as loss of *DCC* and *TP53* seen in 50% of patients (reviewed in Mehlen and Fearon, 2004). Several missense mutations in *UNC-5C* may have a role in CRC susceptibility, some of them (D353N, R603C, and A628K) by reducing wt-UNC-5C proapoptotic activity in the TUNEL and/or caspase-3 assays (Coissieux *et al.*, 2011).

Human genes controlled by the Wnt–APC–β-catenin axis and other signaling components of the Wnt pathways play crucial roles in embryonic development and neurogenesis, genomic stability, cell proliferation, migration, and neoplasia in epithelial cancers (Silhankova and Korswagen, 2007).

Several genes involved in hereditary syndromes, including *APC*, *RAS*, *retinoblastoma* (*RB1*), *TP53*, *HNPCC*, *and patched homologue 1* (*PTCH1*) predispose to the development of tumors in the epithelial and nervous system (reviewed in Ohgaki *et al.*, 2010; Zhang *et al.*, 2011a). Communications between Wnt signals and the dependence–guidance receptor systems are therefore strongly suspected. In amphibians, β-catenin regulates dorsal axis specification and *nodal-related 3* involved in muscle and neural tissue induction during gastrulation (McKendry *et al.*, 1997). In *Drosophila*, the homeotic gene *Ultrabithorax* (*Ubx*) expressed in the central region of the visceral mesoderm is activated by wingless (wg) signaling via the β-catenin homologue armadillo (arm) targeting the lymphocyte enhancer-binding factor-1 (LEF-1) during endoderm induction of the larval midgut epithelium by the adhering visceral mesoderm (Bienz, 1996; Riese *et al.*, 1997). Ubx stimulates the expression of a member of the TGFβ proteins family decapentaplegic (dpp) in the visceral mesoderm and expression of wg in posteriorly adjacent cells. Subsequently, wg signaling at the *Ubx* promoter is efficiently and synergistically coordinated and amplified by additional mechanisms initiated by dpp and targeting the cAMP-response element CRE/CREB transcription factor in close vicinity to the arm/LEF-1 binding site, suggesting that this multimolecular complex is coordinated by LEF-1 (Riese *et al.*, 1997). In the mouse, expression of oncogenic β-catenin leads to the appearance of adenomatous polyps in the intestine and microadenomas in the colon (Harada *et al.*, 1999). Wnt pathway activation is an initiating and early event in colorectal cancerogenesis. APC is mutated in 80% of sporadic cancers (reviewed in Le Floch *et al.*, 2006). The mutations generally produce a truncated APC protein deficient for axin binding. Interestingly, several authors described nuclear-cytoplasmic shuttling of APC in human carcinoma cells (Bienz, 2002; Rosin-Arbesfeld *et al.*, 2003), suggesting that APC controls the β-catenin transcription factor and other mechanisms in transformed cells (reviewed in Narayan and Roy, 2003; Narayan *et al.*, 2005). Epigenetic downregulation of the Wnt antagonists encoded by the *SFRP* genes by promoter methylation may be instrumental in the appearance of ACF as early premalignant changes that predispose to CRC through constitutive Wnt ligand signaling and cell resistance to apoptosis (Deng *et al.*, 2004; Suzuki *et al.*, 2004). Overall, mutations in the Wnt signaling elements cause β-catenin stabilization, and induce and enhance malignant transformation in familial and sporadic colorectal tumors. Sporadic CRC that expresses normal APC frequently display β-catenin mutations (Morin *et al.*, 1997), frequently localized at the GSK-3β phosphorylation site in the N-terminal end. The mutated protein is not recognized by β-TrCP and escapes degradation by the proteasome. In addition, mutations of Axin-2 induced by DNA-repair defects are observed in CRC, in absence of APC dysfunction (Liu *et al.*, 2000). These mutations are also present in cancer-derived cell lines, with mutations in APC or β-catenin, suggesting their additive impact on Wnt pathways. In addition to genetic alterations, overexpression of Wnt-2, -4, -5a, -6, and -7a has been observed in human colorectal tumors (Dimitriadis *et al.*, 2001; Holcombe *et al.*, 2002; Vider *et al.*, 1996). Upregulation of Frizzled-1 and -2 was also detected at the invasive front. In contrast, sFRP1 and sFRP2 inhibit Wnt signaling and are frequently downregulated. The Wnt antagonist sFRP-1 binds Wnt ligands via frizzled-related cysteine-rich domain and comprises a Netrin module (NTR) with unknown functional significance. The NTR module in Netrins is composed of a set of hydrophobic and positively charged amino acids and mediates binding of inhibitors of metalloproteinases (TIMPs) to their protease targets. Recently published data revealed that this module in sFRP-1 interacts with the N-terminal domain of thrombospondin-1 (TSP-1) to abolish carcinoma cell adhesion and migration (Martin-Manso *et al.*, 2011). Again, possible molecular cross talks between sFRP-1, Netrin, TSP-1, and Wnt pathways can be envisaged. Epigenetic inactivation of their genes by promoter methylation has been observed in ACF, an early preneoplastic lesion (Suzuki *et al.*, 2004). In the Wnt signaling cascade, TCF-1 is overexpressed in numerous CRC cell lines. LEF-1 expression has been described in tumoral tissues, whereas normal colonic mucosa was negative (Hovanes *et al.*, 2001; Mayer *et al.*, 1997). Previously unidentified Wnt target genes were identified from a DNA-microarray screen in adenomatous GI tumors of the APC $^{Min/+}$ mouse, including *Igfbp5*, *Lcn2*, *Ly6d*, *Ly6d*, *N4wbp4*, *S100c*, and *Sox4* (Reichling *et al.*, 2005). The APC gene product is a critical component of the β-catenin-degradation complex (Fig. 7). This molecular scaffold functions as a negative regulator of the Wnt/β-catenin pathway (Polakis, 1997). A coiled-coil structure of seven armadillo-type repeats at the N-terminal portion of this large protein (more than 300kDa) is involved in its oligomerization. The central part of APC contains domains of interaction with axin and β-catenin in the degradation complex. Three 15 amino acid repeats and 7 motifs, each containing 20 amino acid residues, participate in the interactions with β-catenin. The 20 amino acid motifs are characterized by the presence of a consensus site for phosphorylation by GSK-3β. Three SAMP domains (Ser-Ala-Met-Pro signature) responsible for the interaction with axin are scattered within the 20-aminoacid repeat region (Behrens *et al.*, 1998; von Kries *et al.*, 2000). Conductin and Axin RGS domains bind the SAMP repeats of APC. The C-terminal APC region contains motifs involved in the interaction with structural proteins: a basic domain involved in the binding to microtubules, a domain of interaction with the microtubule-associated EB1 proteins, and a potential binding site for PDZ-type proteins (Dikovskaya *et al.*, 2001). It is generally accepted that APC needs to bind axin in order to favor β-catenin degradation. Although its role remains unclear, APC may regulate the

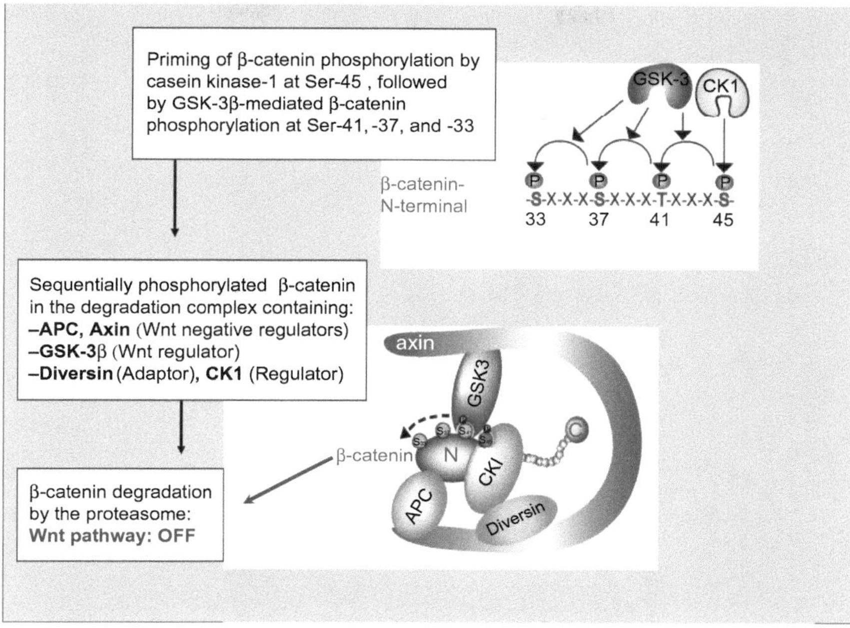

Fig. 7 The APC gene product is a critical component of the APC/β-catenin degradation complex. (See Page 4 in Color Section at the back of the book.)

activity of axin. Overexpression of axin tumor suppressor is sufficient to counteract the accumulation of β-catenin in APC-deficient cells (Hart *et al.*, 1998). APC can be detected in several cellular compartments and structures, including the cytoplasm, nucleus, intercellular junctions at plasma membrane, microtubule ends, mitotic spindles, and F-actin filaments (Bienz, 2002; Mogensen *et al.*, 2002; Näthke, 2004). Actin-dependent pools of APC have been identified at the apical plasma membrane, at lateral membrane interface, and near the basal membrane in microtubule-dependent clusters. As axin, APC has nuclear import and export motifs. APC may facilitate the association of β-catenin and axin at the plasma membrane and E-cadherin/β-catenin interactions (reviewed in Henderson and Fagotto, 2002), suggesting that cell–cell contact associations and dissociations during cell scattering and invasion can supply a new pool of free β-catenin available for nuclear import and canonical Wnt transcription signals (Boissan *et al.*, 2010; Le Floch *et al.*, 2005), as shown in Fig. 8. Accordingly, restoration of wt-APC in APC-deficient human colon cancer cells causes an increase in the translocation of E-cadherin to the plasma membrane, resulting in cell–cell adhesions and maintenance of the epithelial phenotype (Faux *et al.*, 2004). Of note, the Slit–Robo axis (three Slit glycoproteins

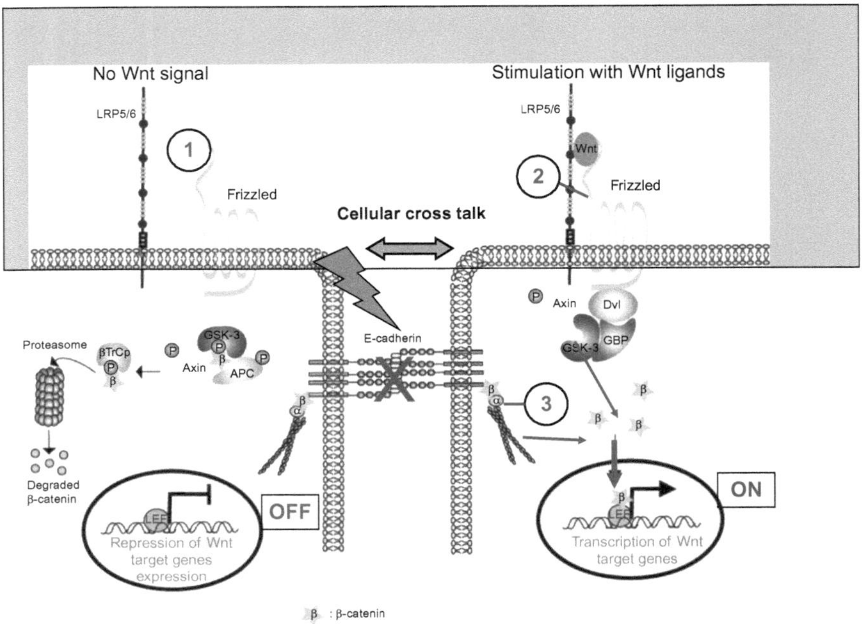

Fig. 8 Dual functions of β-catenin at adherens junctions and in APC/β-catenin degradation complex. Direct and indirect activation of the canonical Wnt transcription pathways: (1) No Wnt signals and maintenance of adherens junctions by E-cadherin: β-catenin is degraded by the proteasome and β-catenin-mediated transcription is OFF! (2) Activation of the Wnt pathways by external Wnt–Frizzled GPCR-mediated inhibition of GSK-3β and β-catenin degradation. The canonical Wnt transcription pathway is ON! (3) Indirect transcriptional activation of β-catenin-dependent genes by the functional invalidation of E-cadherin-mediated adherens junctions via E-Cadherin gene repression by EMT transcription factors (such as Snail, Slug), inactivating mutations in the *E-cadherin* gene, and other mechanisms linked to the metastasis suppressor NM23H1 (Boissan *et al.*, 2010). Immunohistochemical studies show the delocalization of β-catenin from cell–cell adhesions at the plasma membrane to the cytoplasm and its nuclear accumulation. (See Page 4 in Color Section at the back of the book.)

and four Robo receptor members) is involved in CRC progression by inducing E-cadherin degradation, suggesting additional cross talks between Slit–Robo guidance mechanisms and the alternative β-catenin–Wnt transcription pathway (Zhou *et al.*, 2011). Dual roles of β-catenin in cell–cell adhesion and gene expression are further demonstrated by the generation of a β-catenin variant that cannot function in transcription but is proficient in adherens junctions formation during ESC self-renewal and germ layer formation (Lyashenko *et al.*, 2011). Similar to β-catenin, APC is subject to degradation by the proteasome (Choi *et al.*, 2004). Wnt signals slow down APC degradation and favor the nuclear accumulation of APC and β-catenin.

The stabilization of APC by Wnt signaling may constitute a limiting step in which APC exported from the nucleus reduces LEF/TCF-dependent transcription. Wnt targets include several genes involved in transcription, cell positioning and proliferation, survival, adhesion, migration and invasion, inflammation, EMT, and tumorigenesis, namely *TCF-1*, *LEF-1*, *C-JUN*, *Id2*, *RXR*, *SOX9*, *AXIN-2*, *MYC*, *CYCLIN D1*, *BCL2*, *uPAR*, *CD44*, *MMP-7*, *MMP-26*, *MT1-MMP*, *PPAR-δ*, *COX-2*, *WISP-2*, and *MET* (reviewed in Le Floch *et al.*, 2006).

7. COLON CANCER THERAPY

Current chemotherapeutic approaches against CRC are mainly based on combinations of cytotoxic drugs targeting DNA integrity and function. In the past decade, significant improvements have been made in response rates, progression-free survival, and overall survival (André *et al.*, 2004b; Haller *et al.*, 2011). In preclinical models and clinical trials, improvement of the therapeutic activity of several combined regimens was reported for 5-FU and oxaliplatin, 5-FU and folinic acid, FOLFOX-4 (5-FU, leucovorin, and oxaliplatin), FOLFIRI (irinotecan, 5-FU, and leucovorin), intermittent oxaliplatin treatment (FOLFOX-7), and XELOX (oxaliplatin plus the oral fluoropyrimidine, capecitabine) (Tournigand *et al.*, 2004, 2006). More recently, targeted kinase inhibitors and monoclonal antibodies with cytotoxic chemotherapy showed clinical responses and promising improvement in survival among patients with metastatic CRC. Associations between cetuximab/bevacizumab with FOLFOX-4 or FOLFIRI were proposed in previously untreated patients with metastatic CRC (Bokemeyer *et al.*, 2009; Hurwitz *et al.*, 2004; Laurent-Puig *et al.*, 2009; Van Cutsem *et al.*, 2011). Novel targets such as HER-3, protein kinase C (PKC), AKT, mutant B-RAF, galectins, histone deacetylase inhibitor (HDAC) inhibitors, peroxisome proliferator-activated receptor gamma (PPARγ) agonists, and miRs are also challenged. However, anticancer drug resistance, EMT, and increased invasive and metastatic activities are also induced by the classical DNA-damaging agents 5-FU and oxaliplatin in human colon cancer cells (Barbat *et al.*, 1998; Dessein *et al.*, 2010; Yang *et al.*, 2006). Amplification of the *TYMS* gene encoding thymidylate synthetase and induction of cell surface chemokine receptors CXCR4 and transcription factor ASCL2 are observed in 5-FU-resistant HT29 human colon cancer cells and metastatic CRC patients treated by oxaliplatin (André *et al.*, 2004b; Raymond *et al.*, 1997). ASCL2 was recently identified as a critical factor involved in the maintenance of intestinal stem cells (van der Flier *et al.*, 2009). It is now well accepted that the chemokine CXCR2/CXCR4/CXCR7 receptor family and their ligands CXCL12 and MIF mediate a series of transforming functions, including tumor cell invasion, angiogenesis, metastasis, and resistance to chemotherapy. However, the overall survival of metastatic CRC patients and therapeutic efficacy of the anti-HER-1 mAbs cetuximab and panitumumab alone in the treatment of CRC is highly dependent upon the mutational status of the *KRAS* and *PI3KCA* genes (reviewed in De Roock *et al.*, 2010; Lièvre *et al.*, 2008; Saridaki *et al.*, 2011). Of note, activation of the Ki-Ras/B-RAF/mitogen-activated protein kinase/extracellular signal-regulated kinase/kinase (MEK/ERK) and HER-1/Src/PI3K/PTEN/AKT/mTOR signaling cascades is a common oncogenic defect supporting the development of chemoresistances and invasive potential in CRC patients. These transduction pathways are not linear since several alternative pathways and cross talks are connected at different branches and levels in this signal transduction tree. Therapeutic perspectives to induce and restore sensitivity of epithelial tumors presenting intrinsic or acquired resistances to first-line regimens were introduced by preclinical and clinical studies and development of first-generation anticancer drugs targeting the tyrosine kinases Src, HER-1, VEGF-R, PDGF-R and the serine–threonine kinases AKT, mTOR, PKC, and TGFβRI (reviewed in Faivre *et al.*, 2006; reviewed in Giaccone and Wang, 2011; Mayer and Krop, 2010; Nguyen *et al.*, 2006; Rivat *et al.*, 2003; Rodrigues *et al.*, 2003b; reviewed in Winder and Lenz, 2010). Complex mechanisms of tumor resistance are induced during anti-HER-1 and HER-2-targeted therapy, including receptor amplification, mutations, *MET* gene amplification, mutations in the *PI3KCA* gene, conversion of receptor agonist activity into a partial agonist, EMT, and clonal selection of resistant subpopulations with defective cell death responses and metastatic potential (Dessein *et al.*, 2010; Sequist *et al.*, 2011; Yang *et al.*, 2006). Interestingly, resistance to the HER-2 mAb trastuzumab and PTEN inactivation mechanisms may occur independently of PTEN loss or mutational status (Zhang *et al.*, 2011b). This was accomplished through the hyperactivation of the non-RTK Src observed in various HER-2/trastuzumab resistance models. Src activation by trastuzumab resistance or constitutive activation of the IGF-R1, HER-1, and HER-3 pathways has been shown to compromise the activity of the tumor suppressor PTEN through tyrosine phosphorylation and inhibition of its localization at the plasma membrane. Reciprocally, Src was further hyperactivated through PTEN phosphatase-mediated dephosphorylation of Src at Tyr416. Both IGF-R1 and HER-1 are activated by autocrine signaling loops and GPCR-mediated cross talks in CRC and several other epithelial cancers. Oncogenic activation of Src following signal transactivation, mutation, and overexpression correlates with early and late stages of CRC. This tyrosine kinase was shown to modulate several critical targets involved in invasive growth, survival, and metastasis, including STAT3, AP1, c-MYC oncogenes, FAK, PKC family, and MAPK.

New generation of drugs targeting both HER-1 and HER-2 such as the tyrosine kinase inhibitor lapatinib is under preclinical and clinical evaluation in a variety of tumor types including CRC (Labonte *et al.*, 2011). In addition, combining lapatinib with the HDAC panobinostat produced a synergic therapeutic response in CRC models. This interesting response may be explained by the anticancer activity of HDAC inhibitors inducing the hyperacetylation of DNA-bound histones and nonhistone proteins, alterations of the chromatin structure, gene expression, growth arrest, and apoptosis (reviewed in Kim and Bae, 2011; Labonte *et al.*, 2011). Combined lapatinib and panobinostat treatment resulted in apoptosis, DNA double-strand breaks monitored by γH2AX formation (De Wever *et al.*, 2011), caspase-8 activation, and PARP cleavage in CRC cell lines. HDAC inhibitors can also induce the acetylation of heat-shock protein HSP90 involved in the stability of HER-1 and HER-2 and required for matrix metalloproteinase-2 activation (Hendrix *et al.*, 2010; Labonte *et al.*, 2011). It was shown that cancer cell invasion mediated by MMP-2 activation depends on extracellular HSP90α chaperoning. In addition, irreversible inhibitors of the HER family, such as Afatinib (BIBW 2992) with dual activity against HER-1 and HER-2, show interesting potential for the treatment of cancer patients with T790M mutations in exon 20 of HER-1 (reviewed in Giaccone and Wang, 2011). Taken together, it appears that these resistance phenotypes originate from complex and multiple epigenetic and genetic defects and persistent activation of oncogenic and survival signaling pathways. Consistently, these intrinsic and acquired anticancer drug-resistance mechanisms illustrate the clonal diversification, adaptation, plasticity, and pluripotency of cancer cells, a reminiscence of the stem cell progenitor behavior.

As stated above, current practices that personalize CRC therapy are therefore based on the integration of a series of biomarkers and oncogenic pathways in relation with genetic characteristics of a given tumor in a selected CRC patient. Several miRs that are playing important roles in colon cancer cell survival, EMT, angiogenesis, metastasis, and critical signaling pathways targeted by current therapeutic strategies have been recently identified by several laboratories (reviewed in de Krijger *et al.*, 2011; de Planell-Saguer and Rodicio, 2011; Sayed and Abdellatif, 2011). As expected, miRome analysis of primary tumors and metastases should improve the prognosis, therapeutic outcome, and survival of CRC patients. Future "à la carte" molecular profiling practices will include more global analyses to conduct genome-wide sequencing, single-nucleotide polymorphism, copy number, DNA ploidy, and epigenetic platform analyses (mutome, amplicome, CIMP) as well as transcript, protein, and metabolic profiles (transcriptome, miR analysis, phosphoproteome). The costs of personalizing care and global analysis are in balance with the expected benefits linked to the efficacy of the treatments adapted to each individual biomarker in CRC patients (reviewed in Catenacci *et al.*, 2010; Stricker *et al.*, 2011).

B. Inflammation, Colorectal Neoplasia, and Netrin Expression and Signaling

As discussed above, molecular defects targeting Netrin-like molecules and dependence–guidance receptors as well as Integrins and ECM molecules are involved in the restriction and progression of most common types of epithelial cancers including CRC. The Netrin signaling systems constitute a new paradigm of versatile effectors with opposing roles in neural guidance, cancer limitation and promotion during chronic inflammation, and neoplasia. It is widely accepted that inflammatory situations play key roles in the predisposition to CRC development and progression (reviewed in Cho and Brant, 2011; Ullman and Itzkowitz, 2011). Consistently, Netrin/DCC pathways have a major relevance in the limitation of inflammatory diseases, neoplasia, and patient survival during chemotherapy. The majority of CRCs show the loss of both DCC and UNC-5C Netrin-1 receptors: UNC-5C inactivation occurs early, whereas DCC loss occurs in later stages through genetic and epigenetic processes (Bernet *et al.*, 2007; reviewed in Mehlen and Fearon, 2004; Shin *et al.*, 2007). However, other data claim that DCC expression and loss are not related to tumor stage in CRC patients, but represent a dominant predictor of survival in patients with unresectable metastatic disease treated by 5-FU combined with either methotrexate or leucovorin (Aschele *et al.*, 2004).

Differential expression and distribution of Netrin-1 and DCC are detected by *in situ* hybridization and IHC in normal intestinal mucosa, with high-density Netrin signals in the bottom crypts and low Netrin levels at the mucosal/lumen interface (Mazelin *et al.*, 2004). In contrast, DCC was found all along the intestinal epithelium in mice and human gut (Hsu *et al.*, 2001). The progressive disequilibrium in the relative expression of these two conditional guidance molecules, involved in cell survival and cell death states, may support the continuous self-renewal, migration, desquamation, and elimination of damaged normal epithelial cells and their transformed counterparts under premalignant, noninvasive states. Quiescent and dormant cancer cells are considered as the source of tumor recurrence, a reversible cellular state linked to cell survival and cell cycle control. Forced expression of Netrin-1 in mouse GI tract reduced intestinal epithelial cell death associated with the formation of adenomatous lesions in the small intestine and colon, in 17% and 7% of animals, respectively (Mazelin *et al.*, 2004). Of note, transgenic Netrin-1 mice showed focal or diffuse hyperplasia of the colon mucosa. When backcrossed with the $APC^{+/1638N}$ mice, Netrin-1 mice exhibited adenocarcinomas with focal invasion of the *muscularis mucosa*. Thus, a fine balance between guidance and cell death signals controlled by dependence receptors and their ligands determine their opposing roles in inflammation, epithelial cancer cell survival, invasion, and metastasis.

Several factors have been shown to affect the expression of Netrin-like molecules during inflammatory situations and hypoxia (induction by NF-κB and hypoxia-inducible factor HIF-1α), *S. aureus* infection (downregulation) in gut, lung, and CNS (Mirakaj *et al.*, 2010, 2011a,b; Paradisi *et al.*, 2008; Rosenberger *et al.*, 2009). NF-κB interacts directly with the Netrin-1 promoter in human colon cancer cells HCT116. Consistently, Netrin-1 is overexpressed in several human epithelial tumors (Table I) and can be used as a diagnostic biomarker of many human cancers (Ramesh *et al.*, 2011). Netrin-1 in turn attenuates hypoxia-induced inflammation, inflammatory peritonitis, and acute lung injury and therefore coordinates several inflammatory responses through inhibition of leukocyte transmigration (including monocytes, lymphocytes, and granulocytes) and engagement of UNC-5B and A2b-R (Ly *et al.*, 2005; Mirakaj *et al.*, 2010, 2011a,b; Rosenberger *et al.*, 2009). Approaches that prevent inflammatory cells from entering the GI mucosa can use engineered enteric bacteria producing Netrin-like molecules to counteract the early phases of IBD. Several clinical development programs of novel therapeutic agents in IBD (Crohn's disease, ulcerative colitis) are focusing on neutralization of chemokines and proinflammatory mediators, inflammatory cells and stem cells, and adhesion molecules such as Integrins α4 (inflammatory cells), α4β7 (GI tract specific), and β7 (reviewed in Plevy and Targan, 2011). CRCs in IBD patients are expressing high levels of Netrin-1 associated with the retention of the DCC receptor in 28 out of the 30 tumors analyzed (Paradisi *et al.*, 2009). Similar scenario, including Netrin-1 and DCC upregulation, was observed in neoplastic lesions of the adult mice IBD model induced by the DNA alkylating agent azoxymethane and proinflammatory agent DSS. Most interestingly, neutralization of Netrin-1 by intraperitoneal injections of the DCC ectodomain DCC-4Fbn (fourth fibronectin extracellular domain) decreased high-grade adenoma and adenocarcinoma, suggesting that IBD-induced Netrin-1 is causally involved in intestinal neoplasia linked to inflammatory situations, giving another illustration of the dual roles of the DCC–Netrin axis in inflammation and cancer. However, DCC-4Fbn failed to reverse the induction of DCC-4Fbn and mucosal inflammation in the IBD model, suggesting that this Netrin-1 interference system is not fully efficient and accessible to neutralize intratumoral Netrin levels in this drastic model of IBD. It is not excluded that *DCC* expression and protein stability are also induced by NF-κB. NF-κB activation is associated with all hallmarks of cancer predisposition and progression: inflammation, EMT, invasive growth, angiogenesis, survival, metastasis, and resistance to chemotherapy. DCC was also strongly upregulated in OA via induction of the *DCC* promoter by the AP-2 and Sox9 transcription factors (Schubert *et al.*, 2009). Netrin-1 and DCC are implicated in the migration of chondrocytes, induction of MMP-1 and -3 expression, and upregulation of the MIA/CD-RAP marker associated with the OA phenotype. Before and after renal IRI, an inflammatory disease mediated by innate and adoptive immune systems, administration of Netrin-1 attenuates kidney injury, apoptosis, and monocyte and neutrophil infiltration (Tadagavadi *et al.*, 2010). Netrin-1 reduced the production of several chemokines and cytokines (IL-6, IL-1β, and TNF-α) by T-helper lymphocytes Th1/Th2/Th17. Other mechanisms and pathways regulating Netrin expression are identified during development, neuronal connectivity, and plasticity. Human Netrin-4, also called β-Netrin (Koch *et al.*, 2000; Yin *et al.*, 2000), is expressed and secreted as a spliced isoform with distinct expression pattern in several regions of the human brain (Zhang *et al.*, 2004). This Netrin-4 variant lacks the signal peptide region and is truncated at the laminin-N-terminal domain of the peptide, suggesting that this isoform may have some unknown intracellular functions. Netrin-1 overexpression observed in several human cancers (Table I) and after ectopic expression may regulate atypical cell survival and death functions at conditional guidance receptors and lateral signaling (Roperch *et al.*, 2008). Thus, Netrin silencing and overexpression may dictate critical functions during development and neoplasia.

C. Proximal Events in Netrin Signaling

Our current knowledge on the stoichiometry, molecular determinants, and characteristics of the interactions between Netrin-like ligands and their corresponding binding sites, receptors, coreceptors, and modifiers is still very limited. The relative affinity, binding capacity, and molecular specificity of the dependence receptors and their functional interactions with their respective counterparts and ligands that are simultaneously coexpressed or induced in a given cell type are still unknown parameters. As stated above in Fig. 2, a given guidance molecule can bind multiple Netrin receptors, including DCC, as shown for the recently identified axon guidance protein Draxin (Ahmed *et al.*, 2011) interacting with Neogenin, UNC-5 family receptors, and DSCAM. Both Netrins and Netrin receptors exhibit complex structures comprising several modules involved in multiple molecular interactions (e.g., ligand–receptor–ECM) connected with selective signaling pathways and biological significances, in a cell context-dependent manner. Consistently, discrete domains of Netrin/UNC-6 are required for guidance activity in *C. elegans* dorsal migration (V-1, -2, and -3) and ventral migration (V-1 and -3). These domains mediate different guidance and binding activities (Ishii *et al.*, 1992; Kruger *et al.*, 2004; Lim and Wadsworth, 2002). Therefore, parallel or convergent signaling messages and cytoskeletal changes originate from Netrin-like molecules, including cell-ECM-mediated adhesion (V-1 laminin domain), attraction or repulsion (V-2, -3), and blockade of axon branching (C domain) in response to the signals elicited by the N-terminal domains (Lim and Wadsworth, 2002). Similarly, modulation of the proximal Netrin signaling axis messages are orchestrated by the compartmentalization of DCC within microdomains of the plasma membrane comprising signaling molecules, with early signaling events propagated in the soluble, nonraft-associated DCC membrane fraction (Petrie *et al.*, 2009). However, the majority of cell-surface DCC was found in the detergent-resistant membrane fraction up to 60min after the addition of Netrin-1. This situation may reflect an additional static role for DCC molecules anchored at the plasma membrane in the maintenance of cell–cell, cell–matrix, and Netrin substrates adhesions, or their dissociation during cellular motility, invasion, and survival, as suggested earlier (Martin *et al.*, 2006; Shekarabi *et al.*, 2005). Similar apparent constitutive association of guidance receptors with lipid rafts has been described for the axonal guidance receptors Ephrin-A5 and -B1. However, this reservoir of latent DCC in lipid rafts highly enriched in signaling molecules might be recruited, activated, and translocated to soluble plasma membrane fractions through Fyn-dependent DCC Tyr^{1418} phosphorylation required to mediate Rac1 activation and other dynamic signaling responses linked to association and activation of the actin cytoskeleton (Li *et al.*, 2002a,b, 2004, 2006). Consistently, FAK activation and localization of the DCC SH3/SH2 adaptor protein Nck-1 are localized outside lipid rafts (Petrie *et al.*, 2009). Ectopic expression of DCC in DCC-null NG108-15 neuroblastoma cells produced a Netrin-1-dependent increase in the filopodia actin-based membrane extrusions and cell spreading by activating Cdc42 and Rac1 (Shekarabi and Kennedy, 2002). Another pool of latent DCC is also mobilized by exocytosis from of a preexisting intracellular vesicular pool translocated to the plasma membrane through protein kinase A (PKA) activation by forskolin (FK), a direct adenylate cyclase activator (Bouchard *et al.*, 2004). However, the biological relevance of these FK- and cAMP-dependent PKA responses was not validated by physiological agonists of this signaling cascade. The cAMP–PKA axis is considered as a key mediator of chemotaxis and growth cone attraction and repulsion (Han *et al.*, 2007; Ming *et al.*, 1997). The role of cAMP and PKA in guidance receptor signaling was directly linked to molecular interactions of A2b-R with Netrin-1 and DCC (Corset *et al.*, 2000). However, some divergences emerged in data collected on outgrowth-promoting actions of Netrin-1 in dorsal spinal cord axons (Corset *et al.*, 2000) and commissural axon outgrowth or Xenopus spinal axon attraction to Netrin-1 (Stein *et al.*, 2001). These divergent cellular responses are expected to occur following DCC/A2b activation by DCC agonists and adenosine in neural, epithelial, and inflammatory cells, in concert with signaling modulations controlled by the other adenosine receptor subtypes A1, A2a, and A3 (reviewed in Linden, 2011). Alpha2b adrenergic receptors directly interact with α- and β-tubulin and control its transport trafficking from the ER to the cell surface (Duvernay *et al.*, 2011). One can imagine that this tubulin cargo not only recognize GPCR export motifs but may also mediates the cotransport and membrane targeting of other newly synthesized receptors, such as DCC and other guidance cues. In DCC-deficient human colon cancer cells, we have shown that the A2b receptor antagonist enprofylline abolished invasion induced by adenosine and adenosine receptor agonist NECA, whereas higher enprofylline concentrations were required to neutralize the invasive response to Netrin-1. The A2bR was coupled to cAMP generation in HCT8/S11 cells, whereas exogenous Netrin-1 was ineffective, suggesting that DCC is required to initiate this pathway (Rodrigues *et al.*, 2007). Of note, the invasive responses induced by Netrin-1 and NECA were interrupted by the PKA inhibitor KT5720, whereas adenosine-induced invasion was KT5720 resistant. So, direct activation of the cAMP system and permissive roles of PKA may coexist in a given model, pending the status of guidance receptors equipment and signals linked to inflammatory situations, cell-type specificity, and oncogenic events. Another cAMP-dependent mechanism mobilized in axon guidance may also originate from the cAMP–Epac–Rap1 axis connected to PKCα activation, RhoA inhibition, and subsequent permissive activation of Rac1 in the mechanisms driving filopodia-associated cancer cell invasion (Murray *et al.*, 2009). Functions attributed to the small GTPase Rap1 comprise the control of cell proliferation, polarity and secretion, activation of Integrin-mediated cell adhesion to the ECM, and regulation of cell–cell contacts (reviewed in Frische and Zwartkruis, 2010). The Rap1–GEF Epac (exchange protein directly activated by cAMP) binds cAMP, is cAMP activated and PKA phosphorylated, showing that cAMP may affect also Rap1 signaling in a PKA-dependent manner (Altschuler *et al.*, 1995; de Rooij *et al.*, 1998). Accordingly, a developmental switch induced by Netrin-1 distinctly activates Epac in embryonic growth cones but PKA in postnatal neurons and governs gradients of cAMP-dependent guidance cues from attraction to repulsion (Murray *et al.*, 2009).

The cytoplasmic juxtamembrane region ZU5 of UNC-5A (Fig. 9) binds NRAGE/MAGE-D1 (melanoma antigen-D1) to induce apoptosis in native NRAGE-proficient PC12 primary neuronal cells but not in differentiated NRAGE-deficient pheochromocytoma counterparts. NRAGE-dependent apoptosis is also supported by degradation of the survival protein XIAP (X-linked inhibitor of apoptosis), NGF binding to cell death receptor p75NTR, activation of c-Jun N-terminal kinase (JNK)/BAD, and caspases signaling (Bhakar *et al.*, 2003; Jordan *et al.*, 2001; Salehi *et al.*, 2000, 2002; Williams *et al.*, 2003). There is evidence that DCC and UNC-5B multimerize in response to Netrin-1 and that this event is sufficient to inhibit apoptosis (Mille *et al.*, 2009a). In addition, the recombinant soluble fifth fibronectin domain of DCC (FN-5), also designated DCC-5Fbn, specifically binds to

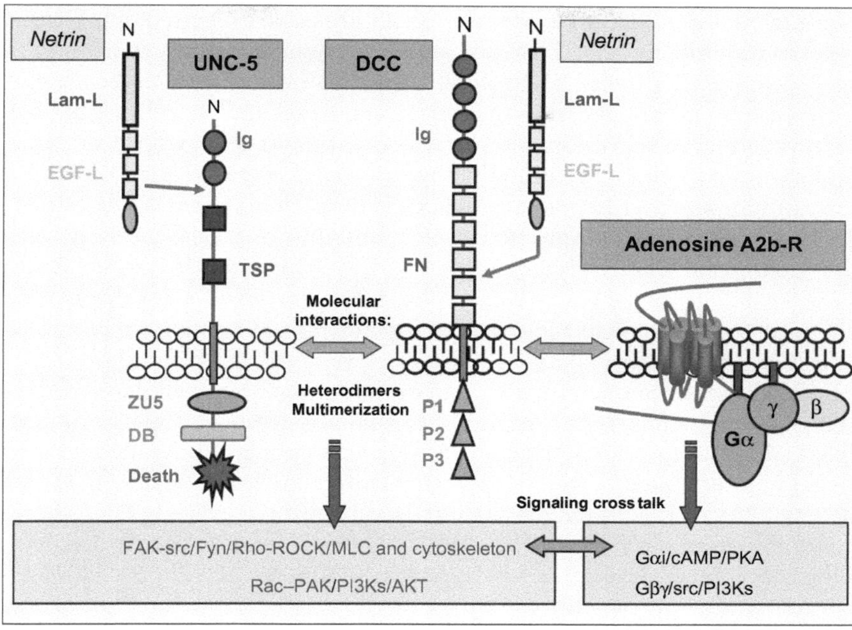

Fig. 9 Molecular structure of the guidance receptors DCC and UNC-5 and their interplay with Netrin-1, adenosine A2b receptors, and GPCR signaling. The adhesion and guidance molecule Netrin-1 is structurally related to laminins according to its laminin-like domains (Lam-L) at the globular N-terminal: Netrin-1 is thought to have derived from the laminin-γ chain. Netrin-1 domains comprise three EGF-like repeats (EGF-L) and a positively charged C-terminal domain. Binding of soluble, extracellular, ECM-bound, and transmembrane forms of Netrin(s) with DCC, Neogenin, UNC-5, and DSCAM receptor family implicates Lam-L and EGF-L repeats. Netrin receptors UNC-5 and DCC are members of the immunoglobulin superfamily (Ig domains) and display extracellular thrombospondin and fibronectin-like domains (TSP, FN) in UNC-5C/DCC, and in DCC, juxtamembrane cytoplasmic modules ZU5, DB (DCC-binding domain), the death domain in UNC-5, and C-terminal conserved regions P1, P2, and P3 (reviewed in Bradford *et al.*, 2009; Sun *et al.*, 2011; Wang *et al.*, 2009b). Molecular interactions between naive and activated Netrin receptors DCC and adenosine A2b coreceptors recruit several molecular signaling scaffolds and adaptors, and trigger complex and interdependent intracellular signaling platforms implicated in cellular guidance and dependence (e.g., the FAK–cytoskeleton axis, Rac–AKT, Gαi/cAMP–PKA, and Gβγ–PI3Ks). (See Page 5 in Color Section at the back of the book.)

Netrin-1 (K_d=5 nM) with similar affinity of the DCC–Netrin-1 interaction but is not competent to inhibit DCC–Netrin-1 interactions, suggesting that other domains of DCC play a critical role in the active conformation required to accomplish high-affinity interactions with Netrin-1 and signal transduction to biological responses. FN-5 has been implicated in DCC–Netrin binding through an interaction of heparin with a Netrin fragment

(Geisbrecht *et al.*, 2003). However, these authors also suggested that FN-4 is the primary domain in DCC necessary for Netrin binding and contributes to the FN-5 binding activity. These conflicting data may arise from differential posttranslational modification of Netrins in normal and cancer cell-derived Netrins and its DCC-binding sites and modulation of their reciprocal multiple binding domains. Interestingly, this truncated extracellular DCC-5Fbn domain inhibits Netrin-1-induced DCC multimerization and induces apoptosis in Netrin-1-expressing cancer cells, and thus restricted tumor growth and metastasis (Delloye-Bourgeois *et al.*, 2009a,b; Fitamant *et al.*, 2008; Mille *et al.*, 2009a). Thus, DCC-5Fbn may also serve as a therapeutic scavenger of Netrin-1 in biological fluids in order to counteract autocrine Netrin-1-induced survival and invasion in several epithelial tumors.

DCC transmembrane receptors are connected with intracellular signaling pathways and subjected to multiple regulations (reviewed in Cirulli and Yebra, 2007; Sun *et al.*, 2011). The ubiquitin–proteasome pathways and the E3 ubiquitin ligase seven in absentia mammalian homologue (SIAH) are playing a major role in the intracellular mechanisms regulating DCC integrity and degradation (Hu *et al.*, 1997). Netrin-1 was shown to decrease cell-surface DCC levels via ubiquitination (Kim *et al.*, 2005). UNC-5B autoinhibition was recently explained by the molecular architecture of the receptor cytoplasmic domain containing three distinct folded domains ZU5, UPA, and death domain (DD) engaged in a structural supramodule in a closed intramolecular conformation blocking its signaling activity on apoptosis and angiogenesis (Wang *et al.*, 2009b). Deletion mutants of DCC, UNC-5, and Netrin revealed that individual domains in DCC and UNC-5, and multiple domains in Netrin, are required for receptor–ligand binding (Kruger *et al.*, 2004). Transduction mechanisms controlled by Y482 tyrosine phosphorylation of UNC-5 by Netrin, in the presence of DCC, are important for repulsive signals (Killeen *et al.*, 2002; Tong *et al.*, 2001). In UNC-5, the second Ig domain is autoinhibitory for this indicator of activity, while the TSP domains are permissive to sustain basal phosphorylation activity even though they are dispensable for Netrin binding activity. This model provides the intramolecular information and rationale for the opposing and convergent biological responses supported by the Netrin and DCC/UNC-5 respective binding domains and interplay in attraction, repulsion, and branching morphogenesis (Kruger *et al.*, 2004). Interestingly, the axon guidance receptor Frazzled/DCC generates a transcriptional response from the commissureless gene *Comm* independently of the known Fra ligands, suggesting that neuronal gene expression is altered by a putative external signal(s) encountered during navigation of the growth cone toward the midline (Kidd, 2009; Yang *et al.*, 2009).

DCC cleavage by metalloprotease-dependent ectodomain shedding was first evidenced in the regulation of axonal migration by the chemoattractant

Netrin-1 (Galko and Tessier-Lavigne, 2000). In rat embryonic cortical neurons, the tumor necrosis factor-α converting enzyme (TACE), also called ADAM-17, cleaves the Netrin receptor Neogenin and desensitizes the neurons to its ligand RGM, a glycosylphosphatidylinositol-anchored glycoprotein (Okamura *et al.*, 2011). It is likely that DCC ectodomain shedding can mimick some molecular interactions controlled by the DCC-5Fbn domain. The cleavage of the DCC intracellular domain by caspase-3 at the major site Asp 1290 is involved in the proapoptotic function of DCC in the absence of Netrin in DCC-defective and -transiently transfected HEK-293T cells (Mehlen *et al.*, 1998). Caspase-3 is rapidly activated by Netrin in a proteasome-dependent manner and is required for the chemotropic responses of Netrin-1 (Campbell and Holt, 2003). This apoptotic protease is unexpectedly incriminated in tumor cell repopulation, rate of tumor recurrence, and death during cancer radiotherapy in human subjects (Huang *et al.*, 2011). These indesirable consequences are associated with markedly increased caspase-3 activity in the tumors, according to its capacity to upregulate prostaglandin E2 as a compensatory survival escape-rescue signaling response loop, in response to persistent cell death signals. Caspase-3 regulates the activation of other signaling molecules such as the cleavage of the Rho-kinase ROCK-I during cell death induced by various agonists and cytotoxic cancer chemotherapy (Coleman *et al.*, 2001; Coudray *et al.*, 2005; Sebbagh *et al.*, 2001). This constitutively active truncated serine–threonine kinase in turn upregulates myosin light chain (MLC) phosphorylation, apoptotic membrane blebbing and cellular fragmentation followed by phagocytic engulfment by neighboring cells or by professional phagocytes in apoptotic cell clearance (Leverrier and Ridley, 2001).

Apoptotic nuclear disintegration requires actin–myosin contractile force and lamin proteolysis, making apoptosis analogous to, but distinct from, mitosis where nuclear disintegration results from microtubule-based forces and from lamin phosphorylation and depolymerization (Croft *et al.*, 2005). Opposing roles of Netrin and dependence receptors on cell death responses are connected with apoptosis via direct activation of the serine/threonine kinase death-associated protein kinase (DAPK) by UNC-5B and its reversion in the presence of Netrin-1 (Llambi *et al.*, 2005). This mechanism of cell death induction and survival is mediated by a protein complex formed of the PR65β subunit of the protein phosphatase 2A (PP2A) and DAPK implicated in UNC-5B-induced and Netrin-1-reversed DAPK phosphorylation in endothelial cell death and angiogenesis regulation (Guenebeaud *et al.*, 2010). The Rho–ROCK–MLC axis also controls the actomyosin filament assembly involved in cancer cell migration and invasion induced by several agonists, including Netrin-1 (Nguyen *et al.*, 2005a,b; Rodrigues *et al.*, 2007). Similarly, the formation of an activated ezrin–radixin–moesin complex binding actin in growth cone filopodia is induced by Netrin-1 during the outgrowth of cortical neurons. Src kinase and RhoA/ROCK axis are required to orchestrate these mechanisms (Antoine-Bertrand *et al.*, 2011a). Convergent reports incriminate a selective, sequential, asymmetric localization and recruitment of Rho-like small GTPases (RhoA, Rac1, Cdc42, and the Rac1 ortholog CED-1) that are linked to the Rac–PI3K–PTEN–AKT axis in Netrin pathways and differential controls of neurite outgrowth, axon guidance, asymmetric distribution of MIG-10/lamellipodin, cellular adhesion and survival, neurodegenerative diseases, cancer cell invasion and metastasis (Adler *et al.*, 2006; Antoine-Bertrand *et al.*, 2011a,b; Bae *et al.*, 2010; Causeret *et al.*, 2004; Kotelevets *et al.*, 1998; Li *et al.*, 2002a,b; Picard *et al.*, 2009; Quinn *et al.*, 2006, 2008; reviewed in Quinn and Wadsworth, 2008; Rodrigues *et al.*, 2007). For example, Rac1 regulates axon guidance and neurite complexity, Cdc42 regulates axon specification and growth, and RhoA inhibits neurite extension (Peng *et al.*, 2010). A signaling complex comprising Cdc42, Rac1, p21-activated kinase PAK-1 and the neuronal Wiskott–Aldrich syndrome protein N-WASP is recruited by DCC during Netrin-1-induced substrate adhesion and growth cone morphology (Shekarabi *et al.*, 2005). Nucleotide exchange factors (GEFs), such as TRIO in the TRIO–Rac–PAK axis, and the molecular complex comprising engulfment and cell motility factor (ELMO1)/DOCK180 (dedicator of cytokinesis) in the UNC-5B–Rac1 axis are both acting as intracellular signaling switches and effectors downstream Netrin pathways in the developing cerebellum, neurite growth, migration and guidance, and vascular morphogenesis (Epting *et al.*, 2010; Peng *et al.*, 2010). This molecular signaling trio is activated by Disrupted-in-Schizophrenia 1 (DISC1) interaction with TRIO, facilitating the recruitment of Rac1 to TRIO during axon guidance (Chen *et al.*, 2011).

Phosphorylation of DCC and multifunctional Neogenin receptors by Netrin-1 has been demonstrated in Netrin signaling and neurite outgrowth and requires FAK signaling and Src family kinases Src, Fyn, and Lck for axon guidance and Netrin signaling (Li *et al.*, 2004; Liu *et al.*, 2004a; Ren *et al.*, 2004, 2008; reviewed in Wilson and Key, 2007). Accordingly, signaling cross talks between Integrins, FAK/JNK, and HER-1/Ras/PI3K/Src have been established (Ayoub *et al.*, 2009; Grijelmo *et al.*, 2007), as shown in Fig. 1. Netrin-1 signaling induces UNC-5B interaction with the brain GTPase and PI3K enhancer PIKE-L, PI3K activation, and neuronal survival through Fyn-dependent mechanisms (Tang *et al.*, 2008). On the other hand, the C-terminal domain of FAK interacts with the P3 domain of DCC, while the SH3 domain of Src interacts with the DCC tyrosine residue in position 1400. Both FAK and Src phophorylates DCC at tyrosine Y1420, and UNC-5, at multiple tyrosine sites (Li *et al.*, 2006). These findings are consistent with the implication of FAK and Src kinases in focal adhesions recruited by Netrin-1-activated Integrins $\alpha 3\beta 1$ to promote the Integrins/FAK/Src and

Rho/Rac signaling platform involved in directional interneural migration, cellular adhesion, survival, and cancer cell invasion (Itoh *et al.*, 2005; Stanco *et al.*, 2009; Staquicini *et al.*, 2009; Yebra *et al.*, 2003). Netrin/DCC-induced axonal migration requires local caspase-3 activity and is linked to extracellular signal-regulated 1/2 (ERK1/2) activation, a MAPK involved in growth cone guidance, protein synthesis, epithelial cell growth and differentiation, and cancer cell invasion (Boissan *et al.*, 2010; Campbell and Holt, 2003; Forcet *et al.*, 2002; Sawai *et al.*, 2005). Coherently, the cytoplasmic P1 domain of DCC (1139–1165 fragment) is a direct MAPK-docking motif for DCC-ERK2 binding and ERK2-mediated phosphorylation of the cell death receptor at five potential Ser/Thr sites in close vicinity with the kinase-active site (Ma *et al.*, 2010). DCC signaling though the ERK1/2–eNOS (endothelial Nitric Oxyde Synthase) pathway was also induced by Netrin-1 in primary isolated cardiomocytes (Zhang and Cai, 2010). In addition, Netrin-1-induced translation and growth cone guidance require the MAPK/stress-activated protein kinase (SAPK) p38/caspase-3 cascade for guidance factor-induced protein degradation and rapid caspase-3-dependent cleavage of PARP involved in the maintenance of genomic stability and activation of proteasome-dependent proteolysis (Campbell and Holt, 2003; Ullrich *et al*, 1999). The JNK/SAPK pathway regulating cellular stress responses during cell death and inflammation is not activated in response of growth cone collapse to Netrin-1 (Campbell and Holt, 2003). Conversely, Netrin-4 was shown to reduce signal activation within the AKT–JNK2–c-Jun pathway (Liu *et al.*, 2004a; Nacht *et al.*, 2009; Park *et al.*, 2004; Yang *et al.*, 2007). α-1 Integrin-associated FAK/Src and p130Cas-JNK molecular scaffold and signaling cascades are linked to the activation of the proinvasive metalloproteinases MMP-2 and MMP-9 in human colon cancer cells (Van Slambrouck *et al.*, 2007).

There is cumulating evidence that the PI3K–AKT–mTOR axis is interconnected with multiple signaling pathways driving cancer cell survival and invasion, EMT, metastasis, and anticancer drug resistances. It was shown above that the PI3K enhancers PIKE-A and -L are integrated as convergent signaling components of the AKT pathways regulated by cell death receptors, Integrins, and guidance molecules, as demonstrated here for several other upstream AKT effectors, such as Rac and Src family kinases/FAK (Fig. 1). Consistently, the ribosomal protein S6, an mTor (mammalian target of rapamycin) target, is activated by Netrin-4, and the use of rapamycin, an mTor inhibitor, suppresses Netrin-4-induced lymphatic endothelial proliferation (Larrieu-Lahargue *et al.*, 2010). Similarly, mTor has been linked to both Wnt and Ephrin signaling in a cancer context (Inoki *et al.*, 2006; Sahin, 2010). Some components of this signaling machinery are well connected with Netrin–DCC receptors responses including Nck/PAK/PI3K/Rac and AKT (Fig. 10). mTOR is a member of the ATM (ataxia

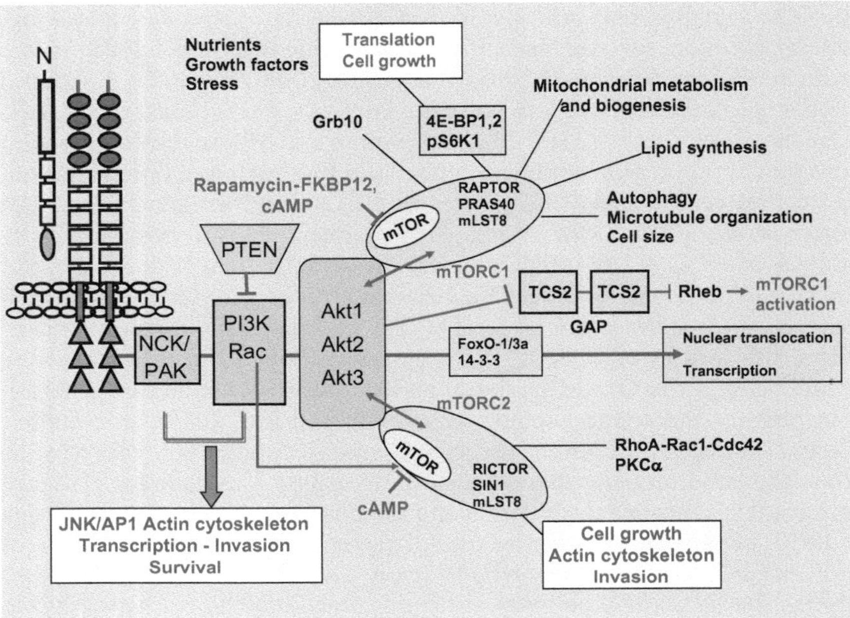

Fig. 10 The AKT–mTOR signaling platform as a target of cellular guidance and dependence regulators. Current advances in the field support the notion that some Netrin signaling pathways are integrated in the NCK/PI3K/Rac and AKT axis (orange-filled signaling boxes). Up to now, no information is available on the possible role of guidance and dependence receptors on proximal and distant elements of the mTOR signaling platforms downstream AKT (blue signaling boxes) (reviewed in Zoncu *et al.*, 2011). The mammalian kinase target of rapamycin mTOR is the catalytic subunit of the two multiprotein complexes mTORC1 and mTORC2. In addition, these molecular scaffolds include [RAPTOR/PRAS-40 and MLST8] in mTORC1, and the rapamycin-insensitive companion of mTOR [(RICTOR), SIN1, and mLST8] in mTORC2. Through its interaction with FKBP12, rapamycin is able to exert its effects by binding the mTORC1 and inhibiting the mTOR signaling pathway. The mTOR kinase has been shown to directly phosphorylate the ribosomal protein S6 kinase p70 (S6k), the translational inhibitors 4E-BP1 and 4E-BP2, and PHAS involved in protein synthesis and cell growth. In contrast, the mTOR2 is rapamycin insensitive. Reciprocal signaling cross talk and phosphorylations are organized between the AKT modules and the mTORC2 and mTORC1–2 signalosomes. AKT is phosphorylated by mTORC2 at Thr450 of the turn motif and at Ser473 of the HM sequence containing Ser/Thr–Tyr/Phe. In addition, AKT was described to phosphorylate and negate the activity of TCS2, a negative mTOR regulator involved in the TSC1–TSC2 complex acting as a GTPase-activating protein (GAP) for Rheb. Thus, AKT-mediated TSC2 phosphorylation is likely to promote mTORC1 activation. The mTORC2 appears to be the substrate of phosphoinositide-dependent protein kinase-2 (PDK2) responsible for AKT activation. Accordingly, signaling dialogs are identified between mTORC2 and several signaling molecules Rho-like GTPases, PKCα, the Rho-GEF P-Rex1, which associate functionally with mTORC2, and leads to pleiotropic regulations at cell adhesions, survival, autophagy, growth, cytoskeletal organization, and invasive phenotypes (reviewed in Oh and Jacinto, 2011). The mTORC2 complex is engaged in a plasma membrane molecular scaffold comprising [PDK1, AKT, Rac-1] and activating preferentially Akt-1 isoform versus Akt-2 (Higuchi *et al.*, 2008). (See Page 6 in Color Section at the back of the book.)

telangiectasia mutated)-related family of kinases and is deregulated in cancer. In addition, overlapping biological functions are noticed between guidance–dependence activities and the Ser/Thr kinase mTOR engaged in the two large molecular complexes mTORC1 and mTORC2 (reviewed in Oh and Jacinto, 2011; Sengupta *et al.*, 2010; Zoncu *et al.*, 2011). The adaptor protein Grb10 was identified as an mTOR1 substrate that mediates the inhibition of PI3K in cells lacking tuberous sclerosis complex 2 (TCS2), a tumor suppressor and negative regulator of mTORC1 (Hsu *et al.*, 2011). Grb10 functions as a multimolecular interacting adaptor and regulator for growth factor receptors stability and internalization (insulin-R, IGF-R), for AKT, Raf-1 and the dependence receptor Ret in cellular responses to nutritional deprivation, stress, migration, proliferation, and survival (reviewed in Holt and Siddle, 2005; Jahn *et al.*, 2002; Kebache *et al.*, 2007; Pandey *et al.*, 1995). In addition, cAMP/PKA signals converge to mTORC1 and mTORC2 to dissociate these complexes, leading to a decrease in mTOR intrinsic catalytic activity (Xie *et al.*, 2011). It is therefore plausible that some elements of these two molecular platforms mTORC1 and -2 are integrated in the guidance and dependence signaling systems during neoplasia.

D. Netrin Cross Talks with pRb1 and p53 Family, Hedgehog, Wnt and TGFβ Signaling

Opposing functions of Netrin-1 are described in attractive and repulsive axonal responses and promotion or restriction of cancer cell proliferation, apoptosis, invasive tumor growth, angiogenesis, and metastasis. These versatile phenotypes are dependent on Netrin receptor equipment (*vide supra*) and interactions with a vast array of signaling networks that are shifted to tumor-suppressive or -promoting functions inherent to environmental factors and neoplasia.

A direct role of the retinoblastoma (Rb1)/E2F3 pathway in the regulation of neuronal cell migration and Neogenin promoter activity was reported in neural precursor cells of the developing forebrain. This interesting data raise the possibility that the Rb1–E2F axis and upstream regulators (cyclin D1, Cdk4, and p16) could contribute to other mechanisms involved in cancer beyond the cell cycle and maintenance of the quiescent G0 arrest in stem cells, such as signaling connections with the dependence receptor pathways and programmed cell death and differentiation (Andrusiak *et al.*, 2011; Ciavarra and Zacksenhaus, 2011; Djelloul *et al.*, 1997). In a mouse model of colitis and inflammation-to-cancer sequence, the NF-κB activator nitric oxide synthase (NOS) induces mitogenic signals via soluble guanylyl cyclase (sGC)/cGMP signaling and Rb1 hyperphosphorylation–inactivation. These signals are dependent on the MEK/ERK/PI3K and AKT pathways (Ying *et al.*, 2007). We have previously shown that the NO/sGC and cGMP cascade drives permissive signals to the proinvasive thrombin PAR-1 pathways in colon cancer cells (Nguyen *et al.*, 2005a,b). LOH and loss of Rb1 function are associated with genomic instability, loss of cellular senescence, paradoxal mitogenic and apoptotic responses, and initiation and progression of common human cancer types through tumor angiogenesis and metastasis (reviewed in Burkhart and Sage, 2008). In colon cancer, loss of pRb1 is not involved in intestinal tumor initiation but cooperates with Wnt pathway activation to influence tumor initiation and location in the proximal colon and cecum in the APC^{1638N} mice (Kucherlapati *et al.*, 2008). The double-mutant mice have higher incidence and multiplicity of villous adenomas and reduced survival. Clinical colorectal adenocarcinomas undergo allelic loss at the *RB1* locus, whereas almost half of CRC show nonrandom chromosomal 13 gains. Approximately two-thirds of intestinal tumors have increased levels of Rb1 expression (Kucherlapati *et al.*, 2008; Vogelstein *et al.*, 1989). In downstream pathways, pRb1 restricts the transcriptional activity of the E2F family regulating cell proliferation (G1 to G0 transition), initiation of DNA replication, and apoptosis (Rowland and Bernards, 2006). Dual oncogenic and tumor suppressor roles have been reported for E2F1 in transgenic mice lacking this transcription factor. Firstly, homozygous mice for a nonfunctional allele develop tissular atrophy and a broad and unusual spectrum of tumors, suggesting that E2F1 has opposing roles in cellular growth and functions as a tumor suppressor in a tissue-specific context (Yamasaki *et al.*, 1996). Secondly, loss of E2F1 reduced pituitary and thyroid tumorigenesis in *RB1* +/− mice and also prolonged the survival of the *RB1* (+/−) and *E2F1* (−/−) animals (Yamasaki *et al.*, 1998). Of note, E2F1 is a negative regulator of Wnt signaling and β-catenin/TCF4-mediated transcription in human CRC cell lines SW480, DLD1, and HCT116 (Morris *et al.*, 2008). Consistently, E2F1 can activate the transcription of the Wnt repressors *AXIN1*, *AXIN2*, and of seven in absentia homologue 1 (*SIAH1* encoding an E3 ubiquitin ligase) which are involved in β-catenin degradation (Morris *et al.*, 2008; Xie *et al.*, 2009). Moreover, this tumor-suppressive function of E2F1 is antagonized by pRb1 and CDK8, a cyclin C-dependent kinase frequently overexpressed in 50% of CRC patients, and involved in the multiprotein Mediator complex coupled to the actions of transcription (Bernards, 2008; Morris *et al.*, 2008). Convergent data published by Firestein *et al.* (2008) support the notion that CDK8 is a CRC oncogene that stabilizes the interaction of β-catenin with the c-MYC promoter, supporting canonical Wnt pathway activity. Thus, a fine dosage of the Rb1 and E2F1 levels, and their partners in CNS and peripheral tissues, and their regulation by posttranscriptional mechanisms, signaling modifiers, and cross talks have significant influence on cancer progression and can explain the paradoxical and dual roles of the Rb1–E2F axis as a molecular commutator for the regulation of cell proliferation, apoptosis, and neoplasia. It is plausible that these pRb1 intersections would converge to signaling and functional regulations at the Netrin receptor and guidance family.

Convergences and negative feedback have been recently described between Netrin UNC-5B receptors, activation of PI3K/AKT signaling, nuclear GTPase PIKE-A isoform phosphorylation, and the *TP53* tumor suppressor gene (He *et al.*, 2011). The cleavage of UNC-5B by caspases seems to be indispensable for the proapoptotic functions of this Netrin receptor mediated by the interaction of DAPK with the UNC-5B DD (Llambi *et al.*, 2005). PIKE-A phosphorylation and AKT activation in turn enhances its association with UNC-5B, blocking UNC-5B fragmentation by caspases, thus inhibiting apoptosis. The PIKE-A protooncogene selectively binds to UNC-5B, but not to UNC-5A/C. PIKE-A is overexpressed in neuroblastomas and many common human epithelial cancers (e.g., colon, kidney, bladder, lung, breast, prostate, ovary) and promotes cellular transformation and invasive growth (Liu *et al.*, 2007). This PI3K enhancer also exerts its prosurvival functions by activating NF-κ pathways and enhancing the AKT–MDM2 axis that triggers p53 ubiquitination, degradation, nuclear export and culminating to suppression of *UNC-5B* transcription and inhibition of apoptosis (Cai *et al.*, 2009; He *et al.*, 2011; Zhou *et al.*, 2001). As the intron-1 of UNC-5B contains a p53-binding sequence, this additional cross talk induced by hypoxia and genotoxic stress is therefore interrupted by the Netrin-1/UNC-5B cascade in wt-*TP53* cells (Tanikawa *et al.*, 2003). In *DCC* and *TP53*-deficient cancer cells, Netrin-1-induced cell death through upregulation of the transcriptionally active TAp73α isoform (Roperch *et al.*, 2008), suggesting the interesting possibility that the other p53 family members p73 and p63, and their multiple N-terminal isoforms and splice variants, are implicated in the transcriptional and multifactorial regulation of UNC-5 family Netrin receptors at the cell death and survival pathways. It is therefore apparent that experimental data performed on cell death and guidance studies using epithelial cancer cells with defective tumor suppressor genes, and expressing viral oncogenes (HEK-293, Hela), should be interpreted with caution, as described for the expression and inhibitory impact of the *E1A* gene on miR-520h expression, cell death, and tumorigenicity (Su *et al.*, 2010).

Morphogenic events controlled by the Wnt/Hedgehog and TGFβ/BMP pathways to orchestrate axonal pathfinding and neurogenesis are also operating in tissue patterning and cell-fate specification (reviewed in Charron and Tessier-Lavigne, 2005). Coherently, emerging evidence support the existence of signaling connections between these classical morphogens and the Netrin guidance systems.

Functional coincidences and convergences between the Netrin and the morphogenic hedgehog (Hh) signaling systems are now well established during vertebrate development and neoplasia in CRC and other epithelial tumors (Charron *et al.*, 2003). In the zebrafish, Netrin-1a expression is differentially regulated by Hh signaling in the ventral neural tube and in a subset of adjacent mesodermal cells, the somites (reviewed in Ingham *et al.*, 2011; Lauderdale *et al.*, 1998). Of note, Netrin-1a expression is induced by ectopic expression of the morphogens *sonic hedgehog* (*shh*), *tiggy-winkle hedgehog* (*twhh*), or *echida hedgehog* (*ehh*), whereas both *shh* and *twhh* (but not *ehh*) are effective in myotomes (Lauderdale *et al.*, 1998). Also, the secreted protein Shh restricts the expression of transcription factors and dependence receptor agonist Ephrin-A5 in the dorsal neural tube and concomitantly upregulates the glioma-associated (Gli-1) transcription factor, as well as cell surface and secreted proteins in the ventral aspect, namely PTCH and Netrin-1, respectively (Hynes *et al.*, 2000; Kennedy *et al.*, 1994; Serafini *et al.*, 1994). The signaling components and mechanisms of this Hh morphogenic pathway include posttranscriptionally and sequentially modified, cleaved, secreted, and "dispatched" Hh ligands binding membrane receptors PTCH1, acting as a tumor suppressor, and PTCH2 (reviewed in Gallet, 2011; Ingham *et al.*, 2011; Li *et al.*, 2011a). Reception of Hh ligands by PTCH in turn alleviates PTCH-mediated repression of the signal transducer Smoothened (SMO), a putative member of the GPCR family presumably coupled to Gαi subunits (reviewed in Ayers and Thérond, 2010; Ogden *et al.*, 2008). Molecular complex or heterodimers between PTCH and the PTCH coreceptors SMO are therefore plausible. This configuration is reminiscent of the DCC/A2b-R dialog and is coherent with the idea that DCC may use, or restrict, the signaling competences of this GPCR that is currently considered as a deleterious signaling system in neoplasia and proinflammatory situations such as colitis (reviewed in Haskó *et al.*, 2008, 2009; Kolachala *et al.*, 2008; Rodrigues *et al.*, 2007; Rosenberger *et al.*, 2009; Zaynagetdinov *et al.*, 2010). Molecular cross signaling has been identified between Gαi subunits and UNC-5B (Komatsuzaki *et al.*, 2002). As shown in Fig. 11, GPCR and their associated downstream signaling pathways are currently considered as therapeutic targets in clinical cancers (Ayoub *et al.*, 2009; Harrison, 2006; Lappano and Maggiolini, 2011; Prévost *et al.*, 2006). Within the PTCH/SMO dialog, PTCH receptors antagonize the pathway by inhibiting SMO in the absence of Hh ligands. Activated PTCH receptors release SMO inhibition, trigger cytoskeleton activation by Src family kinases during guidance of commissural neurons (Yam *et al.*, 2009), SMO-dependent induction of the Gli family of vertebrate transcription factors (Gli1, Gli2, Gli3), and provide further negative and positive feedback for this canonical Hh/PTCH pathway by targeting *PTCH1* and *Gli1*, respectively (reviewed in Li *et al.*, 2011a). Gli target genes include critical regulators of the cell proliferation cycle (*cyclins D* and *B*, *CDC25*, *CDC2*, *myc*) and genes involved in the balance between cell survival (*Bcl-2*, *PDGFRα*)

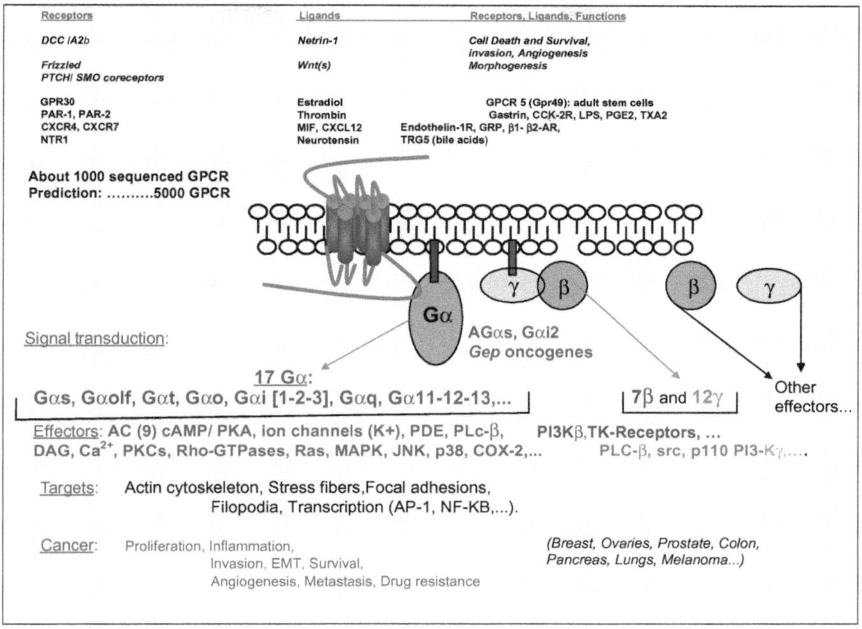

Fig. 11 Signals interconnections between guidance–dependence receptors and G-protein-coupled receptors (GPCR) signaling. GPCR are connected with: (i) the guidance–dependence receptors using DCC/adenosine A2b/Netrin coreceptors; (ii) the Wnt–Fz pathways; (iii) the PTCH coreceptors SMO; and (IV) cAMP and PKA-dependent and -independent pathways (cAMP–EPAC–Rap). Subset elements of the GPCR signalomes are functioning as oncogenic pathways (Gα subunits, Gβγ dimers, and downstream elements) and are now considered as new targets for anticancer therapy (Ayoub *et al.*, 2009; Harrison, 2006; Kimple *et al.*, 2011; Lappano and Maggiolini, 2011; Prévost *et al.*, 2006). GPCR signal transduction is mediated by heterotrimeric G-proteins Gαβγ including multiple combinations between 17 Gα, 7 Gβ, and 12 Gγ subunits. GPCR signals occur through the GαGDP–GαGTP and GαGDP cycle induced by receptor activation and GαGTP–GDP inactivation by GAP (GTPase-activating proteins). Several other proximal GPCR signaling adaptors, modifiers, and reglators are also described, such as the family of 37 regulators of G-protein signaling (RGS) functioning like GAPs for large G-protein Gα subunits. Mutated and oncogenic forms of Gα proteins are discovered in clinical human cancers (Gαi, Gαs) and were identified experimentally as constitutively activated, GTPase-deficient mutants of Gα12/13 and designated as the *Gep* oncogenes in soft tissue sarcoma. GPCR and GαGTP/Gβγ activation and heterotrimer dissociation as Gα+Gβγ subunits, as well as their individual single Gβ and Gγ components, are then missioned to undergo their own signaling activity (Faivre *et al.*, 2006; Nguyen *et al.*, 2006; Régnauld *et al.*, 2002). Thus, the next downstream GPCR effectors include, at least in part, nine adenylyl cyclases (AC) connected with the cAMP–PKA axis, ion channels, phosphodiesterases (PDE), Phospholipase-C (PLC-β)/diacyl glycerol (DAG)-protein kinase C (PKC) cascades, Rho-like GTPases, MAPK/SAPK, PI3K-β, PI3K-γ, cyclooxygenase-2 (COX-2), and multiple other targets included in the GPCR connections. GPCR involved in human epithelial cancers include GPR30 (membrane receptors activated by 17β-estradiol and estrogens), PAR-1 and PAR-2 thrombin receptors, CXCR4 and CXCR7 binding the macrophage inhibiting factor (MIF) and the CXCL12/SDF-1

and cell death (*DR5*, *PDGFRα–Fas* axis) during migration directed by guidance cues. At the ventral midline of the embryonic spinal cord, a subset of axons is attracted from the dorsal neural tube toward the floor plate, where they form axon commissures with a cluster of highly secretory cells. The outgrowth and guidance of these axons is directed by Netrin-1 and persists in *NETRIN-1* mutant floor plate (Charron *et al.*, 2003). Netrin-1 defience was rescued by Hh signaling, suggesting functional redundancy and supply between these two essential pathways. Thus, Shh expressed by floor plate and notochord functions as a collaborator of Netrin-1 in guiding commissural axons to the floor plate (Salinas, 2003). Accordingly, the most severe phenotype was observed in the *Gli2 −/−*, *Netrin-1 −/−* mutant. Connections between Hh and Netrin-1 were further illustrated by therapeutic applications using genetic modification of mesenchymal stem cells producing Shh to improve their survival and angiogenic potential *in vivo* through the promotion of Netrin-1-dependent angiogenesis and preservation of global functions in the ischemic heart (Ahmed *et al.*, 2010).

Deregulated expression and inactivating or activating mutations of signaling elements of the Hh–PTCH–SMO axis play an important role in chronic myeloid leukemia and predisposition for skin and brain cancers (Gorlin's syndrome) and more common tumors of the colon, lung, breast, pancreas, and prostate (Jagani *et al.*, 2010; Moraes *et al.*, 2007). As a negative regulator of the SMO cell death pathway, the Hh/PTCH signaling elements may function transiently or constitutively as a conditional dependence receptor pathway to induce apoptosis in cell types dependent on Hh ligand–PTCH1 interactions and environmental factors for their survival. Accordingly, treatment of human colon cancer cells with a small inhibitor targeting Gli1 and Gli2, the classical inhibitor of SMO cyclopamine, or suppression of Gli1 by shRNA increased the expression of cell death receptors (Fas, DR5) and downregulation of PDGF-Rα and Bcl-2 (Mazumdar *et al.*, 2011). Consistently, cyclopamine treatment induced apoptosis in both adenoma- and carcinoma-derived cell lines, which can be rescued in part by further stimulation of autocrine Hh/PTCH/SMO and Gli1 signaling in this model (Qualtrough *et al.*, 2004). In contrast, PTCH was not detected in colon

chemokines, NTR1 neurotensin receptors, adult stem cell-associated GPCR5, Endothelin-1R receptors, gastrin-releasing peptide (GRP) receptors, and TGR5, binding bile acids (reviewed in Zimber and Gespach, 2008). GPR30 mediates neuroprotective actions. Biological events linked to GPCR include functional regulations at focal adhesions and actin cytoskeleton, nuclear signals and gene expression, and many other essential mechanisms engaged during normal development, environmental stimulants, physiological responses to hormones and neurotransmitters, and pathological states. (For color version of this figure, the reader is referred to the Web version of this book.)

tumor cell lines, although a few of these cell lines (HT29, HCT116, and HCT116-chr3+) expressed the *GLI* transcripts (Berman *et al.*, 2003), suggesting that subsets of clinical colorectal tumors are not driven by this autocrine loop. In addition, activated Hh signaling is also observed in early and late stages of pancreatic cancer tumorigenesis (Thayer *et al.*, 2003) and may trigger the adenoma–carcinoma sequence linked to the invasive process during colorectal carcinogenesis characterized by an increased rate of PTCH and SMO expression (Yoshikawa *et al.*, 2009). Moreover, high Hh/Gli signaling and concurrent low Wnt/TCF signaling hallmark the metastatic transition in CRC patients (Varnat *et al.*, 2010).

Several arguments strongly suggest that the PTCH/SMO pathways are involved in several functions inherent to the neoplastic phenotype at the proliferation/differentiation interface. Constitutive activation of SMO in the mammary glands of transgenic mice leads to increased proliferation, altered differentiation, and ductal hyperplasia distinct from those caused by PTCH1 heterozygosity (Moraes *et al.*, 2007). SMO activation was also shown to promote the mammosphere-forming efficiency of primary mammary epithelial cells by increasing the pool of division-competent cells capable of anchorage-independent growth and survival rather than by increasing the generation of stem cell subtypes. In the colon of Indian Hh (*Ihh*+) mouse embryos, the polarized and differentiated epithelium is organized into crypts and is negative for the proliferation marker PCNA at day 16.5 *post coitum* (van den Brink *et al.*, 2004). In contrast, this epithelium in *Ihh*–/– embryos is a PCNA-positive proliferative multilayer and does not organize into crypts. Cyclopamine-treated rats display an extensive loss of differentiation criteria, notably redistribution of villin and brush-border enzymes, and inappropriate induction of the goblet cell marker intestinal trefoil factor (ITF/TFF3) in the enterocyte lineage. In the mucosa of the lower digestive tract and colon, TFFs are found to be overexpressed and display pejorative functions on invasive growth and tumor angiogenesis during chronic inflammation and CRC (reviewed in Emami *et al.*, 2004; Gespach, 2008). Thus, in the normal mucosa, *Ihh* functions as a differentiation promoter and as a proliferation inhibitor of the colonic epithelium. Conversely, in human CRC cells, high Hh–Gli1 signature of active Hg pathway is coincident to tumor recurrence, metastasis, and EMT (Varnat *et al.*, 2009).

Epithelial cell dedifferentiation at the invasive front of clinical CRC identifies a network of eight genes included in oncogenic pathways controlled by Hg (Gli1 and Gli2), Wnt, and TGFβ (Oku *et al.*, 2008). In addition, van den Brink *et al.* (2004) showed that Ihh abrogates β-catenin/TCF signaling in human DLD-1 colon cells displaying constitutive activation of the Wnt pathway. Of note, *Ihh* expression was lost in sporadic colon adenomas and dysplasic epithelial cells in FAP adenoma, accompanied with loss of β-catenin staining at the plasma membrane, a signature of Wnt pathway activation. Most interestingly, blockade of the canonical Wnt pathway by a dominant-negative mutant of TFC-4 induced a rapid induction of Ihh, indicating that Ihh expression is negatively regulated by β-catenin/TCF signaling and that APC mutations caused the loss of Ihh in colon epithelial cells. Several evidences indicate that the Hh and Wnt pathways are intimately implicated in the patterning and growth of the stem cell niche in the intestinal epithelium (reviewed in Gespach, 2010; Taipale and Beachy, 2001). It seems that the Hh and Wnt pathways are interdependent and reciprocally control their signaling activities during the early and late phases of CRC progression (Akiyoshi *et al.*, 2006; van den Brink and Hardwich, 2011; Watt, 2004). In support with this model, *SMO* depletion and heterozygosity in *APC+/− SMO+/−* mice and human colon cancer cell lines respectively decreased numbers of polyps and restricted cellular proliferation by G1/S arrest (Arimura *et al.*, 2009). Thus, therapeutic interventions targeting one of these two morphogenic pathways in pathological states and epithelial cancers may provide beneficial response in CRC patients (Lauth *et al.*, 2007; reviewed in Peukert and Miller-Moslin, 2010). This interesting assumption is however challenged by several interferences targeting this model integrated in the extreme complexity and multitude of molecular alterations in Hh/Wnt pathways and other mechanisms associated with neoplasia. Recent data however suggest that Wnt–TCF blockade may produce unexpected undesirable responses and instead promotes metastatic growth (Varnat *et al.*, 2010). Thus, the molecular status of these two signaling components and extrapolation of their functional significance and divergence are tightly dependent of their respective roles in normal development and neoplasia. For example, reduced expression of SMO by siRNA suppressed β-catenin-dependent transcription rather than Gli-dependent transcription and *PTCH1/GLI1* expression in SW480 and HCT116 colon cancer cells (Arimura *et al.*, 2009). Other data also indicate that Wnt and Hh signaling interferences initiated by Wnt agonist or β-catenin/TCF transcription factors resulted in the transcriptional activation of the Hh pathway (Noubissi *et al.*, 2009). Conversely, recent data demonstrate that human floor plate specification from ESCs is dependent on early and high levels of Shh commitment signals that overcome the blockade of the Wnt pathway by Dkk-1, a molecular switch that drive their derivation into anterior neurectoderm (Fasano *et al.*, 2010). It will be important to keep in mind that during CRC progression several Hh signaling and interacting elements are differentially up- and downregulated by epigenetic silencing and other mechanisms, as seen for *Ihh*, *Shh*, hedgehog-interacting protein (*HHIP*), *PTCH*, *SMO*, and *Gli1* (Akiyoshi *et al.*, 2006; Fu *et al.*, 2010; Kober *et al.*, 2011; Oniscu *et al.*, 2004; Taniguchi *et al.*, 2007; Yoshikawa *et al.*, 2009).

Molecular alterations in components of the TGFβ superfamily pathway are well described in colorectal carcinomas and other malignancies (reviewed in Levy and Hill, 2006; Papageorgis *et al.*, 2011). Downstream signals include canonic SMAD2/3-dependent transcriptional responses and SMAD-independent signals (TGFβ, Activin), as well as SMAD1/5/8 signals (BMP) and common co-SMAD4 (mutations and deletions found in 30% of LOH patients). Mutations in the TGFβRII found in the MSI pathway are estimated to occur in approximately 30% of CRC. Mutations in SMAD4 and BMPR1A are found in patients with familial juvenile polyposis, an autosomal-dominant condition associated with an increased risk of CRC. The cytokine TGFβ was originally described as a tumor suppressor at the early stages of tumor development by acting through antimitogenic and apoptotic signals in normal and premalignant colon epithelial cells. However, there is growing evidence that TGFβ enhances tumor progression, EMT, invasion, survival, and metastasis at later stages of the neoplasia in CRC and other epithelial tumor types. TGFβ, BMPs subtypes, and downstream signaling components within the SMAD and non-SMAD pathways can promote, suppress, or reverse inflammation, stem cell differentiation, cancer progression and tumor invasion, metastasis, EMT induced by various mechanisms including chemotherapy resistances, and adaptor protein Kindlin-1 in Integrin-containing focal adhesions (Grijelmo *et al.*, 2007; reviewed in Hardwick *et al.*, 2008; Liu *et al.*, 2011; Lombardo *et al.*, 2011; reviewed in Massagué, 2008; Padua and Massagué, 2009; Pohl *et al.*, 2010; Prunier *et al.*, 1999; Sabbah *et al.*, 2011; Sin *et al.*, 2011). Consistently, molecular alterations and neutralization of this morphogenic pathway by additional interference mechanisms, such as Ras activation, may explain these opposing roles of TGFβ in cancer prevention, promotion, and progression.

Our current knowledge on the signaling connections between Netrin and the morphogenic TGFβ/BMP pathways is still very scarce and patchy. BMP-7 appears to act as roof plate-derived chemorepellent that guides the early trajectory of the axons of commissural neurons in the developing spinal cord (Augsburger *et al.*, 1999). Interestingly, unorthodox structure–activity relationship of BMPs (BMP-7 versus BMP-4 and BMP-6) in reorienting growth cones was established. In *C. elegans*, the *UNC-129* gene encodes a BMP-like TGFβ superfamily ligand expressed and secreted in dorsal but not ventral rows of the body wall muscle (Colavita *et al.*, 1988). Axon guidance along the dorsoventral axis involves UNC-6/Netrin and requires UNC-40/DCC and UNC-5 Netrin receptors (Leonardo *et al.*, 1997). Mutations in *UNC-129* cause defects in the dorsally oriented trajectories of motoraxons that are mimicked by *UNC-5*, *UNC-6*, and *UNC-40* mutants (Colavita and Culotti, 1998; Hedgecock *et al.*, 1990). In the nematode, *UNC-129* has been identified as a suppressor of dorsal migration of the touch receptor axons induced by ectopic expression of *UNC-5* in these cells (Colavita *et al.*, 1988). In addition, UNC-129 promotes long-range repulsive guidance of UNC-6 by enhancing both UNC-5 and UNC-40 signaling at the expense of UNC-5 signaling alone, and thus regulates the balance between UNC-40 dependent and independent UNC-5 signaling pathways (MacNeil *et al.*, 2009). Therefore, Netrin dependence receptors are interconnected with a vast array of critical developmental pathways, oncogenic defects, and tumor suppressor genes.

V. PERSPECTIVES AND CONCLUSION

As discussed above, extensive research efforts devoted to academic practice and clinical applications highlight the dimension of the Netrin signaling galaxy. Consistently, accumulating evidence shows that guidance and dependence receptors and pathways play a critical role in development, homeostasis, and several human diseases. Additional lessons and discoveries are expected to come from worms and flies.

New therapeutic strategies are currently based on manipulating several elements of the Netrin axis in regenerative medicine and stem cell engineering, wound repair, tissue regeneration, and angiogenesis to fight several human disorders such as neurodegenerative diseases (Lourenço *et al.*, 2009; Sentürk *et al.*, 2011), inflammation and atherogenesis (Khan *et al.*, 2010; Ly *et al.*, 2005; Mirakaj *et al.*, 2010, 2011a,b; Rosenberger *et al.*, 2009), neoplastic progression, and tumor chemoresistances (Table I). Guidance and dependence signaling systems are very sophisticated and are directly balancing between tumor suppression and promotion. Intersections of the small interfering RNA networks with the Netrin roads have been recently discovered and their role in the modulation of cellular guidance and death opens new research avenues and clinical applications in neuroscience and oncology (Table II). Accordingly, these alternative directions will be presented and summarized in the following paragraphs in order to update this domain.

A. Other Therapeutic Interventions and Biological Markers

Additional reports support the notion that Netrin-1 reduced the cerebral ischemic stroke and apoptosis and attracted intrinsic neuronal stem cells to the infracted area (Wu *et al.*, 2008). Immune and inflammatory responses are the key components of the pathobiology of stroke leading to immunosuppressive effects that promote fatal infections. As regulators of the cardiovascular development and myogenesis, Netrins/Neogenin exerts cardioprotective roles against cardiac ischemia and myocardial infarction (Zhang and Cai, 2010) and promotes skeletal myogenic differentiation and

Table II MiRome Connections at Guidance and Dependence Receptor Pathways

Receptors/ligands/targets	miRs	Functional impact	References
Integrin receptors			
Alpha-1, (RhoA, N-cadherin)	miR-375	Neurite differentiation, neuronal signaling	Abdelmohsen *et al.* (2010)
Alpha-5	miR-92a	Angiogenesis inhibition, myocardial ischemia, and infarction	Bonauer *et al.* (2009)
Alpha-5, (RhoA)	miR-31	Activated in already-formed breast metastases Regression of metastasis, increased patient survival	Valastyan *et al.* (2011)
Beta-1, (extracellular matrix genes)	miR-29b	Protection from renal medullary injury	Liu *et al.* (2010a,b)
Beta-1, (laminin γ1, Ephrin-B1)	miR-124a	Downregulated in Glioblastoma Inhibition of migration and invasion Negatively regulated by Ephrin-B1	Fowler *et al.* (2011)
Beta-1	miR-183	Decreased adhesion to laminin, gelatin, collagen type-I Associated with cellular senescence, antimetastatic role Influence the development of Neurosensory organs	Li *et al.* (2010a)
Beta-3	miR-30, let-7	Tumor and lung metastasis suppressor in breast tumor-initiating cells	Yu *et al.* (2010)
Beta-8	miR-93	Promotion of survival, tumor growth and angiogenesis	Fang *et al.* (2011)
	miR-145	Maintenance of epithelial integrity Suppression of p63	Lee *et al.* (2011)
Neuropilin receptors			
NRP-1	miR-9 miR-181b	Both downregulated by arsenic, inhibition of angiogenesis	Cui *et al.* (2011)
Slit/Robo receptors			
Slit-2/Robo-1 (Survivin)	miR-218	Nasopharyngeal cancer suppression	Alajez *et al.* (2011)
Robo-1, Robo-2	miR-218	Heart tube formation in zebrafish, blockade of VEGF-induced angiogenesis and MAPK in HUVEC	Fish *et al.* (2011)
GLCE (HSPG pathway)	miR-218	Normal vascular patterning of the retina, vascular guidance regulator	Small *et al.* (2010)

(*continues*)

Table II (*continued*)

Receptors/ligands/targets	miRs	Functional impact	References
Robo-1	miR-218	Inhibition of invasion and metastasis in gastric cancer	Tie *et al.* (2010)
Ephrin–Ephrin-RTK			
Ephrin-A3	miR-210	Induced by hypoxia in HUVEC adaptive response. Inhibition of apoptotic cell death. Increased endothelial cell chemotaxis	Fasanaro *et al.*, 2008
Ephrin-B1	miR-124	Induced by Ephrin-B1, repressor of *Ephrin-B1* gene (*EfnB1*) transcription: (Negative reverse signaling) Promotion of neurogenesis and neuronal differentiation Downregulated in several carcinomas (Ephrin-B1 upregulated)	Arvanitis *et al.* (2010)
Sema–Plexin receptors			
Semaphorin-6A	miR-141	Downregulated in renal cell carcinoma; upregulated in lung, ovarian cancers	Liu *et al.* (2010a)
Plexin-B1	miR-214	Downregulated in cervical cancer. Inhibitor of cell proliferation and invasion	

myotube formation (Kang *et al.*, 2004). Netrin-1 is also strongly induced after renal ischemia-reperfusion experiments as its urinary levels increased rapidly as early as 1h and markedly within 3h (30- to 40-fold) and is therefore an early marker of kidney injury as compared to classical clinical markers (Reeves *et al.*, 2008). Accordingly, partial Netrin-1 deficiency worsens renal dysfunction and inflammation during acute kidney and colon injury in Ntn+/− mice (Aherne *et al.*, 2012; Grenz *et al.*, 2011).

Engineering and reprogramming fetal, embryonic, and mesenchymal stem cells, and tissue-specific resident stem cell progenitors, to produce, respectively, survival or cell death factors to injured and neoplastic tissues become a challenging alternative to improve their survival after transplantation and the therapeutic activity of conventional treatments.

These expectations have been supported by preliminary and encouraging experimental reports in Parkinson's disease, myocardial angiogenesis and repair, neoangiogenesis in hind limb ischemia, and manipulation of pancreatic epithelial progenitors in the ductal and endocrine lineages in pancreatitis and diabetes (Ahmed *et al.*, 2010; Li *et al.*, 2011b; Lin and Isacson, 2006; Yebra *et al.*, 2003). However, efficacy and safety limitations of ectopic mesenchymal stem cells and ESC derivatives in clinical therapy are linked to the risk of genomic instability, spontaneous chromosome changes, and malignant transformation after their propagation and amplification in cell culture and following systemic and topical administration, and their long-term persistence in blood and target organs (reviewed in Gespach, 2010). These genetic alterations seem to be associated with relatively rare mutations but the risk must be taken into account. These ethical and biosafety limitations are reminiscent of the natural history of sporadic cancers during aging. Thus, in case of stem cell and mesenchymal stem cell therapies, the intrinsic genomic instability and genetic alterations of mesenchymal stem cell populations in each patient should be reconsidered. Indeed, it was speculated that the probable primordial cells for the emergence of GI stromal tumors might be primitive mesenchymal cells. Other reports support the possibility that transplanted mesenchymal stem cells might contribute to carcinogenesis, metastasis, and elevation of the tumor mass in colon and prostate cancer (Prantl *et al.*, 2010; Shinagawa *et al.*, 2010).

B. Small Interfering RNAs (miRs) at Guidance and Dependence Pathways

Development and integrative analyses of large-scale miRomes revealed the establishment of several connections of these epigenetic mechanisms targeting posttranscriptional mechanisms with key elements involved in guidance and dependence receptor components (Table II). In addition to their use as biomarkers, these small noncoding RNAs are emerging as potential targets or tools in the prevention, diagnosis, and treatment of human diseases (reviewed in de Planell-Saguer and Rodicio, 2011; Sayed and Abdellatif, 2011). They are multitarget regulatory short interference RNAs acting at several RNA sequences via the RNA-inducing signaling complex and gene expression programs during normal development and neoplasia. Thus, during therapeutic applications, the main limitation and risk of miRs inducing transcript degradation and translation repression is related to their ability to target several epigenetic mechanisms, opposing cellular functions and unrelated pathways and diseases. miRs may exert tumor and metastasis suppressive functions, or oncogenic potential. Conversely, a given transcript might be at the convergence of several miRs, such as Integrin β8, a cell–matrix adhesion molecule targeted by miR-93 and miR-145 (Fang *et al.*, 2011; Lee *et al.*, 2011). Similarly, the p53-induced miR-192, miR-194-2, and miR-215 are capable of inducing p21 expression and cell cycle arrest controlled by the p53–p21 axis in CRCs (Braun *et al.*, 2008). Positive and negative cross talks inside and outside miR-target regulatory loops are described (*vide infra*). Coherently, differentiation potential of miRs 9/9* and miR-124 are recently described to induce the conversion of human fibroblasts into neurons, in concert with neurogenic transcription factors (Yoo *et al.*, 2011). Additional complexity is sustained by upstream signals converging to miRs expression (including pathological states, global and local methylation of the genome, hypoxia, stress conditions and inflammation, chemotherapy) and their downstream functional consequences on cell survival and guidance. These mechanisms are integrated in the regulation of gene expression and signaling during normal development and diseases (Sayed and Abdellatif, 2011).

As shown in Table II, several microRNAs are targeting Integrin subunits and cell adhesion molecules, such as Integrin subunit α1 and proteins with key functions in neuronal signaling and cytoskeleton organization (miR-375), Integrin α5 (miR-31, miR-92a), Integrin subunits β1 (miR-29, miR-124a, miR-183), β3 (miR-30, let-7a), β8 (miR-93 and miR-145, *vide supra*), and FAK (miR-205), as reviewed in Valastyan and Weinberg (2011). Other miR elements of the guidance–dependence receptor axis include miR-9 and miR-181b known to downregulate *NRP1* gene expression in arsenic-induced endothelial cell migration and angiogenic tube formation (Cui *et al.*, 2011). These two noncoding RNAs are induced by arsenic exposure, an environmental teratogen and carcinogen pollutant. Arsenic is combined with interferon-α for consolidative targeted therapy in T-cell leukemia/lymphoma and is the most active agent in the treatment of acute promyelocytic leukemia (PML) through induction of differentiation and apoptosis and degradation of the PML–RARα (retinoic acid receptor-α) fusion transcript (Bazarbachi *et al.*, 2011; Hu, 2011). Convergent data identified the intronically encoded miR-218 within the *Slit-2* and *Slit-3* genes as an orchestrator of signaling elements of the guidance Slit/Robo pathways, including Robo-1, Robo-2, Slit-2, Slit-3, and Survivin (Alajez *et al.*, 2011; Fish *et al.*, 2011; Small *et al.*, 2010; Tie *et al.*, 2010). Other components of the HS biosynthetic pathway are direct and indirect targets of miR-218: the miR-218-sensitive HS-modifying enzyme glucuronyl C5-epimerase and its downstream products HS proteoglycans (HSPGs) that are essential for Slit–Robo signaling (Small *et al.*, 2010). Internal miR loop of ligand–receptor interaction and additional miR-218 cross talks are identified with VEGF-R2 autophosphorylation and control of the Slit–Robo axis (Fish *et al.*, 2011). The Ephrin–Ephrin–RTK pathways are controlled, at least in part, by miR-124 (an Ephrin-B1 repressor) and miR-210, an Ephrin-A3 repressor (Arvanitis *et al.*, 2010; Fasanaro *et al.*, 2008; Fowler *et al.*, 2011). Sema-4D–Plexin-B1 and Sema-6A–Plexin-A4 pathways are controlled by the hypoxia-induced miR-210 (Liu *et al.*, 2010a; Qiang *et al.*, 2011). Up to now, miR(s) targeting the DCC/UNC and Neogenin family receptors and the Netrin(s)/RGM ligands are not identified yet. Another quest is to attribute the functional and biological significance of long noncoding RNAs and to establish their possible roles in cellular guidance and dependence (Baker, 2011). Discovery of new territories and "black boxes" filled by surprising novel mechanisms driving integrated genomic and postgenomic developmental and somatic functions is not excluded.

VI. CONCLUSION

As we have understood from astrophysics and space navigation in the suburbs of our planet in the solar system, may be there is no limit to discover the universal dimensions and complexity of the guidance and dependence receptor pathways during development, survival, and death, from microorganisms and worms to humans. Molecular dissection and high-resolution imaging of these signaling platforms and operators, as well as the persistence and spatiotemporal integration of these semaphore signals by global and nanoscale analyses, are the necessary prerequisites to identify pertinent therapeutic targets in relevant biological context. Orientation of our future investigations will provide more precise approaches to establish the reality and validity of guidance molecules and signals in their environment outside and inside the cellular system. Visualization of the dynamic remodeling of the cell architecture in relation to the temporal translocation and activation of signaling elements within the cell compartments will help to define more comprehensive picture of the guidance and dependence scenario. These advances will constitute the basis and rational of novel anticancer drug design and cellular therapy approaches for efficient and safe clinical applications. Because of the pleiotropic, versatile, interdependent, and critical roles of the guidance–dependence axis in several developmental processes and human diseases, therapeutic interventions aimed at this signaling network should be done with caution at the stem cell and individual levels, when translating experimental data to the clinic.

ACKNOWLEDGMENTS

Aided by INSERM (Hospital Saint-Antoine Research Center U938) and IPSEN. Dedicated to the *Virtuoso Antonio Vivaldi*. The author declares no competing financial interests and would like to apologize to all colleagues whose work he was unable to cite in this Review. Most significant and recent papers in Guidance and Colorectal Cancer fields are listed in the Note Added in Proof.

REFERENCES

Abdelmohsen, K., Hutchison, E. R., Lee, E. K., Kuwano, Y., Kim, M. M., Masuda, K., Srikantan, S., Subaran, S. S., Marasa, B. S., Mattson, M. P., and Gorospe, M. (2010). miR-375 inhibits differentiation of neurites by lowering HuD levels. *Mol. Cell. Biol.* **30**, 4197–4210.

Adams, R. H., and Eichmann, A. (2010). Axon guidance molecules in vascular patterning. *Cold Spring Harb. Perspect. Biol.* **2**, a001875.

Adler, C. E., Fetter, R. D., and Bargmann, C. I. (2006). UNC-6/Netrin induces neuronal asymmetry and defines the site of axon formation. *Nat. Neurosci.* **9**, 511–518.

Aherne, C. M., Collins, C. B., Masterson, J. C., Tizzano, M., Boyle, T. A., Westrich, J. A., Parnes, J. A., Furuta, G. T., Rivera-Nieves, J., and Eltzschig, H. K. (2012). Neuronal guidance molecule netrin-1 attenuates inflammatory cell trafficking during acute experimental colitis. *Gut* **61**(5), 695–705.

Ahmed, R. P., Haider, K. H., Shujia, J., Afzal, M. R., and Ashraf, M. (2010). Sonic Hedgehog gene delivery to the rodent heart promotes angiogenesis via iNOS/Netrin-1/PKC pathway. *PLoS One* **5**, e8576.

Ahmed, G., Shinmyo, Y., Ohta, K., Islam, S. M., Hossain, M., Naser, I. B., Riyadh, M. A., Su, Y., Zhang, S., Tessier-Lavigne, M., and Tanaka, H. (2011). Draxin inhibits axonal outgrowth through the netrin receptor DCC. *J. Neurosci.* **31**, 14018–14023.

Akiyoshi, T., Nakamura, M., Koga, K., Nakashima, H., Yao, T., Tsuneyoshi, M., Tanaka, M., and Katano, M. (2006). Gli1, downregulated in colorectal cancers, inhibits proliferation of colon cancer cells involving Wnt signalling activation. *Gut* **55**, 991–999.

Alajez, N. M., Lenarduzzi, M., Ito, E., Hui, A. B., Shi, W., Bruce, J., Yue, S., Huang, S. H., Xu, W., Waldron, J., O'Sullivan, B., and Liu, F. F. (2011). miR-218 suppresses nasopharyngeal cancer progression through downregulation of survivin and the SLIT2-ROBO1 pathway. *Cancer Res.* **71**, 2381–2391.

Al-Tassan, N., Chmiel, N. H., Maynard, J., Fleming, N., Livingston, A. L., Williams, G. T., Hodges, A. K., Davies, D. R., David, S. S., Sampson, J. R., and Cheadle, J. P. (2002). Inherited variants of MYH associated with somatic G:C→T:A mutations in colorectal tumors. *Nat. Genet.* **2**, 227–232.

Altschuler, D. L., Peterson, S. N., Ostrowski, M. C., and Lapetina, E. G. (1995). Cyclic AMP-dependent activation of Rap1b. *J. Biol. Chem.* **270**, 10373–10376.

André, F., Janssens, B., Bruyneel, E., van Roy, F., Gespach, C., Mareel, M., and Bracke, M. (2004a). α-catenin is required for IGF-I induced cellular migration but not invasion in human colonic cancer cells. *Oncogene* **23**, 1177–1186.

André, T., Boni, C., Mounedji-Boudiaf, L., Navarro, M., Tabernero, J., Hickish, T., Topham, C., Zaninelli, M., Clingan, P., Bridgewater, J., Tabah-Fisch, I., and de Gramont, A. (2004b). Multicenter international study of oxaliplatin/5-fluorouracil/leucovorin in the adjuvant treatment of colon cancer (MOSAIC) investigators. Oxaliplatin, fluorouracil, and leucovorin as adjuvant treatment for colon cancer. *N. Engl. J. Med.* **350**, 2343–2351.

Andrusiak, M. G., McClellan, K. A., Dugal-Tessier, D., Julian, L. M., Rodrigues, S. P., Park, D. S., Kennedy, T. E., and Slack, R. S. (2011). Rb/E2F regulates expression of neogenin during neuronal migration. *Mol. Cell. Biol.* **31**, 238–247.

Anselmo, M. A., Dalvin, S., Prodhan, P., Komatsuzaki, K., Aidlen, J. T., Schnitzer, J. J., Wu, J. Y., and Kinane, T. B. (2003). Slit and robo: expression patterns in lung development. *Gene Expr. Patterns* **3**, 13–19.

Antoine-Bertrand, J., Ghogha, A., Luangrath, V., Bedford, F. K., and Lamarche-Vane, N. (2011a). The activation of Ezrin-Radixin-Moesin (ERM) proteins is regulated by netrin-1 through Src kinase and RhoA/Rho kinase activities and mediates netrin-1-induced axon outgrowth. *Mol. Biol. Cell* **22**, 3734–3746.

Antoine-Bertrand, J., Villemure, J. F., and Lamarche-Vane, N. (2011b). Implication of rho GTPases in neurodegenerative diseases. *Curr. Drug Targets* **12**, 1202–1215.

Arakawa, H. (2004). Netrin-1 and its receptors in tumorigenesis. *Nat. Rev. Cancer* **4**, 978–987.

Arimura, S., Matsunaga, A., Kitamura, T., Aoki, K., Aoki, M., and Taketo, M. M. (2009). Reduced level of smoothened suppresses intestinal tumorigenesis by down-regulation of Wnt signaling. *Gastroenterology* **137**, 629–638.

Arumugam, M., Raes, J., Pelletier, E., Le Paslier, D., Yamada, T., Mende, D. R., Fernandes, G. R., Tap, J., Bruls, T., Batto, J. M., Bertalan, M., Borruel, N., *et al.* (2011). Enterotypes of the human gut microbiome. *Nature* **474**, 174–180.

Arvanitis, D. N., Jungas, T., Behar, A., and Davy, A. (2010). Ephrin-B1 reverse signaling controls a posttranscriptional feedback mechanism via miR-124. *Mol. Cell. Biol.* **30**, 2508–2517.

Aschele, C., Debernardis, D., Lonardi, S., Bandelloni, R., Casazza, S., Monfardini, S., and Gallo, L. (2004). Deleted in colon cancer protein expression in colorectal cancer metastases: a major predictor of survival in patients with unresectable metastatic disease receiving palliative fluorouracil-based chemotherapy. *J. Clin. Oncol.* **22**, 3758–3765.

Attoub, S., Veerle, N., Pirola, L., Bruyneel, E., Chastre, E., Mareel, M., Wymann, M., and Gespach, C. (2000). Leptin promotes invasiveness of kidney and colonic epithelial cells via phosphoinositide 3-kinase-, Rho- and Rac-dependent signaling pathways. *FASEB J.* **14**, 2329–2338.

Attoub, S., Rivat, C., Bedin, M., Bruyneel, E., Louvet, C., André, T., Mareel, M., Mester, J., and Gespach, C. (2002). The c-kit tyrosine kinase inhibitor STI571 for colorectal cancer therapy. *Cancer Res.* **62**, 4879–4883.

Augsburger, A., Schuchardt, A., Hoskins, S., Dodd, J., and Butler, S. (1999). BMPs as mediators of roof plate repulsion of commissural neurons. *Neuron* **24**, 127–141.

Ayers, K. L., and Thérond, P. P. (2010). Evaluating Smoothened as a G-protein-coupled receptor for Hedgehog signalling. *Trends Cell Biol.* **20**, 287–298.

Ayoub, M. A., Damian, M., Gespach, C., Ferrandis, E., Lavergne, O., De Wever, O., Banères, J. L., Pin, J. P., and Prévost, G. P. (2009). Small molecule inhibition of heterotrimeric G-protein signalling through action on Gα subunit. *J. Biol. Chem.* **284**, 29136–29145.

Bae, Y. H., Ding, Z., Das, T., Wells, A., Gertler, F., and Roy, P. (2010). Profilin1 regulates PI(3,4)P2 and lamellipodin accumulation at the leading edge thus influencing motility of MDA-MB-231 cells. *Proc. Natl. Acad. Sci. USA* **107**, 21547–21552.

Bai, G., Chivatakarn, O., Bonanomi, D., Lettieri, K., Franco, L., Xia, C., Stein, E., Ma, L., Lewcock, J. W., and Pfaff, S. L. (2011). Presenilin-dependent receptor processing is required for axon guidance. *Cell* **144**, 106–118.

Baker, M. (2011). Long noncoding RNAs: the search for function. *Nat. Methods* **8**, 379–383.

Balkwill, F., and Mantovani, A. (2010). Cancer and inflammation: implications for pharmacology and therapeutics. *Clin. Pharmacol. Ther.* **87**, 401–406.

Barbat, A., Pandrea, I., Cambier, D., Zweibaum, A., and Lesuffleur, T. (1998). Resistance of the human colon carcinoma cell line HCT-8 to methotrexate results in selection of cells with features of enterocytic differentiation. *Int. J. Cancer* **75**, 731–737.

Bazarbachi, A., Suarez, F., Fields, P., and Hermine, O. (2011). How I treat adult T-cell leukemia/lymphoma. *Blood* **118**, 1736–1745.

Behrens, J., Jerchow, B. A., Wurtele, M., Grimm, J., Asbrand, C., Wirtz, R., Kuhl, M., Wedlich, D., and Birchmeier, W. (1998). Functional interaction of an axin homolog, conductin, with beta-catenin, APC, and GSK3β. *Science* **280**, 596–599.

Berger, A. H., Knudson, A. G., and Pandolfi, P. P. (2011). A continuum model for tumor suppression. *Nature* **476**, 163–1169.

Berman, D. M., Karhadkar, S. S., Maitra, A., Montes De Oca, R., Gerstenblith, M. R., Briggs, K., Parker, A. R., Shimada, Y., Eshleman, J. R., Watkins, D. N., and Beachy, P. A. (2003). Widespread requirement for Hedgehog ligand stimulation in growth of digestive tract tumours. *Nature* **425**, 846–851.

Bernards, R. (2008). Cancer: entangled pathways. *Nature* **455**, 479–480.

Bernet, A., and Fitamant, J. (2008). Netrin-1 and its receptors in tumour growth promotion. *Expert Opin. Ther. Targets* **12**, 995–1007.

Bernet, A., Mazelin, L., Coissieux, M. M., Gadot, N., Ackerman, S. L., Scoazec, J. Y., and Mehlen, P. (2007). Inactivation of the UNC5C Netrin-1 receptor is associated with tumor progression in colorectal malignancies. *Gastroenterology* **133**, 1840–1848.

Bhakar, A. L., Howell, J. L., Paul, C. E., Salehi, A. H., Becker, E. B., Said, F., Bonni, A., and Barker, P. A. (2003). Apoptosis induced by p75NTR overexpression requires Jun kinase-dependent phosphorylation of Bad. *J. Neurosci.* **23**, 11373–11381.

Bienz, M. (1996). Induction of the endoderm in *Drosophila*. *Semin. Cell Dev. Biol.* **7**, 113–119.

Bienz, M. (2002). The subcellular destinations of APC proteins. *Nat. Rev. Mol. Cell Biol.* **3**, 328–338.

Bogan, C., Chen, J., O'Sullivan, M. G., and Cormier, R. T. (2009). Loss of EphA2 receptor tyrosine kinase reduces ApcMin/+ tumorigenesis. *Int. J. Cancer* **124**, 1366–1371.

Boissan, M., de Wever, O., Lizarraga, F., Wendum, D., Poincloux, R., Chignard, N., Desbois-Mouthon, C., Dufour, S., Nawrocki-Raby, B., Birembaut, P., Bracke, M., Chavrier, P., *et al.* (2010). Implication of metastasis suppressor NM23-H1 in maintaining adherens junctions and limiting the invasive potential of human cancer cells. *Cancer Res.* **70**, 7710–7722.

Boivin, G. P., Washington, K., Yang, K., Ward, J. M., Pretlow, T. P., Russell, R., Besselsen, D. G., Godfrey, V. L., Doetschman, T., Dove, W. F., Pitot, H. C., Halberg, R. B., *et al.* (2003). Pathology of mouse models of intestinal cancer: consensus report and recommendations. *Gastroenterology* **124**, 762–777.

Bokemeyer, C., Bondarenko, I., Makhson, A., Hartmann, J. T., Aparicio, J., De Braud, F., Donea, S., Ludwig, H., Schuch, G., Stroh, C., Loos, A. H., Zubel, A., *et al.* (2009). Fluorouracil, leucovorin, and oxaliplatin with and without cetuximab in the first-line treatment of metastatic colorectal cancer. *J. Clin. Oncol.* **27**, 663–671.

Boland, C. R., and Goel, A. (2010). Microsatellite instability in colorectal cancer. *Gastroenterology* **138**, 2073–2087.

Bonauer, A., Carmona, G., Iwasaki, M., Mione, M., Koyanagi, M., Fischer, A., Burchfield, J., Fox, H., Doebele, C., Ohtani, K., Chavakis, E., Potente, M., *et al.* (2009). MicroRNA-92a controls angiogenesis and functional recovery of ischemic tissues in mice. *Science* **324**, 1710–1713.

Bouchard, J. F., Moore, S. W., Tritsch, N. X., Roux, P. P., Shekarabi, M., Barker, P. A., and Kennedy, T. E. (2004). Protein kinase A activation promotes plasma membrane insertion of DCC from an intracellular pool: a novel mechanism regulating commissural axon extension. *J. Neurosci.* **24,** 3040–3050.

Bradford, D., Cole, S. J., and Cooper, H. M. (2009). Netrin-1: diversity in development. *Int. J. Biochem. Cell Biol.* **41,** 487–493.

Braun, C. J., Zhang, X., Savelyeva, I., Wolff, S., Moll, U. M., Schepeler, T., Ørntoft, T. F., Andersen, C. L., and Dobbelstein, M. (2008). p53-Responsive micrornas 192 and 215 are capable of inducing cell cycle arrest. *Cancer Res.* **68,** 10094–10104.

Brenner, S. (1974). The genetics of *Caenorhabditis elegans*. *Genetics* **77,** 71–94.

Brouns, I., Adriaensen, D., Burnstock, G., and Timmermans, J. P. (2000). Intraepithelial vagal sensory nerve terminals in rat pulmonary neuroepithelial bodies express P2X(3) receptors. *Am. J. Respir. Cell Mol. Biol.* **23,** 52–61.

Browning, K. N. (2010). Glucose and the vagus: sensory cells savour sweet substances. *J. Physiol.* **588,** 749–750.

Buhner, S., Li, Q., Vignali, S., Barbara, G., De Giorgio, R., Stanghellini, V., Cremon, C., Zeller, F., Langer, R., Daniel, H., Michel, K., and Schemann, M. (2009). Activation of human enteric neurons by supernatants of colonic biopsy specimens from patients with irritable bowel syndrome. *Gastroenterology* **137,** 1425–1434.

Burkhart, D. L., and Sage, J. (2008). Cellular mechanisms of tumour suppression by the retinoblastoma gene. *Nat. Rev. Cancer* **8,** 671–682.

Cabodi, S., del Pilar Camacho-Leal, M., Di Stefano, P., and Defilippi, P. (2010). Integrin signalling adaptors: not only figurants in the cancer story. *Nat. Rev. Cancer* **10,** 858–870.

Cabrera-Vásquez, S., Navarro-Tableros, V., Sánchez-Soto, C., Gutiérrez-Ospina, G., and Hiriart, M. (2009). Remodelling sympathetic innervation in rat pancreatic islets ontogeny. *BMC Dev. Biol.* **9,** 34.

Cai, Y., Wang, J., Li, R., Ayala, G., Ittmann, M., and Liu, M. (2009). GGAP2/PIKE-A directly activates both the Akt and nuclear factor-kappaB pathways and promotes prostate cancer progression. *Cancer Res.* **69,** 819–827.

Campbell, D. S., and Holt, C. E. (2003). Apoptotic pathway and MAPKs differentially regulate chemotropic responses of retinal growth cones. *Neuron* **37,** 939–952.

Carvalho, A. L., Chuang, A., Jiang, W. W., Lee, J., Begum, S., Poeta, L., Zhao, M., Jerónimo, C., Henrique, R., Nayak, C. S., Park, H. L., Brait, M. R., *et al.* (2006). Deleted in colorectal cancer is a putative conditional tumor-suppressor gene inactivated by promoter hypermethylation in head and neck squamous cell carcinoma. *Cancer Res.* **66,** 9401–9407.

Castets, M., Coissieux, M. M., Delloye-Bourgeois, C., Bernard, L., Delcros, J. G., Bernet, A., Laudet, V., and Mehlen, P. (2009). Inhibition of endothelial cell apoptosis by Netrin-1 during angiogenesis. *Dev. Cell* **16,** 614–620.

Catenacci, D. V., Kozloff, M., Kindler, H. L., and Polite, B. (2010). Personalized colon cancer care in 2010. *Semin. Oncol.* **38,** 284–308.

Causeret, F., Hidalgo-Sanchez, M., Fort, P., Backer, S., Popoff, M. R., Gauthier-Rouvière, C., and Bloch-Gallego, E. (2004). Distinct roles of Rac1/Cdc42 and Rho/Rock for axon outgrowth and nucleokinesis of precerebellar neurons toward Netrin 1. *Development* **131,** 2841–2852.

Charron, F., and Tessier-Lavigne, M. (2005). Novel brain wiring functions for classical morphogens: a role as graded positional cues in axon guidance. *Development* **132,** 2251–2262.

Charron, F., Stein, E., Jeong, J., McMahon, A. P., and Tessier-Lavigne, M. (2003). The morphogen sonic hedgehog is an axonal chemoattractant that collaborates with Netrin-1 in midline axon guidance. *Cell* **113,** 11–23.

Chastre, E., Empereur, S., Di Gioia, Y., El Mahdani, N., Mareel, M., Vleminckx, K., Van Roy, F., Bex, V., Emami, S., Spandidos, D., and Gespach, C. (1993). Neoplastic progression of human and rat intestinal cell lines after transfer of the ras and polyoma middle T oncogenes. *Gastroenterology* **105,** 1776–1789.

Chen, S. Y., Huang, P. H., and Cheng, H. J. (2011). Disrupted-in-Schizophrenia 1-mediated axon guidance involves TRIO-RAC-PAK small GTPase pathway signaling. *Proc. Natl. Acad. Sci. USA* **108,** 5861–5866.

Cho, J. H., and Brant, S. R. (2011). Recent insights into the genetics of inflammatory bowel disease. *Gastroenterology* **140,** 1704–1712.

Choi, J., Park, S. Y., Costantini, F., Jho, E. H., and Joo, C. K. (2004). APC is down-regulated by the ubiquitin-proteasome pathway in a process facilitated by Axin. *J. Biol. Chem.* **279,** 49188–49198.

Christofori, G. (2006). New signals from the invasive front. *Nature* **441,** 444–450.

Ciavarra, G., and Zacksenhaus, E. (2011). Multiple pathways counteract cell death induced by RB1 loss: implications for cancer. *Cell Cycle* **10,** 1533–1539.

Cirulli, V., and Yebra, M. (2007). Netrins: beyond the brain. *Nat. Rev. Mol. Cell Biol.* **8,** 296–306.

Coissieux, M. M., Tomsic, J., Castets, M., Hampel, H., Tuupanen, S., Andrieu, N., Comeras, I., Drouet, Y., Lasset, C., Liyanarachchi, S., Mazelin, L., Puisieux, A., *et al.* (2011). Variants in the netrin-1 receptor UNC5C prevent apoptosis and increase risk for familial colorectal cancer. *Gastroenterology* **141,** 2039–2046.

Colavita, A., and Culotti, J. G. (1998). Suppressors of ectopic UNC-5 growth cone steering identify eight genes involved in axon guidance in *Caenorhabditis elegans. Dev. Biol.* **194,** 72–85.

Colavita, A., Krishna, S., Zheng, H., Padgett, R. W., and Culotti, J. G. (1988). Pioneer axon guidance by UNC-129, a *C. elegans* TGF-β. *Science* **281,** 706–709.

Coleman, M. L., Sahai, E. A., Yeo, M., Bosch, M., Dewar, A., and Olson, M. F. (2001). Membrane blebbing during apoptosis results from caspase-mediated activation of ROCK I. *Nat. Cell Biol.* **3,** 339–345.

Conery, A. R., Sever, S., and Harlow, E. (2010). Nucleoside diphosphate kinase Nm23-H1 regulates chromosomal stability by activating the GTPase dynamin during cytokinesis. *Proc. Natl. Acad. Sci. USA* **107,** 15461–15466.

Corset, V., Nguyen-Ba-Charvet, K. T., Forcet, C., Moyse, E., Chédotal, A., and Mehlen, P. (2000). Netrin-1-mediated axon outgrowth and cAMP production requires interaction with adenosine A2b receptor. *Nature* **407,** 747–750.

Coudray, A. M., Louvet, C., Raymond, E., André, T., Tournigand, C., Kornprobst, M., Faivre, S., de Gramont, A., Larsen, A., and Gespach, C. (2005). Increased anticancer activity of the thymidylate synthase inhibitor BGC9331 combined with the topoisomerase I inhibitor SN-38/CPT-11 in human colorectal and breast cancer cells: induction of apoptosis and ROCK cleavage through caspase-3 -dependent and -independent mechanisms. *Int. J. Oncol.* **27,** 553–561.

Croft, D. R., Coleman, M. L., Li, S., Robertson, D., Sullivan, T., Stewart, C. L., and Olson, M. F. (2005). Actin-myosin-based contraction is responsible for apoptotic nuclear disintegration. *J. Cell Biol.* **168,** 245–255.

Crosnier, C., Stamataki, D., and Lewis, J. (2006). Organizing cell renewal in the intestine: stem cells, signals and combinatorial control. *Nat. Rev. Genet.* **7,** 349–359.

Cui, H., Onyango, P., Brandenburg, S., Wu, Y., Hsieh, C. L., and Feinberg, A. P. (2002). Loss of imprinting in colorectal cancer linked to hypomethylation of H19 and IGF2. *Cancer Res.* **62,** 6442–6446.

Cui, H., Cruz-Correa, M., Giardiello, F. M., Hutcheon, D. F., Kafonek, D. R., Brandenburg, S., Wu, Y., He, X., Powe, N. R., and Feinberg, A. P. (2003). Loss of IGF2 imprinting: a potential marker of colorectal cancer risk. *Science* **299,** 1753–1755.

Cui, Y., Han, Z., Yi, H., Song, G., Hao, C., Xia, H., and Ma, X. (2011). MicroRNA-181b and microRNA-9 mediate arsenic-induced angiogenesis via NRP1. *J. Cell. Physiol.* **227,** 772–783.

Dakouane-Giudicelli, M., Duboucher, C., Fortemps, J., Missey-Kolb, H., Brulé, D., Giudicelli, Y., and de Mazancourt, P. (2010). Characterization and expression of Netrin-1 and its receptors UNC5B and DCC in human placenta. *J. Histochem. Cytochem.* **58,** 73–82.

Dallol, A., Morton, D., Maher, E. R., and Latif, F. (2003). SLIT2 axon guidance molecule is frequently inactivated in colorectal cancer and suppresses growth of colorectal carcinoma cells. *Cancer Res.* **63**, 1054–1058.

Dalvin, S., Anselmo, M. A., Prodhan, P., Komatsuzaki, K., Schnitzer, J. J., and Kinane, T. B. (2003). Expression of Netrin-1 and its two receptors DCC and UNC5H2 in the developing mouse lung. *Gene Expr. Patterns* **3**, 279–283.

De Breuck, S., Lardon, J., Rooman, I., and Bouwens, L. (2003). Netrin-1 expression in fetal and regenerating rat pancreas and its effect on the migration of human pancreatic duct and porcine islet precursor cells. *Diabetologia* **46**, 926–933.

de Krijger, I., Mekenkamp, L. J., Punt, C. J., and Nagtegaal, I. D. (2011). MicroRNAs in colorectal cancer metastasis. *J. Pathol.* **224**, 438–447.

de la Chapelle, A. (2004). Genetic predisposition to colorectal cancer. *Nat. Rev. Cancer* **4**, 769–780.

de Planell-Saguer, M., and Rodicio, M. C. (2011). Analytical aspects of microRNA in diagnostics: a review. *Anal. Chim. Acta* **699**, 134–152.

De Roock, W. D., Vriendt, V. D., Normanno, N., Ciardiello, F., and Tejpar, S. (2010). KRAS, BRAF, PIK3CA, and PTEN mutations: implications for targeted therapies in metastatic colorectal cancer. *Lancet Oncol.* **12**, 594–603.

de Rooij, J., Zwartkruis, F. J., Verheijen, M. H., Cool, R. H., Nijman, S. M., Wittinghofer, A., and Bos, J. L. (1998). Epac is a Rap1 guanine-nucleotide-exchange factor directly activated by cyclic AMP. *Nature* **396**, 474–477.

De Vries, M., and Cooper, H. M. (2008). Emerging roles for neogenin and its ligands in CNS development. *J. Neurochem.* **106**, 1483–1492.

De Wever, O., Nguyen, Q. D., Van Hoorde, L., Brake, M., Bruyneel, E., Gespach, C., and Mareel, M. (2004a). Tenascin-C and SF/HGF produced by myofibroblasts in vitro provide convergent pro-invasive signals to human colon cancer cells through RhoA and Rac. *FASEB J.* **18**, 1016–1028.

De Wever, O., Westbroek, W., Verloes, A., Bloemen, N., Bracke, M., Gespach, C., Bruyneel, E., and Mareel, M. (2004b). Critical role of N-cadherin in myofibroblast invasion and migration *in vitro* stimulated by colon-cancer-cell-derived TGF-β or wounding. *J. Cell Sci.* **117**, 4691–4703.

De Wever, O., Sobczak-Thépot, J., Vercoutter-Edouart, A. S., Michalski, J. C., Ouelaa-Benslama, R., Stupack, D. G., Bracke, M., Wang, J. Y., Gespach, C., and Emami, S. (2011). Priming and potentiation of DNA damage response by fibronectin in human colon cancer cells and tumor-derived myofibroblasts. *Int. J. Oncol.* **39**, 393–400.

Debruyne, P., Bruyneel, E., Karaguni, I. M., Li, X. D., Flatau, G., Müller, O., Zimber, A., Gespach, C., and Mareel, M. (2002). Bile acids stimulate invasion and haptotaxis in human colorectal cancer cells through activation of multiple oncogenic pathways. *Oncogene* **21**, 6740–6750.

Degirolamo, C., Modica, S., Palasciano, G., and Moschetta, A. (2011). Bile acids and colon cancer: solving the puzzle with nuclear receptors. *Trends Mol. Med.* **17**, 564–572.

Dejana, E., Orsenigo, F., and Lampugnani, M. G. (2008). The role of adherens junctions and VE-cadherin in the control of vascular permeability. *J. Cell Sci.* **121**, 2115–2122.

Delage, S., Chastre, E., Empereur, S., Wicek, D., Veyssiere, D., Capeau, J., Gespach, C., and Cherqui, G. (1993). Increased expression and activity of PKC-α in human colonic Caco-2 cells after insertion of activated human Ha-ras or polyoma middle T oncogenes. *Cancer Res.* **53**, 2762–2770.

Delloye-Bourgeois, C., Brambilla, E., Coissieux, M. M., Guenebeaud, C., Pedeux, R., Firlej, V., Cabon, F., Brambilla, C., Mehlen, P., and Bernet, A. (2009a). Interference with Netrin-1 and tumor cell death in non-small cell lung cancer. *J. Natl. Cancer Inst.* **101**, 237–247.

Delloye-Bourgeois, C., Fitamant, J., Paradisi, A., Cappellen, D., Douc-Rasy, S., Raquin, M. A., Stupack, D., Nakagawara, A., Rousseau, R., Combaret, V., Puisieux, A., Valteau-Couanet, D., *et al.* (2009b). Netrin-1 acts as a survival factor for aggressive neuroblastoma. *J. Exp. Med.* **206**, 833–847.

Deng, G., Song, G. A., Pong, E., Sleisenger, M., and Kim, Y. S. (2004). Promoter methylation inhibits APC gene expression by causing changes in chromatin conformation and interfering with the binding of transcription factor CCAAT-binding factor. *Cancer Res.* **64**, 2692–2698.

Denys, H., Derycke, L., Hendrix, A., Westbroek, W., Gheldof, A., Narine, K., Pauwels, P., Gespach, C., Bracke, M., and De Wever, O. (2008). Differential impact of TGF-beta and EGF on fibroblast differentiation and invasion reciprocally promotes colon cancer cell invasion. *Cancer Lett.* **266**, 263–274.

Dessein, A. F., Stechly, L., Jonckheere, N., Dumont, P., Monté, D., Leteurtre, E., Truant, S., Kerckaert, J. P., Lecellier, C. H., Lesuffleur, T., Dessein, R., Grard, G., *et al.* (2010). Autocrine induction of invasive and metastatic phenotypes by the MIF-CXCR4 axis in drug-resistant human colon cancer cells. *Cancer Res.* **70**, 4644–4654.

Di Renzo, F., Olivero, M., Giacomini, A., Porte, H., Chastre, E., Mirossay, L., Nordlinger, B., Bretti, S., Bottardi, S., Giordano, S., Plebani, M., Gespach, C., *et al.* (1995). Overexpression and amplification of the Met/HGF receptor gene during the progression of colorectal cancer. *Clin. Cancer Res.* **1**, 147–154.

Dickson, B. J. (2002). Molecular mechanisms of axon guidance. *Science* **298**, 1959–1964.

Dikovskaya, D., Zumbrunn, J., Penman, G. A., and Nathke, I. S. (2001). The adenomatous polyposis coli protein: in the limelight out at the edge. *Trends Cell Biol.* **11**, 378–384.

Dimitriadis, A., Vincan, E., Mohammed, I. M., Roczo, N., Phillips, W. A., and Baindur-Hudson, S. (2001). Expression of Wnt genes in human colon cancers. *Cancer Lett.* **166**, 185–191.

Djelloul, S., Forgue-Lafitte, M. E., Hermelin, B., Mareel, M., Baldi, A., Giordano, A., Chastre, E., and Gespach, C. (1997). Enterocyte differentiation is compatible with SV40 large T expression and loss of p53 function in human colonic Caco-2 cells. Status of the pRb1 and pRb2 tumor suppressor gene products. *FEBS Lett.* **406**, 234–242.

Duman-Scheel, M. (2009). Netrin and DCC: axon guidance regulators at the intersection of nervous system development and cancer. *Curr. Drug Targets* **10**, 602–610.

Dumartin, L., Quemener, C., Laklai, H., Herbert, J., Bicknell, R., Bousquet, C., Pyronnet, S., Castronovo, V., Schilling, M. K., Bikfalvi, A., and Hagedorn, M. (2010). Netrin-1 mediates early events in pancreatic adenocarcinoma progression, acting on tumor and endothelial cells. *Gastroenterology* **138**, 1595–1606.

Duval, A., Iacopetta, B., Ranzani, G. N., Lothe, R. A., Thomas, G., and Hamelin, R. (1999). Variable mutation frequencies in coding repeats of TCF-4 and other target genes in colon, gastric and endometrial carcinoma showing microsatellite instability. *Oncogene* **18**, 6806–6809.

Duvernay, M. T., Wang, H., Dong, C., Guidry, J. J., Sackett, D. L., and Wu, G. (2011). Alpha2B-adrenergic receptor interaction with tubulin controls its transport from the endoplasmic reticulum to the cell surface. *J. Biol. Chem.* **286**, 14080–14089.

Emami, S., Mir, L., Gespach, C., and Rosselin, G. (1989). Transfection of fetal rat intestinal epithelial cells by viral oncogenes: establishment and characterization of the E1A-immortalized SLC-11 cell line. *Proc. Natl. Acad. Sci. USA* **86**, 3194–3198.

Emami, S., Le Floch, N., Bruyneel, E., Thim, L., May, F., Westley, B., Rio, M. C., Mareel, M., and Gespach, C. (2001). Induction of scattering and cellular invasion by trefoil peptides in src- and RhoA-transformed kidney and colonic epithelial cells. *FASEB J.* **15**, 351–361.

Emami, S., Rodrigues, S., Rodrigue, C. M., Floch, N., Rivat, C., Attoub, S., Bruyneel, E., and Gespach, C. (2004). Trefoil factors (TFFs) and cancer progression. *Peptides* **25**, 885–898.

Empereur, S., Djelloul, S., Di Gioia, Y., Bruyneel, E., Mareel, M., Van Hengel, J., Van Roy, F., Comoglio, P., Courtneidge, S., Paraskeva, C., Chastre, E., and Gespach, C. (1997). Progression of familial adenomatous polyposis (FAP) colonic cells after transfer of the src or polyoma middle T oncogenes. Cooperation between src and HGF/Met in invasion. *Br. J. Cancer* **75**, 241–250.

Endo, H., Hosono, K., Uchiyama, T., Sakai, E., Sugiyama, M., Takahashi, H., Nakajima, N., Wada, K., Takeda, K., Nakagama, H., and Nakajima, A. (2011). Leptin acts as a growth factor for colorectal tumours at stages subsequent to tumour initiation in murine colon carcinogenesis. *Gut* **60**, 1363–1371.

Epting, D., Wendik, B., Bennewitz, K., Dietz, C. T., Driever, W., and Kroll, J. (2010). The Rac1 regulator ELMO1 controls vascular morphogenesis in zebrafish. *Circ. Res.* **107**, 45–55.

Esseghir, S., Kennedy, A., Seedhar, P., Nerurkar, A., Poulsom, R., Reis-Filho, J. S., and Isacke, C. M. (2007). Identification of NTN4, TRA1, and STC2 as prognostic markers in breast cancer in a screen for signal sequence encoding proteins. *Clin. Cancer Res.* **13**, 3164–3173.

Eveno, C., Broqueres-You, D., Feron, J. G., Rampanou, A., Tijeras-Raballand, A., Ropert, S., Leconte, L., Levy, B. I., and Pocard, M. (2011). Netrin-4 delays colorectal cancer carcinomatosis by inhibiting tumor angiogenesis. *Am. J. Pathol.* **178**, 1861–1869.

Faivre, S., Kroemer, G., and Raymond, E. (2006). Current development of mTOR inhibitors as anticancer agents. *Nat. Rev. Drug Discov.* **5**, 671–688.

Fang, L., Deng, Z., Shatseva, T., Yang, J., Peng, C., Du, W. W., Yee, A. J., Ang, L. C., He, C., Shan, S. W., and Yang, B. B. (2011). MicroRNA miR-93 promotes tumor growth and angiogenesis by targeting integrin-β8. *Oncogene* **30**, 806–821.

Fasanaro, P., D'Alessandra, Y., Di Stefano, V., Melchionna, R., Romani, S., Pompilio, G., Capogrossi, M. C., and Martelli, F. (2008). MicroRNA-210 modulates endothelial cell response to hypoxia and inhibits the receptor tyrosine kinase ligand Ephrin-A3. *J. Biol. Chem.* **283**, 15878–15883.

Fasano, C. A., Chambers, S. M., Lee, G., Tomishima, M. J., and Studer, L. (2010). Efficient derivation of functional floor plate tissue from human embryonic stem cells. *Cell Stem Cell* **6**, 336–347.

Faux, M. C., Ross, J. L., Meeker, C., Johns, T., Ji, H., Simpson, R. J., Layton, M. J., and Burgess, A. W. (2004). Restoration of full-length adenomatous polyposis coli (APC) protein in a colon cancer cell line enhances cell adhesion. *J. Cell Sci.* **117**, 427–439.

Fazeli, A., Dickinson, S. L., Hermiston, M. L., Tighe, R. V., Steen, R. G., Small, C. G., Stoeckli, E. T., Keino-Masu, K., Masu, M., Rayburn, H., Simons, J., Bronson, R. T., *et al.* (1997). Phenotype of mice lacking functional Deleted in colorectal cancer (Dcc) gene. *Nature* **386**, 796–804.

Fearon, E. R., Cho, K. R., Nigro, J. M., Kern, S. E., Simons, J. W., Ruppert, J. M., Hamilton, S. R., Preisinger, A. C., Thomas, G., Kinzler, K. W., *et al.* (1990). Identification of a chromosome 18q gene that is altered in colorectal cancers. *Science* **247**, 49–56.

Firestein, R., Bass, A. J., Kim, S. Y., Dunn, I. F., Silver, S. J., Guney, I., Freed, E., Ligon, A. H., Vena, N., Ogino, S., Chheda, M. G., Tamayo, P., *et al.* (2008). CDK8 is a colorectal cancer oncogene that regulates β-catenin activity. *Nature* **455**, 547–551.

Fish, J. E., Wythe, J. D., Xiao, T., Bruneau, B. G., Stainier, D. Y., Srivastava, D., and Woo, S. (2011). A Slit/miR-218/Robo regulatory loop is required during heart tube formation in zebrafish. *Development* **138**, 1409–1419.

Fitamant, J., Guenebeaud, C., Coissieux, M. M., Guix, C., Treilleux, I., Scoazec, J. Y., Bachelot, T., Bernet, A., and Mehlen, P. (2008). Netrin-1 expression confers a selective advantage for tumor cell survival in metastatic breast cancer. *Proc. Natl. Acad. Sci. USA* **105**, 4850–4855.

Fodde, R., Edelmann, W., Yang, K., van Leeuwen, C., Carlson, C., Renault, B., Breukel, C., Alt, E., Lipkin, M., Khan, P. M., and Kucherlapati, R. (1994). A targeted chain-termination mutation in the mouse Apc gene results in multiple intestinal tumors. *Proc. Natl. Acad. Sci. USA* **91**, 8969–8973.

Fodde, R., Kuipers, J., Rosenberg, C., Smits, R., Kielman, M., Gaspar, C., van Es, J. H., Breukel, C., Wiegant, J., Giles, R. H., and Clevers, H. (2001). Mutations in the APC tumour suppressor gene cause chromosomal instability. *Nat. Cell Biol.* **3**, 433–438.

Forcet, C., Stein, E., Pays, L., Corset, V., Llambi, F., Tessier-Lavigne, M., and Mehlen, P. (2002). Netrin-1-mediated axon outgrowth requires deleted in colorectal cancer-dependent MAPK activation. *Nature* **417**, 443–447.

Fowler, A., Thomson, D., Giles, K., Maleki, S., Mreich, E., Wheeler, H., Leedman, P., Biggs, M., Cook, R., Little, N., Robinson, B., and McDonald, K. (2011). miR-124a is frequently down-regulated in glioblastoma and is involved in migration and invasion. *Eur. J. Cancer* **47**, 953–963.

Frisch, S. M. (2008). Caspase-8: fly or die. *Cancer Res.* **68**, 4491–4493.

Frisch, S. M., and Francis, H. (1994). Disruption of epithelial cell-matrix interactions induces apoptosis. *J. Cell Biol.* **124**, 619–626.

Frische, E. W., and Zwartkruis, F. J. (2010). Rap1, a mercenary among the Ras-like GTPases. *Dev. Biol.* **340**, 1–9.

Fu, V. X., Schwarze, S. R., Kenowski, M. L., Leblanc, S., Svaren, J., and Jarrard, D. F. (2004). A loss of insulin-like growth factor-2 imprinting is modulated by CCCTC-binding factor down-regulation at senescence in human epithelial cells. *J. Biol. Chem.* **279**, 52218–52226.

Fu, X., Yang, X., Li, J., Tian, X., Cai, J., and Zhang, Y. (2010). Opposite expression patterns of Sonic hedgehog and Indian hedgehog are associated with aberrant methylation status of their promoters in colorectal cancers. *Pathology* **42**, 553–559.

Furne, C., Rama, N., Corset, V., Chédotal, A., and Mehlen, P. (2008). Netrin-1 is a survival factor during commissural neuron navigation. *Proc. Natl. Acad. Sci. USA* **105**, 14465–14470.

Galko, M. J., and Tessier-Lavigne, M. (2000). Function of an axonal chemoattractant modulated by metalloprotease activity. *Science* **289**, 1365–1367.

Gallet, A. (2011). Hedgehog morphogen: from secretion to reception. *Trends Cell Biol.* **21**, 238–246.

Geisbrecht, B. V., Dowd, K. A., Barfield, R. W., Longo, P. A., and Leahy, D. J. (2003). Netrin binds discrete subdomains of DCC and UNC5 and mediates interactions between DCC and heparin. *J. Biol. Chem.* **278**, 32561–32568.

Gellad, Z. F., and Provenzale, D. (2010). Colorectal cancer: national and international perspective on the burden of disease and public health impact. *Gastroenterology* **138**, 2177–2190.

Gershon, M. D. (2010). Developmental determinants of the independence and complexity of the enteric nervous system. *Trends Neurosci.* **33**, 446–456.

Gespach, C. (2008). Trefoil Factors in heath and cancer disease. *In* Encyclopedia of Cancer, (Manfred Schwab, Ed.). 2nd ed. Springer, New York.

Gespach, C. (2010). Stem cells and colon cancer: the questionable cancer stem cell hypothesis. L'hypothèse ambiguë des cellules souches cancéreuses. *Gastroenterol. Clin. Biol.* **34**, 653–661.

Giaccone, G., and Wang, Y. (2011). Strategies for overcoming resistance to EGFR family tyrosine kinase inhibitors. *Cancer Treat. Rev.* **37**, 456–464.

Goichberg, P., Bai, Y., D'Amario, D., Ferreira-Martins, J., Fiorini, C., Zheng, H., Signore, S., Del Monte, F., Ottolenghi, S., D'Alessandro, D., Michler, R. E., Hosoda, T., *et al.* (2011). The Ephrin A1-EphA2 system promotes cardiac stem cell migration after infarction. *Circ. Res.* **108**, 1071–1083.

Goyal, R. K., and Hirano, I. (1996). The enteric nervous system. *N. Engl. J. Med.* **334**, 1106–1115.

Grabauskas, G., Song, I., Zhou, S., and Owyang, C. (2010). Electrophysiological identification of glucose-sensing neurons in rat nodose ganglia. *J. Physiol.* **588**, 617–632.

Grenz, A., Dalton, J. H., Bauerle, J. D., Badulak, A., Ridyard, D., Gandjeva, A., Aherne, C. M., Brodsky, K. S., Kim, J. H., Tuder, R. M., and Eltzschig, H. K. (2011). Partial Netrin-1 deficiency aggravates acute kidney injury. *PLoS One* **6**, e14812.

Grijelmo, C., Rodrigue, C., Svrcek, M., Bruyneel, E., Hendrix, A., de Wever, O., and Gespach, C. (2007). Proinvasive activity of BMP-7 through SMAD4/src-independent and ERK/Rac/JNK-dependent signaling pathways in colon cancer cells. *Cell. Signal.* **19**, 1722–1732.

Guenebeaud, C., Goldschneider, D., Castets, M., Guix, C., Chazot, G., Delloye-Bourgeois, C., Eisenberg-Lerner, A., Shohat, G., Zhang, M., Laudet, V., Kimchi, A., Bernet, A., *et al.* (2010). The dependence receptor UNC5H2/B triggers apoptosis *via* PP2A-mediated dephosphorylation of DAP kinase. *Mol. Cell* **40**, 863–876.

Gupta, R. A., Wang, D., Katkuri, S., Wang, H., Dey, S. K., and DuBois, R. N. (2004). Activation of nuclear hormone receptor peroxisome proliferator-activated receptor-delta accelerates intestinal adenoma growth. *Nat. Med.* **10**, 245–247.

Hagedorn, E. J., Yashiro, H., Ziel, J. W., Ihara, S., Wang, Z., and Sherwood, D. R. (2009). Integrin acts upstream of Netrin signaling to regulate formation of the anchor cell's invasive membrane in *C. elegans*. *Dev. Cell* **17**, 187–198.

Hagihara, M., Endo, M., Hata, K., Higuchi, C., Takaoka, K., Yoshikawa, H., and Yamashita, T. (2011). Neogenin, a receptor for bone morphogenetic proteins. *J. Biol. Chem.* **286**, 5157–5165.

Haller, D. G., Tabernero, J., Maroun, J., de Braud, F., Price, T., Van Cutsem, E., Hill, M., Gilberg, F., Rittweger, K., and Schmoll, H. J. (2011). Capecitabine plus oxaliplatin compared with fluorouracil and folinic acid as adjuvant therapy for stage III colon cancer. *J. Clin. Oncol.* **29**, 1465–1471.

Han, J., Han, L., Tiwari, P., Wen, Z., and Zheng, J. Q. (2007). Spatial targeting of type II protein kinase A to filopodia mediates the regulation of growth cone guidance by cAMP. *J. Cell Biol.* **176**, 101–111.

Hanahan, D., and Weinberg, R. A. (2011). Hallmarks of cancer: the next generation. *Cell* **144**, 646–674.

Harada, N., Tamai, Y., Ishikawa, T., Sauer, B., Takaku, K., Oshima, M., and Taketo, M. M. (1999). Intestinal polyposis in mice with a dominant stable mutation of the β-catenin gene. *EMBO J.* **18**, 5931–5942.

Hardwick, J. C., Kodach, L. L., Offerhaus, G. J., and van den Brink, G. R. (2008). Bone morphogenetic protein signalling in colorectal cancer. *Nat. Rev. Cancer* **8**, 806–812.

Harrison, C. (2006). Anticancer drugs. Downstream GPCR inhibition. *Nat. Rev. Drug Discov.* **5**, 894–895.

Hart, M. J., de los Santos, R., Albert, I. N., Rubinfeld, B., and Polakis, P. (1998). Downregulation of β-catenin by human Axin and its association with the APC tumor suppressor, β-catenin and GSK3β. *Curr. Biol.* **8**, 573–581.

Haskó, G., Linden, J., Cronstein, B., and Pacher, P. (2008). Adenosine receptors: therapeutic aspects for inflammatory and immune diseases. *Nat. Rev. Drug Discov.* **7**, 759–770.

Haskó, G., Csóka, B., Németh, Z. H., Vizi, E. S., and Pacher, P. (2009). A(2B) adenosine receptors in immunity and inflammation. *Trends Immunol.* **30**, 263–270.

Hata, K., Fujitani, M., Yasuda, Y., Doya, H., Saito, T., Yamagishi, S., Mueller, B. K., and Yamashita, T. (2006). RGMa inhibition promotes axonal growth and recovery after spinal cord injury. *J. Cell Biol.* **173**, 47–58.

He, K., Jang, S. W., Joshi, J., Yoo, M. H., and Ye, K. (2011). Akt-phosphorylated PIKE-A inhibits UNC5B-induced apoptosis in cancer cell lines in a p53-dependent manner. *Mol. Biol. Cell* **22,** 1943–1954.

Hedgecock, E. M., Culotti, J. G., and Hall, D. H. (1990). The unc-5, unc-6, and unc-40 genes guide circumferential migrations of pioneer axons and mesodermal cells on the epidermis in *C. elegans. Neuron* **4,** 61–85.

Henderson, B. R., and Fagotto, F. (2002). The ins and outs of APC and β-catenin nuclear transport. *EMBO Rep.* **3,** 834–839.

Hendrix, A., Maynard, D., Pauwels, P., Braems, G., Denys, H., Van den Broecke, R., Lambert, J., Van Belle, S., Cocquyt, V., Gespach, C., Bracke, M., Seabra, M. C., *et al.* (2010). Effect of the secretory small GTPase Rab27B on breast cancer growth, invasion, and metastasis. *J. Natl. Cancer Inst.* **102,** 866–880.

Hendrix, A., Gespach, C., Bracke, M., and De Wever, O. (2011). The tumor ecosystem regulates the roads for invasion and metastasis. *Clin. Res. Hepatol. Gastroenterol* **35,** 714–719.

Herz, J., and Chen, Y. (2006). Reelin, lipoprotein receptors and synaptic plasticity. *Nat. Rev. Neurosci.* **7,** 850–859.

Hibi, K., Mizukami, H., Shirahata, A., Goto, T., Sakata, M., and Sanada, Y. (2009). Aberrant methylation of the Netrin-1 receptor genes UNC5C and DCC detected in advanced colorectal cancer. *World J. Surg.* **33,** 1053–1057.

Higuchi, M., Onishi, K., Kikuchi, C., and Gotoh, Y. (2008). Scaffolding function of PAK in the PDK1-Akt pathway. *Nat. Cell Biol.* **10,** 1356–1364.

Hilgers, W., Song, J. J., Haye, M., Hruban, R. R., Kern, S. E., and Fearon, E. R. (2000). Homozygous deletions inactivate DCC, but not MADH4/DPC4/SMAD4, in a subset of pancreatic and biliary cancers. *Genes Chromosomes Cancer* **27,** 353–357.

Hinoue, T., Weisenberger, D. J., Pan, F., Campan, M., Kim, M., Young, J., Whitehall, V. L., Leggett, B. A., and Laird, P. W. (2009). Analysis of the association between CIMP and BRAF in colorectal cancer by DNA methylation profiling. *PLoS One* **4,** e8357.

Hintermann, E., Yang, N., O'Sullivan, D., Higgins, J. M., and Quaranta, V. (2005). Integrin α6β4-erbB2 complex inhibits haptotaxis by up-regulating E-cadherin cell-cell junctions in keratinocytes. *J. Biol. Chem.* **280,** 8004–8015.

Hiramoto, M., Hiromi, Y., Giniger, E., and Hotta, Y. (2000). The *Drosophila* Netrin receptor Frazzled guides axons by controlling Netrin distribution. *Nature* **406,** 886–889.

Holcombe, R. F., Marsh, J. L., Waterman, M. L., Lin, F., Milovanovic, T., and Truong, T. (2002). Expression of Wnt ligands and Frizzled receptors in colonic mucosa and in colon carcinoma. *Mol. Pathol.* **55,** 220–226.

Holt, L. J., and Siddle, K. (2005). Grb10 and Grb14: enigmatic regulators of insulin action—and more? *Biochem. J.* **388,** 393–406.

Hovanes, K., Li, T. W., Munguia, J. E., Truong, T., Milovanovic, T., Lawrence Marsh, J., Holcombe, R. F., and Waterman, M. L. (2001). Beta-catenin-sensitive isoforms of lymphoid enhancer factor-1 are selectively expressed in colon cancer. *Nat. Genet.* **28,** 53–57.

Hsu, Y. H., Shaw, C. K., and Chuong, C. M. (2001). Immunohistochemical localization of deleted-in-colon-cancer (DCC) protein in human epithelial, neural, and neuro-endocrine tissues in paraffin sections with antigen retrieval. *Kaohsiung J. Med. Sci.* **17,** 351–357.

Hsu, P. P., Kang, S. A., Rameseder, J., Zhang, Y., Ottina, K. A., Lim, D., Peterson, T. R., Choi, Y., Gray, N. S., Yaffe, M. B., Marto, J. A., and Sabatini, D. M. (2011). The mTOR-regulated phosphoproteome reveals a mechanism of mTORC1-mediated inhibition of growth factor signaling. *Science* **332,** 1317–1322.

Hu, J. (2011). Arsenic in the treatment of newly diagnosed acute promyelocytic leukemia: current status and future research direction. *Front. Med.* **5,** 45–52.

Hu, G., Zhang, S., Vidal, M., Baer, J. L., Xu, T., and Fearon, E. R. (1997). Mammalian homologs of seven in absentia regulate DCC via the ubiquitin-proteasome pathway. *Genes Dev.* **11**, 2701–2714.

Huang, Q., Li, F., Liu, X., Li, W., Shi, W., Liu, F. F., O'Sullivan, B., He, Z., Peng, Y., Tan, A. C., Zhou, L., Shen, J., *et al.* (2011). Caspase 3-mediated stimulation of tumor cell repopulation during cancer radiotherapy. *Nat. Med.* **17**, 860–866.

Hurwitz, H., Fehrenbacher, L., Novotny, W., Cartwright, T., Hainsworth, J., Heim, W., Berlin, J., Baron, A., Griffing, S., Holmgren, E., Ferrara, N., Fyfe, G., *et al.* (2004). Bevacizumab plus irinotecan, fluorouracil, and leucovorin for metastatic colorectal cancer. *N. Engl. J. Med.* **350**, 2335–2342.

Huusko, P., Ponciano-Jackson, D., Wolf, M., Kiefer, J. A., Azorsa, D. O., Tuzmen, S., Weaver, D., Robbins, C., Moses, T., Allinen, M., Hautaniemi, S., Chen, Y., *et al.* (2004). Nonsense-mediated decay microarray analysis identifies mutations of EPHB2 in human prostate cancer. *Nat. Genet.* **36**, 979–983.

Hynes, M., Ye, W., Wang, K., Stone, D., Murone, M., Sauvage, F., and Rosenthal, A. (2000). The seven-transmembrane receptor smoothened cell-autonomously induces multiple ventral cell types. *Nat. Neurosci.* **3**, 41–46.

Ingham, P. W., Nakano, Y., and Seger, C. (2011). Mechanisms and functions of Hedgehog signalling across the metazoa. *Nat. Rev. Genet.* **12**, 393–406.

Inoda, S., Hirohashi, Y., Torigoe, T., Morita, R., Takahashi, A., Asanuma, H., Nakatsugawa, M., Nishizawa, S., Tamura, Y., Tsuruma, T., Terui, T., Kondo, T., *et al.* (2011). Cytotoxic T lymphocytes efficiently recognize human colon cancer stem-like cells. *Am. J. Pathol.* **178**, 1805–1813.

Inoki, K., Ouyang, H., Zhu, T., Lindvall, C., Wang, Y., Zhang, X., Yang, Q., Bennett, C., Harada, Y., Stankunas, K., Wang, C. Y., He, X., *et al.* (2006). TSC2 integrates Wnt and energy signals via a coordinated phosphorylation by AMPK and GSK3 to regulate cell growth. *Cell* **126**, 955–968.

Ishii, N., Wadsworth, W. G., Stern, B. D., Culotti, J. G., and Hedgecock, E. M. (1992). UNC-6, a laminin-related protein, guides cell and pioneer axon migrations in *C. elegans*. *Neuron* **9**, 873–881.

Itoh, B., Hirose, T., Takata, N., Nishiwaki, K., Koga, M., Ohshima, Y., and Okada, M. (2005). SRC-1, a non-receptor type of protein tyrosine kinase, controls the direction of cell and growth cone migration in *C. elegans*. *Development* **132**, 5161–5172.

Jagani, Z., Dorsch, M., and Warmuth, M. (2010). Hedgehog pathway activation in chronic myeloid leukemia. *Cell Cycle* **9**, 3449–3456.

Jahn, T., Seipel, P., Urschel, S., Peschel, C., and Duyster, J. (2002). Role for the adaptor protein Grb10 in the activation of Akt. *Mol. Cell. Biol.* **22**, 979–991.

Jasperson, K. W., Tuohy, T. M., Neklason, D. W., and Burt, R. W. (2010). Hereditary and familial colon cancer. *Gastroenterology* **138**, 2044–2058.

Jen, J. C., Chan, W. M., Bosley, T. M., Wan, J., Carr, J. R., Rüb, U., Shattuck, D., Salamon, G., Kudo, L. C., Ou, J., Lin, D. D., Salih, M. A., *et al.* (2004). Mutations in a human ROBO gene disrupt hindbrain axon pathway crossing and morphogenesis. *Science* **304**, 1509–1513.

Jiang, Y., Liu, M. T., and Gershon, M. D. (2003). Netrins and DCC in the guidance of migrating neural crest-derived cells in the developing bowel and pancreas. *Dev. Biol.* **258**, 364–384.

Jordan, P., Brazao, R., Boavida, M. G., Gespach, C., and Chastre, E. (1999). Cloning of a novel human Rac1b splice variant with increased expression in human colorectal cancers. *Oncogene* **18**, 6835–6939.

Jordan, B. W., Dinev, D., LeMellay, V., Troppmair, J., Gotz, R., Wixler, L., Sendtner, M., Ludwig, S., and Rapp, U. R. (2001). Neurotrophin receptor-interacting mage homologue is an inducible inhibitor of apoptosis protein-interacting protein that augments cell death. *J. Biol. Chem.* **276**, 39985–39989.

Kanchanawong, P., Shtengel, G., Pasapera, A. M., Ramko, E. B., Davidson, M. W., Hess, H. F., and Waterman, C. M. (2010). Nanoscale architecture of integrin-based cell adhesions. *Nature* **468**, 580–584.

Kang, J. S., Yi, M. J., Zhang, W., Feinleib, J. L., Cole, F., and Krauss, R. S. (2004). Netrins and neogenin promote myotube formation. *J. Cell Biol.* **167**, 493–504.

Kaufmann, S., Kuphal, S., Schubert, T., and Bosserhoff, A. K. (2009). Functional implication of Netrin expression in malignant melanoma. *Cell. Oncol.* **31**, 415–422.

Kebache, S., Ash, J., Annis, M. G., Hagan, J., Huber, M., Hassard, J., Stewart, C. L., Whiteway, M., and Nantel, A. (2007). Grb10 and active Raf-1 kinase promote Bad-dependent cell survival. *J. Biol. Chem.* **282**, 21873–21883.

Keino-Masu, K., Masu, M., Hinck, L., Leonardo, E. D., Chan, S. S., Culotti, J. G., and Tessier-Lavigne, M. (1996). Deleted in Colorectal Cancer (DCC) encodes a netrin receptor. *Cell* **87**, 175–185.

Kennedy, T. E., Serafini, T., de la Torre, J. R., and Tessier-Lavigne, M. (1994). Netrins are diffusible chemotropic factors for commissural axons in the embryonic spinal cord. *Cell* **78**, 425–435.

Khan, J. A., Cao, M., Kang, B. Y., Liu, Y., Mehta, J. L., and Hermonat, P. L. (2010). Systemic human Netrin-1 gene delivery by adeno-associated virus type 8 alters leukocyte accumulation and atherogenesis in vivo. *Gene Ther.* **18**, 437–444.

Kidd, T. (2009). Crossing the line. *Science* **324**, 893–894.

Killeen, M. T., and Sybingco, S. S. (2008). Netrin, Slit and Wnt receptors allow axons to choose the axis of migration. *Dev. Biol.* **323**, 143–151.

Killeen, M., Tong, J., Krizus, A., Steven, R., Scott, I., Pawson, T., and Culotti, J. (2002). UNC-5 function requires phosphorylation of cytoplasmic tyrosine 482, but its UNC-40-independent functions also require a region between the ZU-5 and death domains. *Dev. Biol.* **251**, 348–366.

Kim, D., and Ackerman, S. L. (2011). The UNC5C Netrin receptor regulates dorsal guidance of mouse hindbrain axons. *J. Neurosci.* **31**, 2167–2179.

Kim, H. J., and Bae, S. C. (2011). Histone deacetylase inhibitors: molecular mechanisms of action and clinical trials as anti-cancer drugs. *Am. J. Transl. Res.* **3**, 166–179.

Kim, S., Ren, X. C., Fox, E., and Wadsworth, W. G. (1999). SDQR migrations in *Caenorhabditis elegans* are controlled by multiple guidance cues and changing responses to Netrin UNC-6. *Development* **126**, 3881–3890.

Kim, T. H., Lee, H. K., Seo, I. A., Bae, H. R., Suh, D. J., Wu, J., Rao, Y., Hwang, K. G., and Park, H. T. (2005). Netrin induces down-regulation of its receptor, Deleted in Colorectal Cancer, through the ubiquitin-proteasome pathway in the embryonic cortical neuron. *J. Neurochem.* **95**, 1–8.

Kimple, A. J., Bosch, D. E., Giguère, P. M., and Siderovski, D. P. (2011). Regulators of G-protein signaling and their Gα substrates: promises and challenges in their use as drug discovery targets. *Pharmacol. Rev.* **63**, 728–749.

Kobayashi, S., Kohda, T., Ichikawa, H., Ogura, A., Ohki, M., Kaneko-Ishino, T., and Ishino, F. (2002). Paternal expression of a novel imprinted gene, Peg12/Frat3, in the mouse 7C region homologous to the Prader-Willi syndrome region. *Biochem. Biophys. Res. Commun.* **290**, 403–408.

Kober, P., Bujko, M., Olędzki, J., Tysarowski, A., and Siedlecki, J. A. (2011). Methyl-CpG binding column-based identification of nine genes hypermethylated in colorectal cancer. *Mol. Carcinog.* **50**, 846–856.

Koch, M., Murrell, J. R., Hunter, D. D., Olson, P. F., Jin, W., Keene, D. R., Brunken, W. J., and Burgeson, R. E. (2000). A novel member of the Netrin family, beta-Netrin, shares homology with the beta chain of laminin: identification, expression, and functional characterization. *J. Cell Biol.* **151**, 221–234.

Koch, A. W., Mathivet, T., Larrivée, B., Tong, R. K., Kowalski, J., Pibouin-Fragner, L., Bouvrée, K., Stawicki, S., Nicholes, K., Rathore, N., Scales, S. J., Luis, E., *et al.* (2011). Robo4 maintains vessel integrity and inhibits angiogenesis by interacting with UNC5B. *Dev. Cell* **20**, 33–46.

Kolachala, V., Ruble, B., Vijay-Kumar, M., Wang, L., Mwangi, S., Figler, H., Figler, R., Srinivasan, S., Gewirtz, A., Linden, J., Merlin, D., and Sitaraman, S. (2008). Blockade of adenosine A2B receptors ameliorates murine colitis. *Br. J. Pharmacol.* **155**, 127–137.

Kolesnikov, T., and Beckendorf, S. K. (2005). Netrin and Slit guide salivary gland migration. *Dev. Biol.* **284**, 102–111.

Komatsuzaki, K., Dalvin, S., and Kinane, T. B. (2002). Modulation of G(ialpha(2)) signaling by the axonal guidance molecule UNC5H2. *Biochem. Biophys. Res. Commun.* **297**, 898–905.

Kotelevets, L., Noë, V., Bruyneel, E., Myssiakine, E., Chastre, E., Mareel, M., and Gespach, C. (1998). Inhibition by platelet-activating factor of src- and hepatocyte growth factor-dependent invasiveness of intestinal and kidney epithelial cells. Phosphatidylinositol-3′-kinase is a critical mediator of tumor invasion. *J. Biol. Chem.* **273**, 14138–14145.

Kruger, R., Lee, J., Li, W., and Guan, K. L. (2004). Mapping Netrin receptor binding reveals domains of Unc5 regulating its tyrosine phosphorylation. *J. Neurosci.* **24**, 10826–10834.

Kucherlapati, M. H., Yang, K., Fan, K., Kuraguchi, M., Sonkin, D., Rosulek, A., Lipkin, M., Bronson, R. T., Aronow, B. J., and Kucherlapati, R. (2008). Loss of Rb1 in the gastrointestinal tract of Apc1638N mice promotes tumors of the cecum and proximal colon. *Proc. Natl. Acad. Sci. USA* **105**, 15493–15498.

Kuroki, T., Trapasso, F., Yendamuri, S., Matsuyama, A., Alder, H., Williams, N. N., Kaiser, L. R., and Croce, C. M. (2003). Allelic loss on chromosome 3p21.3 and promoter hypermethylation of semaphorin 3B in non-small cell lung cancer. *Cancer Res.* **63**, 3352–3355.

Labonte, M. J., Wilson, P. M., Fazzone, W., Russell, J., Louie, S. G., El-Khoueiry, A., Lenz, H. J., and Ladner, R. D. (2011). The dual EGFR/HER2 inhibitor lapatinib synergistically enhances the antitumor activity of the histone deacetylase inhibitor panobinostat in colorectal cancer models. *Cancer Res.* **71**, 3635–3648.

Ladle, D. R., Pecho-Vrieseling, E., and Arber, S. (2007). Assembly of motor circuits in the spinal cord: driven to function by genetic and experience-dependent mechanisms. *Neuron* **56**, 270–283.

Lappano, R., and Maggiolini, M. (2011). G protein-coupled receptors: novel targets for drug discovery in cancer. *Nat. Rev. Drug Discov.* **10**, 47–60.

Larrieu-Lahargue, F., Welm, A. L., Thomas, K. R., and Li, D. Y. (2010). Netrin-4 induces lymphangiogenesis in vivo. *Blood* **115**, 5418–5426.

Larrieu-Lahargue, F., Welm, A. L., Thomas, K. R., and Li, D. Y. (2011). Netrin-4 activates endothelial integrin $\alpha6\beta1$. *Circ. Res.* **109**, 770–774.

Larrivée, B., Freitas, C., Trombe, M., Lv, X., Delafarge, B., Yuan, L., Bouvrée, K., Bréant, C., Del Toro, R., Bréchot, N., Germain, S., Bono, F., *et al.* (2007). Activation of the UNC5B receptor by Netrin-1 inhibits sprouting angiogenesis. *Genes Dev.* **21**, 2433–2447.

Latil, A., Chêne, L., Cochant-Priollet, B., Mangin, P., Fournier, G., Berthon, P., and Cussenot, O. (2003). Quantification of expression of Netrins, slits and their receptors in human prostate tumors. *Int. J. Cancer* **103**, 306–315.

Lauderdale, J. D., Pasquali, S. K., Fazel, R., van Eeden, F. J., Schauerte, H. E., Haffter, P., and Kuwada, J. Y. (1998). Regulation of Netrin-1a expression by hedgehog proteins. *Mol. Cell. Neurosci.* **11**, 194–205.

Laurent-Puig, P., Cayre, A., Manceau, G., Buc, E., Bachet, J. B., Lecomte, T., Rougier, P., Lievre, A., Landi, B., Boige, V., Ducreux, M., Ychou, M., *et al.* (2009). Analysis of PTEN, BRAF, and EGFR status in determining benefit from cetuximab therapy in wild-type KRAS metastatic colon cancer. *J. Clin. Oncol.* **27**, 5924–5930.

Lauth, M., Bergström, A., Shimokawa, T., and Toftgård, R. (2007). Inhibition of GLI-mediated transcription and tumor cell growth by small-molecule antagonists. *Proc. Natl. Acad. Sci. USA* **104,** 8455–8460.

Le Douarin, N. M., and Teillet, M. A. (1973). The migration of neural crest cells to the wall of the digestive tract in avian embryo. *J. Embryol. Exp. Morphol.* **30,** 31–48.

Le Floch, N., Rivat, C., De Wever, O., Bruyneel, E., Mareel, M., Dale, T., and Gespach, C. (2005). The proinvasive activity of Wnt-2 is mediated through a noncanonical Wnt pathway coupled to GSK-3β and c-jun/AP-1 signaling. *FASEB J.* **19,** 144–146.

Le Floch, N., Forgez, P., Souaze, F., Bruyneel, E., Bracke, M., Sabbah, M., Redeuilh, G., Zimber, A., Mester, J., and Gespach, C. (2006). Wnt signaling in normal development and cancer progression. *In* Trends in Cellular Signaling, (Dave E. Caplin, Ed.), pp. 143–227. Nova Science Publishers, New YorkChapter 5.

Leanne Jones, D., and Rando, T. (2011). Emerging models and paradigms for stem cell ageing. *Nat. Cell Biol.* **23,** 506–512.

Lee, S. K., Teng, Y., Wong, H. K., Ng, T. K., Huang, L., Lei, P., Choy, K. W., Liu, Y., Zhang, M., Lam, D. S., Yam, G. H., and Pang, C. P. (2011). MicroRNA-145 regulates human corneal epithelial differentiation. *PLoS One* **6,** e21249.

Leggett, B., and Whitehall, V. (2010). Role of the serrated pathway in colorectal cancer pathogenesis. *Gastroenterology* **138,** 2088–2100.

Lejmi, E., Leconte, L., Pédron-Mazoyer, S., Ropert, S., Raoul, W., Lavalette, S., Bouras, I., Feron, J. G., Maitre-Boube, M., Assayag, F., Feumi, C., Alemany, M., *et al.* (2008). Netrin-4 inhibits angiogenesis via binding to neogenin and recruitment of Unc5B. *Proc. Natl. Acad. Sci. USA* **105,** 12491–12496.

Leonardo, E. D., Hinck, L., Masu, M., Keino-Masu, K., Ackerman, S. L., and Tessier-Lavigne, M. (1997). Vertebrate homologues of *C. elegans* UNC-5 are candidate Netrin receptors. *Nature* **386,** 833–838.

Leverrier, Y., and Ridley, A. J. (2001). Apoptosis: caspases orchestrate the ROCK "n" bleb. *Nat. Cell Biol.* **3,** E91–E93.

Levy, L., and Hill, C. S. (2006). Alterations in components of the TGF-β superfamily signaling pathways in human cancer. *Cytokine Growth Factor Rev.* **17,** 41–58.

Lévy, P., Emami, S., Cherqui, G., Chastre, E., Gespach, C., and Picard, J. (1990). Altered expression of proteoglycans in E1A-immortalized rat fetal intestinal epithelial cells in culture. *Cancer Res.* **50,** 6716–6722.

Li, X., Meriane, M., Triki, I., Shekarabi, M., Kennedy, T., Larose, L., and Lamarche-Vane, N. (2002a). The adaptor protein Nck-1 couples the Netrin-1 receptor DCC (deleted in colorectal cancer) to the activation of the small GTPase Rac1 through an atypical mechanism. *J. Biol. Chem.* **277,** 37788–37797.

Li, X., Saint-Cyr-Proulx, E., Aktories, K., and Lamarche-Vane, N. (2002b). Rac1 and Cdc42 but not RhoA or Rho kinase activities are required for neurite outgrowth induced by the Netrin-1 receptor DCC (deleted in colorectal cancer) in N1E-115 neuroblastoma cells. *J. Biol. Chem.* **277,** 15207–15214.

Li, W., Lee, J., Vikis, H. G., Lee, S. H., Liu, G., Aurandt, J., Shen, T. L., Fearon, E. R., Guan, J. L., Han, M., Rao, Y., Hong, K., *et al.* (2004). Activation of FAK and Src are receptor-proximal events required for Netrin signaling. *Nat. Neurosci.* **7,** 1213–1221.

Li, W., Aurandt, J., Jürgensen, C., Rao, Y., and Guan, K. L. (2006). FAK and Src kinases are required for Netrin-induced tyrosine phosphorylation of UNC5. *J. Cell Sci.* **119,** 47–55.

Li, G., Luna, C., Qiu, J., Epstein, D. L., and Gonzalez, P. (2010a). Targeting of integrin beta1 and kinesin 2alpha by microRNA 183. *J. Biol. Chem.* **285,** 5461–5471.

Li, Y., Nie, C. J., Hu, L., Qin, Y., Liu, H. B., Zeng, T. T., Chen, L., Fu, L., Deng, W., Chen, S. P., Jia, W. H., Zhang, C., *et al.* (2010b). Characterization of a novel mechanism of genomic instability involving the SEI1/SET/NM23H1 pathway in esophageal cancers. *Cancer Res.* **70** (14), 5695–57052010 Jul 15.

Li, C., Chi, S., and Xie, J. (2011a). Hedgehog signaling in skin cancers. *Cell. Signal.* **23,** 1235–1243.

Li, Q., Yao, D., Ma, J., Zhu, J., Xu, X., Ren, Y., Ding, X., and Mao, X. (2011b). Transplantation of MSCs in combination with netrin-1 improves neoangiogenesis in a rat model of hind limb ischemia. *J. Surg. Res.* **166,** 162–169.

Lièvre, A., Bachet, J. B., Boige, V., Cayre, A., Le Corre, D., Buc, E., Ychou, M., Bouché, O., Landi, B., Louvet, C., André, T., Bibeau, F., *et al.* (2008). KRAS mutations as an independent prognostic factor in patients with advanced colorectal cancer treated with cetuximab. *J. Clin. Oncol.* **26,** 374–379.

Lim, Y. S., and Wadsworth, W. G. (2002). Identification of domains of Netrin UNC-6 that mediate attractive and repulsive guidance and responses from cells and growth cones. *J. Neurosci.* **22,** 7080–7087.

Lim, A. H., Suli, A., Yaniv, K., Weinstein, B., Li, D. Y., and Chien, C. B. (2011). Motoneurons are essential for vascular pathfinding. *Development* **138,** 3847–3857.

Lin, L., and Isacson, O. (2006). Axonal growth regulation of fetal and embryonic stem cell-derived dopaminergic neurons by Netrin-1 and Slits. *Stem Cells* **24,** 2504–2513.

Linden, J. (2011). Regulation of leukocyte function by adenosine receptors. *Adv. Pharmacol.* **61,** 95–114.

Link, B. C., Reichelt, U., Schreiber, M., Kaifi, J. T., Wachowiak, R., Bogoevski, D., Bubenheim, M., Cataldegirmen, G., Gawad, K. A., Issa, R., Koops, S., Izbicki, J. R., *et al.* (2007). Prognostic implications of Netrin-1 expression and its receptors in patients with adenocarcinoma of the pancreas. *Ann. Surg. Oncol.* **14,** 2591–2599.

Lipton, L., Halford, S. E., Johnson, V., Novelli, M. R., Jones, A., Cummings, C., Barclay, E., Sieber, O., Sadat, A., Bisgaard, M. L., Hodgson, S. V., Aaltonen, L. A., *et al.* (2003). Carcinogenesis in MYH-associated polyposis follows a distinct genetic pathway. *Cancer Res.* **63,** 7595–7599.

Liu, W., Dong, X., Mai, M., Seelan, R. S., Taniguchi, K., Krishnadath, K. K., Halling, K. C., Cunningham, J. M., Boardman, L. A., Qian, C., Christensen, E., Schmidt, S. S., *et al.* (2000). Mutations in AXIN2 cause colorectal cancer with defective mismatch repair by activating β-catenin/TCF signalling. *Nat. Genet.* **26,** 146–147.

Liu, G., Beggs, H., Jürgensen, C., Park, H. T., Tang, H., Gorski, J., Jones, K. R., Reichardt, L. F., Wu, J., and Rao, Y. (2004a). Netrin requires focal adhesion kinase and Src family kinases for axon outgrowth and attraction. *Nat. Neurosci.* **7,** 1222–1232.

Liu, Y., Stein, E., Oliver, T., Li, Y., Brunken, W. J., Koch, M., Tessier-Lavigne, M., and Hogan, B. L. (2004b). Novel role for Netrins in regulating epithelial behavior during lung branching morphogenesis. *Curr. Biol.* **14,** 897–905.

Liu, X., Hu, Y., Hao, C., Rempel, S. A., and Ye, K. (2007). PIKE-A is a proto-oncogene promoting cell growth, transformation and invasion. *Oncogene* **26,** 4918–4927.

Liu, H., Brannon, A. R., Reddy, A. R., Alexe, G., Seiler, M. W., Arreola, A., Oza, J. H., Yao, M., Juan, D., Liou, L. S., Ganesan, S., Levine, A. J., *et al.* (2010a). Identifying mRNA targets of microRNA dysregulated in cancer: with application to clear cell Renal Cell Carcinoma. *BMC Syst. Biol.* **4,** 51.

Liu, Y., Taylor, N. E., Lu, L., Usa, K., Cowley, A. W., Jr., Ferreri, N. R., Yeo, N. C., and Liang, M. (2010b). Renal medullary microRNAs in Dahl salt-sensitive rats: miR-29b regulates several collagens and related genes. *Hypertension* **55,** 974–982.

Liu, X. Q., Rajput, A., Geng, L., Ongchin, M., Chaudhuri, A., and Wang, J. (2011). Restoration of TGF-β type II receptor expression in colon cancer cells with microsatellite instability increases metastatic potential in vivo. *J. Biol. Chem.* **286,** 16082–16090.

Llambi, F., Causeret, F., Bloch-Gallego, E., and Mehlen, P. (2001). Netrin-1 acts as a survival factor *via* its receptors UNC5H and DCC. *EMBO J.* **20,** 2715–2722.

Llambi, F., Lourenço, F. C., Gozuacik, D., Guix, C., Pays, L., Del Rio, G., Kimchi, A., and Mehlen, P. (2005). The dependence receptor UNC5H2 mediates apoptosis through DAP-kinase. *EMBO J.* **24,** 1192–1201.

Lombardo, Y., Scopelliti, A., Cammareri, P., Todaro, M., Iovino, F., Ricci-Vitiani, L., Gulotta, G., Dieli, F., de Maria, R., and Stassi, G. (2011). Bone morphogenetic protein 4 induces differentiation of colorectal cancer stem cells and increases their response to chemotherapy in mice. *Gastroenterology* **140,** 297–309.

Lourenço, F. C., Galvan, V., Fombonne, J., Corset, V., Llambi, F., Müller, U., Bredesen, D. E., and Mehlen, P. (2009). Netrin-1 interacts with amyloid precursor protein and regulates amyloid-beta production. *Cell Death Differ.* **16,** 655–663.

Lu, Q., Sun, E. E., Klein, R. S., and Flanagan, J. G. (2001). Ephrin-B reverse signaling is mediated by a novel PDZ-RGS protein and selectively inhibits G protein-coupled chemoattraction. *Cell* **105,** 69–79.

Lu, X., Le Noble, F., Yuan, L., Jiang, Q., De Lafarge, B., Sugiyama, D., Bréant, C., Claes, F., De Smet, F., Thomas, J. L., Autiero, M., Carmeliet, P., *et al.* (2004). The Netrin receptor UNC5B mediates guidance events controlling morphogenesis of the vascular system. *Nature* **432,** 179–186.

Ly, N. P., Komatsuzaki, K., Fraser, I. P., Tseng, A. A., Prodhan, P., Moore, K. J., and Kinane, T. B. (2005). Netrin-1 inhibits leukocyte migration in vitro and in vivo. *Proc. Natl. Acad. Sci. USA* **102,** 14729–14734.

Lyashenko, N., Winter, M., Migliorini, D., Biechele, T., Moon, R. T., and Hartmann, C. (2011). Differential requirement for the dual functions of β-catenin in embryonic stem cell self-renewal and germ layer formation. *Nat. Cell Biol.* **13,** 753–761.

Ma, W., Shang, Y., Wei, Z., Wen, W., Wang, W., and Zhang, M. (2010). Phosphorylation of DCC by ERK2 is facilitated by direct docking of the receptor P1 domain to the kinase. *Structure* **18,** 1502–1511.

MacNeil, L. T., Hardy, W. R., Pawson, T., Wrana, J. L., and Culotti, J. G. (2009). UNC-129 regulates the balance between UNC-40 dependent and independent UNC-5 signaling pathways. *Nat. Neurosci.* **12,** 150–155.

Mancino, M., Esposito, C., Watanabe, K., Nagaoka, T., Gonzales, M., Bianco, C., Normanno, N., Salomon, D. S., and Strizzi, L. (2009). Neuronal guidance protein Netrin-1 induces differentiation in human embryonal carcinoma cells. *Cancer Res.* **69,** 1717–1721.

Mantovani, A., Allavena, P., Sica, A., and Balkwill, F. (2008). Cancer-related inflammation. *Nature* **454,** 436–444.

Markowitz, S. D., and Bertagnolli, M. M. (2009). Molecular origins of cancer: molecular basis of colorectal cancer. *N. Engl. J. Med.* **361,** 2449–2460.

Martín, M., Simon-Assmann, P., Kedinger, M., Martin, M., Mangeat, P., Real, F. X., and Fabre, M. (2006). DCC regulates cell adhesion in human colon cancer derived HT-29 cells and associates with ezrin. *Eur. J. Cell Biol.* **85,** 769–783.

Martin-Manso, G., Calzada, M. J., Chuman, Y., Sipes, J. M., Xavier, C. P., Wolf, V., Kuznetsova, S. A., Rubin, J. S., and Roberts, D. D. (2011). sFRP-1 binds via its Netrin-related motif to the N-module of thrombospondin-1 and blocks thrombospondin-1 stimulation of MDA-MB-231 breast carcinoma cell adhesion and migration. *Arch. Biochem. Biophys.* **509,** 147–156.

Massagué, J. (2008). TGFβ in cancer. *Cell* **134,** 215–230.

Mayer, E. L., and Krop, I. E. (2010). Advances in targeting SRC in the treatment of breast cancer and other solid malignancies. *Clin. Cancer Res.* **16,** 3526–3532.

Mayer, K., Hieronymus, T., Castrop, J., Clevers, H., and Ballhausen, W. G. (1997). Ectopic activation of lymphoid high mobility group-box transcription factor TCF-1 and overexpression in colorectal cancer cells. *Int. J. Cancer* **72,** 625–630.

Mazelin, L., Bernet, A., Bonod-Bidaud, C., Pays, L., Arnaud, S., Gespach, C., Bredesen, D. E., Scoazec, J. Y., and Mehlen, P. (2004). Netrin-1 controls colorectal tumorigenesis by regulating apoptosis. *Nature* **431,** 80–84.

Mazumdar, T., DeVecchio, J., Shi, T., Jones, J., Agyeman, A., and Houghton, J. A. (2011). Hedgehog signaling drives cellular survival in human colon carcinoma cells. *Cancer Res.* **71,** 1092–1102.

McKendry, R., Hsu, S. C., Harland, R. M., and Grosschedl, R. (1997). LEF-1/TCF proteins mediate wnt-inducible transcription from the Xenopus nodal-related 3 promoter. *Dev. Biol.* **192,** 420–431.

Mehlen, P., and Bredesen, D. E. (2004). The dependence receptor hypothesis. *Apoptosis* **9,** 37–49.

Mehlen, P., and Fearon, E. R. (2004). Role of the dependence receptor DCC in colorectal cancer pathogenesis. *J. Clin. Oncol.* **22,** 3420–3428.

Mehlen, P., Rabizadeh, S., Snipas, S. J., Assa-Munt, N., Salvesen, G. S., and Bredesen, D. E. (1998). The DCC gene product induces apoptosis by a mechanism requiring receptor proteolysis. *Nature* **395,** 801–804.

Mehlen, P., Delloye-Bourgeois, C., and Chédotal, A. (2011). Novel roles for Slits and Netrins: axon guidance cues as anticancer targets ? *Nat. Rev. Cancer* **11,** 188–197.

Meyerhardt, J. A., Caca, K., Eckstrand, B. C., Hu, G., Lengauer, C., Banavali, S., Look, A. T., and Fearon, E. R. (1999). Netrin-1: interaction with deleted in colorectal cancer (DCC) and alterations in brain tumors and neuroblastomas. *Cell Growth Differ.* **10,** 35–42.

Mille, F., Llambi, F., Guix, C., Delloye-Bourgeois, C., Guenebeaud, C., Castro-Obregon, S., Bredesen, D. E., Thibert, C., and Mehlen, P. (2009a). Interfering with multimerization of Netrin-1 receptors triggers tumor cell death. *Cell Death Differ.* **16,** 1344–1351.

Mille, F., Thibert, C., Fombonne, J., Rama, N., Guix, C., Hayashi, H., Corset, V., Reed, J. C., and Mehlen, P. (2009b). The Patched dependence receptor triggers apoptosis through a DRAL-caspase-9 complex. *Nat. Cell Biol.* **11,** 739–746.

Ming, G. L., Song, H. J., Berninger, B., Holt, C. E., Tessier-Lavigne, M., and Poo, M. M. (1997). cAMP-dependent growth cone guidance by Netrin-1. *Neuron* **19,** 1225–1235.

Mirakaj, V., Thix, C. A., Laucher, S., Mielke, C., Morote-Garcia, J. C., Schmit, M. A., Henes, J., Unertl, K. E., Köhler, D., and Rosenberger, P. (2010). Netrin-1 dampens pulmonary inflammation during acute lung injury. *Am. J. Respir. Crit. Care Med.* **181,** 815–824.

Mirakaj, V., Gatidou, D., Pötzsch, C., König, K., and Rosenberger, P. (2011a). Netrin-1 signaling dampens inflammatory peritonitis. *J. Immunol.* **186,** 549–555.

Mirakaj, V., Brown, S., Laucher, S., Steinl, C., Klein, G., Köhler, D., Skutella, T., Meisel, C., Brommer, B., Rosenberger, P., and Schwab, J. M. (2011b). Repulsive guidance molecule-A (RGM-A) inhibits leukocyte migration and mitigates inflammation. *Proc. Natl. Acad. Sci. USA* **108,** 6555–6560.

Mogensen, M. M., Tucker, J. B., Mackie, J. B., Prescott, A. R., and Nathke, I. S. (2002). The adenomatous polyposis coli protein unambiguously localizes to microtubule plus ends and is involved in establishing parallel arrays of microtubule bundles in highly polarized epithelial cells. *J. Cell Biol.* **157,** 1041–1048.

Montell, D. J. (1999). The genetics of cell migration in Drosophila melanogaster and *Caenorhabditis elegans* development. *Development* **126,** 3035–3046.

Moore, S. W., Tessier-Lavigne, M., and Kennedy, T. E. (2007). Netrins and their receptors. *Adv. Exp. Med. Biol.* **621,** 17–31.

Moraes, R. C., Zhang, X., Harrington, N., Fung, J. Y., Wu, M. F., Hilsenbeck, S. G., Allred, D. C., and Lewis, M. T. (2007). Constitutive activation of smoothened (SMO) in mammary glands of transgenic mice leads to increased proliferation, altered differentiation and ductal dysplasia. *Development* **134,** 1231–1242.

Morin, P. J., Sparks, A. B., Korinek, V., Barker, N., Clevers, H., Vogelstein, B., and Kinzler, K. W. (1997). Activation of β-catenin-Tcf signaling in colon cancer by mutations in β-catenin or APC. *Science* **275,** 1787–1790.

Morris, E. J., Ji, J. Y., Yang, F., Di Stefano, L., Herr, A., Moon, N. S., Kwon, E. J., Haigis, K. M., Näär, A. M., and Dyson, N. J. (2008). E2F1 represses β-catenin transcription and is antagonized by both pRB and CDK8. *Nature* **455,** 552–556.

Murase, S., and Horwitz, A. F. (2002). Deleted in colorectal carcinoma and differentially expressed integrins mediate the directional migration of neural precursors in the rostral migratory stream. *J. Neurosci.* **22,** 3568–3579.

Murray, A. J., Tucker, S. J., and Shewan, D. A. (2009). cAMP-dependent axon guidance is distinctly regulated by Epac and protein kinase A. *J. Neurosci.* **29,** 15434–15444.

Nacht, M., St Martin, T. B., Byrne, A., Klinger, K. W., Teicher, B. A., Madden, S. L., and Jiang, Y. (2009). Netrin-4 regulates angiogenic responses and tumor cell growth. *Exp. Cell Res.* **315,** 784–794.

Nagaprashantha, L. D., Vatsyayan, R., Lelsani, P. C., Awasthi, S., and Singhal, S. S. (2011). The sensors and regulators of cell-matrix surveillance in anoikis resistance of tumors. *Int. J. Cancer* **128,** 743–752.

Narayan, S., and Roy, D. (2003). Role of APC and DNA mismatch repair genes in the development of colorectal cancers. *Mol. Cancer* **2,** 41–56.

Narayan, S., Jaiswal, A. S., and Balusu, R. (2005). Tumor suppressor APC blocks DNA polymerase β-dependent strand displacement synthesis during long patch but not short patch base excision repair and increases sensitivity to methylmethane sulfonate. *J. Biol. Chem.* **280,** 6942–6949.

Nasarre, P., Potiron, V., Drabkin, H., and Roche, J. (2010). Guidance molecules in lung cancer. *Cell Adh. Migr.* **4,** 130–145.

Näthke, I. S. (2004). The adenomatous polyposis coli protein: the Achilles heel of the gut epithelium. *Annu. Rev. Cell Dev. Biol.* **20,** 337–366.

Nguyen, Q. D., De Wever, O., Bruyneel, E., Hendrix, A., Xie, W. Z., Lombet, A., Leibl, M., Mareel, M., Gieseler, F., Bracke, M., and Gespach, C. (2005a). Commutators of PAR-1 signaling in cancer cell invasion reveal an essential role of the Rho-ROK axis and tumor microenvironment. *Oncogene* **24,** 8240–8251.

Nguyen, Q. D., Faivre, S., Bruyneel, E., Rivat, C., Seto, M., Endo, T., Mareel, M., Emami, S., and Gespach, C. (2005b). RhoA- and RhoD-dependent regulatory switch of Gα subunit signaling by PAR-1 receptors in cellular invasion. *FASEB J.* **16,** 565–576.

Nguyen, Q. D., Rodrigues, S., Rodrigue, C., Rivat, C., Grijelmo, C., Bruyneel, E., Emami, S., Attoub, S., and Gespach, C. (2006). Inhibition of VEGF-165 and Semaphorin 3A-mediated cellular invasion and tumor growth by the VEGF signaling inhibitor ZD4190 in human colon cancer cells and xenografts. *Mol. Cancer Ther.* **5,** 2070–2071.

Nicol, X., Hong, K. P., and Spitzer, N. C. (2011). Spatial and temporal second messenger codes for growth cone turning. *Proc. Natl. Acad. Sci. USA* **108,** 13776–13781.

Nikolopoulos, S. N., and Giancotti, F. G. (2005). Netrin-integrin signaling in epithelial morphogenesis, axon guidance and vascular patterning. *Cell Cycle* **4,** e131–e135.

Noubissi, F. K., Goswami, S., Sanek, N. A., Kawakami, K., Minamoto, T., Moser, A., Grinblat, Y., and Spiegelman, V. S. (2009). Wnt signaling stimulates transcriptional outcome of the Hedgehog pathway by stabilizing GLI1 mRNA. *Cancer Res.* **69,** 8572–8578.

Oberst, A., Dillon, C. P., Weinlich, R., McCormick, L. L., Fitzgerald, P., Pop, C., Hakem, R., Salvesen, G. S., and Green, D. R. (2011). Catalytic activity of the caspase-8-FLIP(L) complex inhibits RIPK3-dependent necrosis. *Nature* **471,** 363–367.

Ogden, S. K., Fei, D. L., Schilling, N. S., Ahmed, Y. F., Hwa, J., and Robbins, D. J. (2008). G protein Gαi functions immediately downstream of Smoothened in Hedgehog signalling. *Nature* **456,** 967–970.

Oh, W. J., and Jacinto, E. (2011). mTOR complex 2 signaling and functions. *Cell Cycle* **10,** 20305–22316.

Ohgaki, H., Kim, Y. H., and Steinbach, J. P. (2010). Nervous system tumors associated with familial tumor syndromes. *Curr. Opin. Neurol.* **23,** 583–591.

Okamura, Y., Kohmura, E., and Yamashita, T. (2011). TACE cleaves neogenin to desensitize cortical neurons to the repulsive guidance molecule. *Neurosci. Res.* **71,** 63–70.

Oku, Y., Shimoji, T., Takifuji, K., Hotta, T., Yokoyama, S., Matsuda, K., Higashiguchi, T., Tominaga, T., Nasu, T., Tamura, K., Matsuura, M., Miyata, S., *et al.* (2008). Identification of the molecular mechanisms for dedifferentiation at the invasion front of colorectal cancer by a gene expression analysis. *Clin. Cancer Res.* **14,** 7215–7722.

Oniscu, A., Jame, R. M., Morris, R. G., Bader, S., Malcomson, R. D., and Harrison, D. J. (2004). Expression of Sonic hedgehog pathway genes is altered in colonic neoplasia. *J. Pathol.* **203,** 909–917.

Osorio, N., and Delmas, P. (2010). Patch clamp recording from enteric neurons *in situ. Nat. Protoc.* **6,** 15–27.

Padua, D., and Massagué, J. (2009). Roles of TGFβ in metastasis. *Cell Res.* **19,** 89–102.

Palu, E., and Liesi, P. (2002). Differential distribution of laminins in Alzheimer disease and normal human brain tissue. *J. Neurosci. Res.* **69,** 243–256.

Pan, Y., Liu, G., Fang, M., Shen, L., Wang, L., Han, Y., Shen, D., and Wang, X. (2010). Abnormal expression of Netrin-G2 in temporal lobe epilepsy neurons in humans and a rat model. *Exp. Neurol.* **224,** 340–346.

Pandey, A., Duan, H., Di Fiore, P. P., and Dixit, V. M. (1995). The Ret receptor protein tyrosine kinase associates with the SH2-containing adapter protein Grb10. *J. Biol. Chem.* **270,** 21461–21463.

Papageorgis, P., Cheng, K., Ozturk, S., Gong, Y., Lambert, A. W., Abdolmaleky, H. M., Zhou, J. R., and Thiagalingam, S. (2011). Smad4 inactivation promotes malignancy and drug resistance of colon cancer. *Cancer Res.* **71,** 998–1008.

Papanastasiou, A. D., Pampalakis, G., Katsaros, D., and Sotiropoulou, G. (2011). Netrin-1 overexpression is predictive of ovarian malignancies. *Oncotarget* **2,** 363–367.

Paradisi, A., Maisse, C., Bernet, A., Coissieux, M. M., Maccarrone, M., Scoazec, J. Y., and Mehlen, P. (2008). NF-kappaB regulates Netrin-1 expression and affects the conditional tumor suppressive activity of the Netrin-1 receptors. *Gastroenterology* **135,** 1248–1257.

Paradisi, A., Maisse, C., Coissieux, M. M., Gadot, N., Lépinasse, F., Delloye-Bourgeois, C., Delcros, J. G., Svrcek, M., Neufert, C., Fléjou, J. F., Scoazec, J. Y., and Mehlen, P. (2009). Netrin-1 up-regulation in inflammatory bowel diseases is required for colorectal cancer progression. *Proc. Natl. Acad. Sci. USA* **106,** 17146–17151.

Park, K. W., Crouse, D., Lee, M., Karnik, S. K., Sorensen, L. K., Murphy, K. J., Kuo, C. J., and Li, D. Y. (2004). The axonal attractant Netrin-1 is an angiogenic factor. *Proc. Natl. Acad. Sci. USA* **101,** 16210–16215.

Park, H. R., Kim, S. H., Lee, S. Y., Sung, J. M., Park, A. R., Kang, S., Kim, B. G., Choi, Y. P., Kim, Y. B., and Cho, N. H. (2011). Nuclear localization of Nm23-H1 in head and neck squamous cell carcinoma is associated with radiation resistance. *Cancer* **117,** 1864–1873.

Peng, Y. J., He, W. Q., Tang, J., Tao, T., Chen, C., Gao, Y. Q., Zhang, W. C., He, X. Y., Dai, Y. Y., Zhu, N. C., Lv, N., Zhang, C. H., *et al.* (2010). Trio is a key guanine nucleotide exchange factor coordinating regulation of the migration and morphogenesis of granule cells in the developing cerebellum. *J. Biol. Chem.* **285,** 24834–24844.

Petrie, R. J., Zhao, B., Bedford, F., and Lamarche-Vane, N. (2009). Compartmentalized DCC signalling is distinct from DCC localized to lipid rafts. *Biol. Cell* **101,** 77–90.

Peukert, S., and Miller-Moslin, K. (2010). Small-molecule inhibitors of the hedgehog signaling pathway as cancer therapeutics. *ChemMedChem* **5,** 500–512.

Phan, H., Boissard, C., Pessah, M., Regnauld, K., Emami, S., Gespach, C., and Rosselin, G. (2000). Decreased ADP-Ribosylation of the Gαolf and Gαs subunits by high glucose in pancreatic B-cells. *Biochem. Biophys. Res. Commun.* **271,** 86–90.

Philipp-Staheli, J., Kim, K. H., Payne, S. R., Gurley, K. E., Liggitt, D., Longton, G., and Kemp, C. J. (2002). Pathway-specific tumor suppression. Reduction of p27 accelerates gastrointestinal tumorigenesis in Apc mutant mice, but not in Smad3 mutant mice. *Cancer Cell* **1,** 355–368.

Picard, M., Petrie, R. J., Antoine-Bertrand, J., Saint-Cyr-Proulx, E., Villemure, J. F., and Lamarche-Vane, N. (2009). Spatial and temporal activation of the small GTPases RhoA and Rac1 by the Netrin-1 receptor UNC5a during neurite outgrowth. *Cell. Signal.* **21,** 1961–1973.

Pino, M. S., and Chung, D. C. (2010). The chromosomal instability pathway in colon cancer. *Gastroenterology* **138,** 2059–2072.

Plevy, S. E., and Targan, S. R. (2011). Future therapeutic approaches for inflammatory bowel diseases. *Gastroenterology* **140,** 1838–1846.

Pohl, M., Radacz, Y., Pawlik, N., Schoeneck, A., Baldus, S. E., Munding, J., Schmiegel, W., Schwarte-Waldhoff, I., and Reinacher-Schick, A. (2010). SMAD4 mediates mesenchymal-epithelial reversion in SW480 colon carcinoma cells. *Anticancer Res.* **30,** 2603–2613.

Polakis, P. (1997). The adenomatous polyposis coli (APC) tumor suppressor. *Biochim. Biophys. Acta* **1332,** F127–F147.

Porcionatto, M. A. (2006). The extracellular matrix provides directional cues for neuronal migration during cerebellar development. *Braz. J. Med. Biol. Res.* **39,** 313–320.

Porte, H., Chastre, E., Prévot, S., Nordlinger, B., Empereur, S., Basset, P., Chambon, P., and Gespach, C. (1995). Neoplastic progression of human colorectal cancer is associated with overexpression of the ST3 and BM-40/SPARC genes. *Int. J. Cancer* **64,** 1–6.

Prantl, L., Muehlberg, F., Navone, N. M., Song, Y. H., Vykoukal, J., Logothetis, C. J., and Alt, E. U. (2010). Adipose tissue-derived stem cells promote prostate tumor growth. *Prostate* **70,** 1709–1715.

Prawitt, D., Enklaar, T., Gartner-Rupprecht, B., Spangenberg, C., Oswald, M., Lausch, E., Schmidtke, P., Reutzel, D., Fees, S., Lucito, R., Korzon, M., Brozek, I., *et al.* (2005). Microdeletion of target sites for insulator protein CTCF in a chromosome 11p15 imprinting center in Beckwith-Wiedemann syndrome and Wilms' tumor. *Proc. Natl. Acad. Sci. USA* **102,** 4085–4090.

Prévost, G. P., Lonchampt, M. O., Holbeck, S., Attoub, S., Zaharevitz, D., Alley, M., Wright, J., Brezak, M. C., Coulomb, H., Savola, A., Huchet, M., Chaumeron, S., *et al.* (2006). Anticancer activity of BIM-46174, a new inhibitor of the heterotrimeric Gα/Gβγ complex. *Cancer Res.* **66,** 9227–9234.

Prunier, C., Mazars, A., Noë, V., Bruyneel, E., Mareel, M., Gespach, C., and Atfi, A. (1999). Evidence that Smad2 is a tumor suppressor implicated in the control of cellular invasion. *J. Biol. Chem.* **274,** 22919–22922.

Qiang, R., Wang, F., Shi, L. Y., Liu, M., Chen, S., Wan, H. Y., Li, Y. X., Li, X., Gao, S. Y., Sun, B. C., and Tang, H. (2011). Plexin-B1 is a target of miR-214 in cervical cancer and promotes the growth and invasion of HeLa cells. *Int. J. Biochem. Cell Biol.* **43,** 632–641.

Qian-Hua, W., Shao-Ping, Z., Jian-Wen, Z., Yun, Y., and Li, Z. (2011). Reduced expression of Netrin-1 is associated with fetal growth restriction. *Mol. Cell. Biochem.* **350,** 81–87.

Qualtrough, D., Buda, A., Gaffield, W., Williams, A. C., and Paraskeva, C. (2004). Hedgehog signalling in colorectal tumour cells: induction of apoptosis with cyclopamine treatment. *Int. J. Cancer* **110,** 831–837.

Quinn, C. C., and Wadsworth, W. G. (2008). Axon guidance: asymmetric signaling orients polarized outgrowth. *Trends Cell Biol.* **18,** 597–603.

Quinn, C. C., Pfeil, D. S., Chen, E., Stovall, E. L., Harden, M. V., Gavin, M. K., Forrester, W. C., Ryder, E. F., Soto, M. C., and Wadsworth, W. G. (2006). UNC-6/Netrin and SLT-1/slit guidance cues orient axon outgrowth mediated by MIG-10/RIAM/lamellipodin. *Curr. Biol.* **16,** 845–853.

Quinn, C. C., Pfeil, D. S., and Wadsworth, W. G. (2008). CED-10/Rac1 mediates axon guidance by regulating the asymmetric distribution of MIG-10/lamellipodin. *Curr. Biol.* **18,** 808–813.

Rajasekharan, S., and Kennedy, T. E. (2009). The Netrin protein family. *Genome Biol.* **10,** 239.

Ramesh, G., Berg, A., and Jayakumar, C. (2011). Plasma Netrin-1 is a diagnostic biomarker of human cancers. *Biomarkers* **16,** 172–180.

Ratcliffe, E. M. (2011). Molecular development of the extrinsic sensory innervation of the gastrointestinal tract. *Auton. Neurosci.* **161,** 1–5.

Ratcliffe, E. M., Setru, S. U., Chen, J. J., Li, Z. S., D'Autréaux, F., and Gershon, M. D. (2006). Netrin/DCC-mediated attraction of vagal sensory axons to the fetal mouse gut. *J. Comp. Neurol.* **498,** 567–580.

Ratcliffe, E. M., D'Autréaux, F., and Gershon, M. D. (2008). Laminin terminates the Netrin/DCC mediated attraction of vagal sensory axons. *Dev. Neurobiol.* **68,** 960–971.

Ratcliffe, E. M., Fan, L., Mohammed, T. J., Anderson, M., Chalazonitis, A., and Gershon, M. D. (2011a). Enteric neurons synthesize Netrins and are essential for the development of the vagal sensory innervation of the fetal gut. *Dev. Neurobiol.* **71,** 362–373.

Ratcliffe, E. M., Farrar, N. R., and Fox, E. A. (2011b). Development of the vagal innervation of the gut: steering the wandering nerve. *Neurogastroenterol. Motil.* **23,** 898–911.

Raymond, E., Buquet-Fagot, C., Djelloul, S., Mester, J., Cvitkovic, E., Allain, P., Louvet, C., and Gespach, C. (1997). Antitumor activity of oxaliplatin in combination with 5-fluorouracil and the thymidylate synthase inhibitor AG337 in human colon, breast and ovarian cancers. *Anticancer Drugs* **8,** 876–885.

Reale, M. A., Hu, G., Zafar, A. I., Getzenberg, R. H., Levine, S. M., and Fearon, E. R. (1994). Expression and alternative splicing of the deleted in colorectal cancer (DCC) gene in normal and malignant tissues. *Cancer Res.* **54,** 4493–4501.

Reeves, W. B., Kwon, O., and Ramesh, G. (2008). Netrin-1 and kidney injury. II. Netrin-1 is an early biomarker of acute kidney injury. *Am. J. Physiol. Renal Physiol.* **294,** F731–F738.

Régnauld, K., Nguyen, Q. D., Vakaet, L., Bruyneel, E., Launay, J. M., Endo, T., Mareel, M., Gespach, C., and Emami, S. (2002). G-protein α olf subunit promotes cellular invasion, survival, and neuroendocrine differentiation in digestive and urogenital epithelial cells. *Oncogene* **21,** 4020–4031.

Reichling, T., Goss, K. H., Carson, D. J., Holdcraft, R. W., Ley-Ebert, C., Witte, D., Aronow, B. J., and Groden, J. (2005). Transcriptional profiles of intestinal tumors in Apc (Min) mice are unique from those of embryonic intestine and identify novel gene targets dysregulated in human colorectal tumors. *Cancer Res.* **65,** 166–176.

Ren, X. R., Ming, G. L., Xie, Y., Hong, Y., Sun, D. M., Zhao, Z. Q., Feng, Z., Wang, Q., Shim, S., Chen, Z. F., Song, H. J., Mei, L., *et al.* (2004). Focal adhesion kinase in Netrin-1 signaling. *Nat. Neurosci.* **7,** 1204–1212.

Ren, X. R., Hong, Y., Feng, Z., Yang, H. M., Mei, L., and Xiong, W. C. (2008). Tyrosine phosphorylation of Netrin receptors in Netrin-1 signaling. *Neurosignals* **16,** 235–245.

Riese, J., Yu, X., Munnerlyn, A., Eresh, S., Hsu, S. C., Grosschedl, R., and Bienz, M. (1997). LEF-1, a nuclear factor coordinating signaling inputs from wingless and decapentaplegic. *Cell* **88,** 777–787.

Rivat, C., LeFloch, N., Sabbah, M., Redeuilh, G., Bruyneel, E., Mareel, M., Matrisian, L., Crowford, H., Gespach, C., and Attoub, S. (2003). Synergistic cooperation between the AP-1 and LEF-1 transcription factors in the activation of the matrilysin promoter by the src oncogene. Consequences on tumor cell invasion. *FASEB J.* **17,** 1721–1723.

Rivat, C., Rodrigues, S., Bruyneel, E., Piétu, G., Robert, A., Redeuilh, G., Brake, M., Gespach, C., and Attoub, S. (2005). Implication of STAT3 signaling in human colonic cancer cells during Intestinal trefoil factor (TFF3) and VEGF-mediated cellular invasion and tumor growth. *Cancer Res.* **65,** 195–202.

Rodrigues, S., Nguyen, Q. D., Faivre, S., Bruyneel, E., Thim, L., Westley, B., May, F., Flatau, G., Mareel, M., Gespach, C., and Emami, S. (2001). Activation of cellular invasion by trefoil peptides and src is mediated by cyclooxygenase- and thromboxane A2 receptor-dependent signaling pathways. *FASEB J.* **15,** 1517–1528.

Rodrigues, S., Van Aken, E., Van Bocxlaer, S., Attoub, S., Nguyen, Q. D., Bruyneel, E., Westley, B., May, F., Thim, L., Mareel, M., Gespach, C., and Emami, S. (2003a). Trefoil peptides as proangiogenic factors *in vivo* and *in vitro*: implication of cyclooxygenase-2 and EGF receptor signaling. *FASEB J.* **17,** 7–16.

Rodrigues, S., Attoub, S., Nguyen, Q. D., Bruyneel, E., Rodrigue, C., Westley, B., May, F. E. B., Thim, L., Mareel, M., Emami, S., and Gespach, C. (2003b). Selective abrogation of the proinvasive activity of the trefoil peptides pS2 and spasmolytic polypeptide by disruption of the EGF receptor signaling pathways in kidney and colonic cancer cells. *Oncogene* **222,** 4488–4497.

Rodrigues, S., Bruyneel, E., Rodrigue, C., Emami, S., and et Gespach, C. (2004). COX-2 et carcinogenèse. *Bull. Cancer* **91,** S59–S76.

Rodrigues, S., De Wever, O., Bruyneel, E., Rooney, R. J., and Gespach, C. (2007). Opposing roles of Netrin-1 and the dependence receptor DCC in cancer cell invasion, tumor growth and metastasis. *Oncogene* **26,** 5615–5625.

Roig, A. I., Eskiocak, U., Hight, S. K., Kim, S. B., Delgado, O., Souza, R. F., Spechler, S. J., Wright, W. E., and Shay, J. W. (2010). Immortalized epithelial cells derived from human colon biopsies express stem cell markers and differentiate in vitro. *Gastroenterology* **138,** 1012–1021.

Roperch, J. P., El Ouadrani, K., Hendrix, A., Emami, S., De Wever, O., Melino, G., and Gespach, C. (2008). Netrin-1 induces apoptosis in human cervical tumor cells via the TAp73alpha tumor suppressor. *Cancer Res.* **68,** 8231–8239.

Rosenberger, P., Schwab, J. M., Mirakaj, V., Masekowsky, E., Mager, A., Morote-Garcia, J. C., Unertl, K., and Eltzschig, H. K. (2009). Hypoxia-inducible factor-dependent induction of Netrin-1 dampens inflammation caused by hypoxia. *Nat. Immunol.* **10,** 195–202.

Rosin-Arbesfeld, R., Cliffe, A., Brabletz, T., and Bienz, M. (2003). Nuclear export of the APC tumour suppressor controls β-catenin function in transcription. *EMBO J.* **22,** 1101–1113.

Rowland, B. D., and Bernards, R. (2006). Re-evaluating cell-cycle regulation by E2Fs. *Cell* **127,** 871–874.

Sabatier, C., Plump, A. S., Le, M.a., Brose, K., Tamada, A., Murakami, F., Lee, E. Y., and Tessier-Lavigne, M. (2004). The divergent Robo family protein rig-1/Robo3 is a negative regulator of slit responsiveness required for midline crossing by commissural axons. *Cell* **117,** 157–169.

Sabbah, M., Emami, S., Redeuilh, G., Julien, S., Prévost, G., Zimber, A., Ouelaa, R., Bracke, M., De Wever, O., and Gespach, C. (2008). Molecular signatures and therapeutic perspective of the epithelial-mesenchymal transitions in human epithelial cancers. *Drug Resist. Updat.* **11,** 152–163.

Sabbah, M., Prunier, C., Ferrand, N., Megalophonos, V., Lambein, K., De Wever, O., Nazaret, N., Lachuer, J., Dumont, S., and Redeuilh, G. (2011). CCN5, a novel transcriptional repressor of the transforming growth factor β signaling pathway. *Mol. Cell. Biol.* **31,** 1459–1469.

Sahin, M. (2010). Eph receptor and mTOR pathway crosstalk: implications for cancer. *Cell Cycle* **9,** 2053–2054.

Sakatani, T., Kaneda, A., Iacobuzio-Donahue, C. A., Carter, M. G., Witzel, S. D., Okano, H., Ko, M. S., Ohlsson, R., Longo, D. L., and Feinberg, A. P. (2005). Loss of imprinting of Igf2 alters intestinal maturation and tumorigenesis in mice. *Science* **307,** 1976–1978.

Salehi, A. H., Roux, P. P., Kubu, C. J., Zeindler, C., Bhakar, A., Tannis, L. L., Verdi, J. M., and Barker, P. A. (2000). NRAGE, a novel MAGE protein, interacts with the p75 neurotrophin receptor and facilitates nerve growth factor-dependent apoptosis. *Neuron* **27,** 279–288.

Salehi, A. H., Xanthoudakis, S., and Barker, P. A. (2002). NRAGE, a p75 neurotrophin receptor-interacting protein, induces caspase activation and cell death through a JNK-dependent mitochondrial pathway. *J. Biol. Chem.* **277,** 48043–48050.

Salinas, P. C. (2003). The morphogen sonic hedgehog collaborates with Netrin-1 to guide axons in the spinal cord. *Trends Neurosci.* **26,** 641–643.

Sanchez-Ruderisch, H., Detjen, K. M., Welzel, M., André, S., Fischer, C., Gabius, H. J., and Rosewicz, S. (2010). Galectin-1 sensitizes carcinoma cells to anoikis *via* the fibronectin receptor $\alpha(5)\beta(1)$-integrin. *Cell Death Differ.* **18,** 806–816.

Sancho, E., Batlle, E., and Clevers, H. (2004). Signaling pathways in intestinal development and cancer. *Annu. Rev. Cell Dev. Biol.* **20,** 695–723.

Saridaki, Z., Tzardi, M., Papadaki, C., Sfakianaki, M., Pega, F., Kalikaki, A., Tsakalaki, E., Trypaki, M., Messaritakis, I., Stathopoulos, E., Mavroudis, D., Georgoulias, V., *et al.* (2011). Impact of KRAS, BRAF, PIK3CA mutations, PTEN, AREG, EREG expression and skin rash in $\geq$2 line cetuximab-based therapy of colorectal cancer patients. *PLoS One* **6,** e15980.

Sawai, H., Okada, Y., Funahashi, H., Matsuo, Y., Takahashi, H., Takeyama, H., and Manabe, T. (2005). Activation of focal adhesion kinase enhances the adhesion and invasion of pancreatic cancer cells via extracellular signal-regulated kinase-1/2 signaling pathway activation. *Mol. Cancer* **4,** 37.

Sayed, D., and Abdellatif, M. (2011). MicroRNAs in development and disease. *Physiol. Rev.* **91,** 827–887.

Schneiders, F. I., Maertens, B., Böse, K., Li, Y., Brunken, W. J., Paulsson, M., Smyth, N., and Koch, M. (2007). Binding of Netrin-4 to laminin short arms regulates basement membrane assembly. *J. Biol. Chem.* **282,** 23750–23758.

Schubert, T., Kaufmann, S., Wenke, A. K., Grässel, S., and Bosserhoff, A. K. (2009). Role of deleted in colon carcinoma in osteoarthritis and in chondrocyte migration. *Rheumatology (Oxford)* **48,** 1435–1441.

Schuchardt, A., D'Agati, V., Larsson-Blomberg, L., Costantini, F., and Pachnis, V. (1994). Defects in the kidney and enteric nervous system of mice lacking the tyrosine kinase receptor Ret. *Nature* **367,** 380–383.

Schwarting, G. A., Raitcheva, D., Bless, E. P., Ackerman, S. L., and Tobet, S. (2004). Netrin 1-mediated chemoattraction regulates the migratory pathway of LHRH neurons. *Eur. J. Neurosci.* **19,** 11–20.

Seaman, C., Anderson, R., Emery, B., and Cooper, H. M. (2001). Localization of the Netrin guidance receptor, DCC, in the developing peripheral and enteric nervous systems. *Mech. Dev.* **103,** 173–175.

Sebbagh, M., Renvoizé, C., Hamelin, J., Riché, N., Bertoglio, J., and Bréard, J. (2001). Caspase-3-mediated cleavage of ROCK I induces MLC phosphorylation and apoptotic membrane blebbing. *Nat. Cell Biol.* **3,** 346–352.

Sengupta, S., Peterson, T. R., and Sabatini, D. M. (2010). Regulation of the mTOR complex 1 pathway by nutrients, growth factors, and stress. *Mol. Cell* **40,** 310–322.

Sentürk, A., Pfennig, S., Weiss, A., Burk, K., and Acker-Palmer, A. (2011). Ephrin Bs are essential components of the Reelin pathway to regulate neuronal migration. *Nature* **472,** 356–360.

Sequist, L. V., Waltman, B. A., Dias-Santagata, D., Digumarthy, S., Turke, A. B., Fidias, P., Bergethon, K., Shaw, A. T., Gettinger, S., Cosper, A. K., Akhavanfard, S., Heist, R. S., *et al.* (2011). Genotypic and histological evolution of lung cancers acquiring resistance to EGFR inhibitors. *Sci. Transl. Med.* **3,** 75ra26.

Serafini, T., Kennedy, T. E., Galko, M. J., Mirzayan, C., Jessell, T. M., and Tessier-Lavigne, M. (1994). The etrins define a family of axon outgrowth-promoting proteins homologous to *C. elegans* UNC-6. *Cell* **78,** 409–424.

Shekarabi, M., and Kennedy, T. E. (2002). The Netrin-1 receptor DCC promotes filopodia formation and cell spreading by activating Cdc42 and Rac1. *Mol. Cell. Neurosci.* **19**, 1–17.

Shekarabi, M., Moore, S. W., Tritsch, N. X., Morris, S. J., Bouchard, J. F., and Kennedy, T. E. (2005). Deleted in colorectal cancer binding Netrin-1 mediates cell substrate adhesion and recruits Cdc42, Rac1, Pak1, and N-WASP into an intracellular signaling complex that promotes growth cone expansion. *J. Neurosci.* **25**, 3132–3141.

Shin, S. K., Nagasaka, T., Jung, B. H., Matsubara, N., Kim, W. H., Carethers, J. M., Boland, C. R., and Goel, A. (2007). Epigenetic and genetic alterations in Netrin-1 receptors UNC5C and DCC in human colon cancer. *Gastroenterology* **133**, 1849–1857.

Shinagawa, K., Kitadai, Y., Tanaka, M., Sumida, T., Kodama, M., Higashi, Y., Tanaka, S., Yasui, W., and Chayama, K. (2010). Mesenchymal stem cells enhance growth and metastasis of colon cancer. *Int. J. Cancer* **127**, 2323–2333.

Sieber, O. M., Howarth, K. M., Thirlwell, C., Rowan, A., Mandir, N., Goodlad, R. A., Gilkar, A., Spencer-Dene, B., Stamp, G., Johnson, V., Silver, A., Yang, H., *et al.* (2004). Myh deficiency enhances intestinal tumorigenesis in multiple intestinal neoplasia (ApcMin/+) mice. *Cancer Res.* **64**, 8876–8881.

Silhankova, M., and Korswagen, H. C. (2007). Migration of neuronal cells along the anterior-posterior body axis of *C. elegans*: Wnts are in control. *Curr. Opin. Genet. Dev.* **17**, 320–325.

Sin, S., Bonin, F., Petit, V., Meseure, D., Lallemand, F., Bièche, I., Bellahcène, A., Castronovo, V., De Wever, O., Gespach, C., Lidereau, R., and Driouch, K. (2011). Role of the focal adhesion protein Kindlin-1 in breast cancer growth and lung metastasis. *J. Natl. Cancer Inst.* **103**, 1323–1337.

Skarnes, W. C., Moss, J. E., Hurtley, S. M., and Beddington, R. S. (1995). Capturing genes encoding membrane and secreted proteins important for mouse development. *Proc. Natl. Acad. Sci. USA* **92**, 6592–6596.

Slorach, E. M., and Werb, Z. (2003). Epithelial morphogenesis: Netrin comes to a sticky and terminal end. *Curr. Biol.* **13**, R491–R493.

Small, E. M., Sutherland, L. B., Rajagopalan, K. N., Wang, S., and Olson, E. N. (2010). MicroRNA-218 regulates vascular patterning by modulation of Slit-Robo signaling. *Circ. Res.* **107**, 1336–1344.

Sodir, N. M., Chen, X., Park, R., Nickel, A. E., Conti, P. S., Moats, R., Bading, J. R., Shibata, D., and Laird, P. W. (2006). Smad3 deficiency promotes tumorigenesis in the distal colon of ApcMin/+ mice. *Cancer Res.* **66**, 8430–8438.

Sonoshita, M., Takaku, K., Sasaki, N., Sugimoto, Y., Ushikubi, F., Narumiya, S., Oshima, M., and Taketo, M. M. (2001). Acceleration of intestinal polyposis through prostaglandin receptor EP2 in Apc(Delta 716) knockout mice. *Nat. Med.* **7**, 1048–1051.

Sonoshita, M., Aoki, M., Fuwa, H., Aoki, K., Hosogi, H., Sakai, Y., Hashida, H., Takabayashi, A., Sasaki, M., Robine, S., Itoh, K., Yoshioka, K., *et al.* (2011). Suppression of colon cancer metastasis by Aes through inhibition of Notch signaling. *Cancer Cell* **19**, 125–137.

Souazé, F., Toy-Miou-Leong, M., Viarddot-Foucault, V., Roullet, N., Bruyneel, E., Compera, E., Gompel, A., Rostène, W., Fléjou, J. F., Mareel, M., Gespach, C., and Forgez, P. (2006). Activation of the neurotensin NT1 receptor gene by the Tcf/β-catenin pathway as a proinvasive signal in human colonic tumors. *Carcinogenesis* **27**, 708–716.

Srinivasan, K., Strickland, P., Valdes, A., Shin, G. C., and Hinck, L. (2003). Netrin-1/neogenin interaction stabilizes multipotent progenitor cap cells during mammary gland morphogenesis. *Dev. Cell* **4**, 371–382.

Stanco, A., Szekeres, C., Patel, N., Rao, S., Campbell, K., Kreidberg, J. A., Polleux, F., and Anton, E. S. (2009). Netrin-1-alpha3beta1 integrin interactions regulate the migration of interneurons through the cortical marginal zone. *Proc. Natl. Acad. Sci. USA* **106**, 7595–7600.

Staquicini, F. I., Dias-Neto, E., Li, J., Snyder, E. Y., Sidman, R. L., Pasqualini, R., and Arap, W. (2009). Discovery of a functional protein complex of Netrin-4, laminin gamma1 chain, and integrin alpha6beta1 in mouse neural stem cells. *Proc. Natl. Acad. Sci. USA* **106,** 2903–2908.

Steeg, P. S., Bevilacqua, G., Kopper, L., Thorgeirsson, U. P., Talmadge, J. E., Liotta, L. A., and Sobel, M. E. (1988). Evidence for a novel gene associated with low tumor metastatic potential. *J. Natl. Cancer Inst.* **80,** 200–204.

Stein, E., Zou, Y., Poo, M., and Tessier-Lavigne, M. (2001). Binding of DCC by Netrin-1 to mediate axon guidance independent of adenosine A2B receptor activation. *Science* **291,** 1976–1982.

Sträter, J., Wedding, U., Barth, T. F., Koretz, K., Elsing, C., and Möller, P. (1996). Rapid onset of apoptosis in vitro follows disruption of beta 1-integrin/matrix interactions in human colonic crypt cells. *Gastroenterology* **110,** 1776–1784.

Stricker, T., Catenacci, D. V., and Seiwert, T. Y. (2011). Molecular profiling of cancer-the future of personalized cancer medicine: a primer on cancer biology and the tools necessary to bring molecular testing to the clinic. *Semin. Oncol.* **38,** 173–185.

Strickland, P., Shin, G. C., Plump, A., Tessier-Lavigne, M., and Hinck, L. (2006). Slit2 and Netrin 1 act synergistically as adhesive cues to generate tubular bi-layers during ductal morphogenesis. *Development* **133,** 823–832.

Strizzi, L., Bianco, C., Raafat, A., Abdallah, W., Chang, C., Raafat, D., Hirota, M., Hamada, S., Sun, Y., Normanno, N., Callahan, R., Hinck, L., *et al.* (2005). Netrin-1 regulates invasion and migration of mouse mammary epithelial cells overexpressing Cripto-1 in vitro and in vivo. *J. Cell Sci.* **118,** 4633–4643.

Su, J. L., Chen, P. B., Chen, Y. H., Chen, S. C., Chang, Y. W., Jan, Y. H., Cheng, X., Hsiao, M., and Hung, M. C. (2010). Downregulation of microRNA miR-520h by E1A contributes to anticancer activity. *Cancer Res.* **70,** 5096–5108.

Sun, K. L., Correia, J. P., and Kennedy, T. E. (2011). Netrins: versatile extracellular cues with diverse functions. *Development* **138,** 2153–2169.

Suzuki, H., Watkins, D. N., Jair, K. W., Schuebel, K. E., Markowitz, S. D., Dong Chen, W., Pretlow, T. P., Yang, B., Akiyama, Y., Van Engeland, M., Toyota, M., Tokino, T., *et al.* (2004). Epigenetic inactivation of SFRP genes allows constitutive WNT signaling in colorectal cancer. *Nat. Genet.* **36,** 417–422.

Tadagavadi, R. K., Wang, W., and Ramesh, G. (2010). Netrin-1 regulates Th1/Th2/Th17 cytokine production and inflammation through UNC5B receptor and protects kidney against ischemia-reperfusion injury. *J. Immunol.* **185,** 3750–3758.

Taipale, J., and Beachy, P. A. (2001). The Hedgehog and Wnt signalilng pathways in cancer. *Nature* **411,** 349–354.

Taketo, M. M., and Edelmann, W. (2009). Mouse models of colon cancer. *Gastroenterology* **136,** 780–798.

Tang, X., Jang, S. W., Okada, M., Chan, C. B., Feng, Y., Liu, Y., Luo, S. W., Hong, Y., Rama, N., Xiong, W. C., Mehlen, P., and Ye, K. (2008). Netrin-1 mediates neuronal survival through PIKE-L interaction with the dependence receptor UNC5B. *Nat. Cell Biol.* **10,** 698–706.

Taniguchi, H., Yamamoto, H., Akutsu, N., Nosho, K., Adachi, Y., Imai, K., and Shinomura, Y. (2007). Transcriptional silencing of hedgehog-interacting protein by CpG hypermethylation and chromatic structure in human gastrointestinal cancer. *J. Pathol.* **213,** 131–139.

Tanikawa, C., Matsuda, K., Fukuda, S., Nakamura, Y., and Arakawa, H. (2003). p53RDL1 regulates p53-dependent apoptosis. *Nat. Cell Biol.* **5,** 216–223.

Teichmann, H. M., and Shen, K. (2011). UNC-6 and UNC-40 promote dendritic growth through PAR-4 in *Caenorhabditis elegans* neurons. *Nat. Neurosci.* **14,** 165–172.

Terzi, J., Grivennikov, S., Karin, E., and Karin, M. (2010). Inflammation and colon cancer. *Gastroenterology* **138,** 2101–2114.

Tessier-Lavigne, M., and Goodman, C. S. (1996). The molecular biology of axon guidance. *Science* **274,** 1123–1133.

Thayer, S. P., di Magliano, M. P., Heiser, P. W., Nielsen, C. M., Roberts, D. J., Lauwers, G. Y., Qi, Y. P., Gysin, S., Fernández-del Castillo, C., Yajnik, V., Antoniu, B., McMahon, M., *et al.* (2003). Hedgehog is an early and late mediator of pancreatic cancer tumorigenesis. *Nature* **425,** 851–856.

Theocharis, A. D., Skandalis, S. S., Tzanakakis, G. N., and Karamanos, N. K. (2010). Proteoglycans in health and disease: novel roles for proteoglycans in malignancy and their pharmacological targeting. *FEBS J.* **277,** 3904–3923.

Thiagalingam, S., Laken, S., Willson, J. K., Markowitz, S. D., Kinzler, K. W., Vogelstein, B., and Lengauer, C. (2001). Mechanisms underlying losses of heterozygosity in human colorectal cancers. *Proc. Natl. Acad. Sci. USA* **98,** 2698–2702.

Thiebault, K., Mazelin, L., Pays, L., Llambi, F., Joly, M. O., Scoazec, J. Y., Saurin, J. C., Romeo, G., and Mehlen, P. (2003). The Netrin-1 receptors UNC5H are putative tumor suppressors controlling cell death commitment. *Proc. Natl. Acad. Sci. USA* **100,** 4173–4178.

Thiery, J. P., Acloque, H., Huang, R. Y., and Nieto, M. A. (2009). Epithelial-mesenchymal transitions in development and disease. *Cell* **139,** 871–890.

Tie, J., Pan, Y., Zhao, L., Wu, K., Liu, J., Sun, S., Guo, X., Wang, B., Gang, Y., Zhang, Y., Li, Q., Qiao, T., *et al.* (2010). MiR-218 inhibits invasion and metastasis of gastric cancer by targeting the Robo1 receptor. *PLoS Genet.* **6,** e1000879.

Tomizawa, Y., Sekido, Y., Kondo, M., Gao, B., Yokota, J., Roche, J., Drabkin, H., Lerman, M. I., Gazdar, A. F., and Minna, J. D. (2001). Inhibition of lung cancer cell growth and induction of apoptosis after reexpression of 3p213 candidate tumor suppressor gene SEMA3B. *Proc. Natl. Acad. Sci. USA* **98,** 13954–13959.

Tong, J., Killeen, M., Steven, R., Binns, K. L., Culotti, J., and Pawson, T. (2001). Netrin stimulates tyrosine phosphorylation of the UNC-5 family of Netrin receptors and induces Shp2 binding to the RCM cytodomain. *J. Biol. Chem.* **276,** 40917–40925.

Tournigand, C., André, T., Achille, E., Lledo, G., Flesh, M., Mery-Mignard, D., Quinaux, E., Couteau, C., Buyse, M., Ganem, G., Landi, B., Colin, P., *et al.* (2004). FOLFIRI followed by FOLFOX6 or the reverse sequence in advanced colorectal cancer: a randomized GERCOR study. *J. Clin. Oncol.* **22,** 229–237.

Tournigand, C., Cervantes, A., Figer, A., Lledo, G., Flesch, M., Buyse, M., Mineur, L., Carola, E., Etienne, P. L., Rivera, F., Chirivella, I., Perez-Staub, N., *et al.* (2006). OPTIMOX1: a randomized study of FOLFOX4 or FOLFOX7 with oxaliplatin in a stop-and-Go fashion in advanced colorectal cancer-a GERCOR study. *J. Clin. Oncol.* **24,** 394–400.

Trusolino, L., and Comoglio, P. M. (2002). Scatter-factor and semaphorin receptors: cell signalling for invasive growth. *Nat. Rev. Cancer* **2,** 289–300.

Tse, C., Xiang, R. H., Bracht, T., and Naylor, S. L. (2002). Human Semaphorin 3B (SEMA3B) located at chromosome 3p21.3 suppresses tumor formation in an adenocarcinoma cell line. *Cancer Res.* **62,** 542–546.

Ullman, T. A., and Itzkowitz, S. H. (2011). Intestinal inflammation and cancer. *Gastroenterology* **140,** 1807–1816.

Ullrich, O., Reinheckel, T., Sitte, N., Hass, R., Grune, T., and Davies, K. J. (1999). Poly-ADP ribose polymerase activates nuclear proteasome to degrade oxidatively damaged histones. *Proc. Natl. Acad. Sci. USA* **96,** 6223–6228.

Valastyan, S., and Weinberg, R. A. (2011). Roles for microRNAs in the regulation of cell adhesion molecules. *J. Cell Sci.* **124,** 999–1006.

Valastyan, S., Chang, A., Benaich, N., Reinhardt, F., and Weinberg, R. A. (2011). Activation of miR-31 function in already-established metastases elicits metastatic regression. *Genes Dev.* **25,** 646–659.

Van Cutsem, E., Köhne, C. H., Láng, I., Folprecht, G., Nowacki, M. P., Cascinu, S., Shchepotin, I., Maurel, J., Cunningham, D., Tejpar, S., Schlichting, M., Zubel, A., *et al.* (2011). Cetuximab plus irinotecan, fluorouracil, and leucovorin as first-line treatment for metastatic colorectal cancer: updated analysis of overall survival according to tumor KRAS and BRAF mutation status. *J. Clin. Oncol.* **29,** 2011–2019.

Van den Brink, G. R., and Hardwick, J. C. H. (2011). Hedgehog Wnteraction in colorectal cancer. *Gut* **55,** 912–914.

van den Brink, G. R., Bleuming, S. A., Hardwick, J. C., Schepman, B. L., Offerhaus, G. J., Keller, J. J., Nielsen, C., Gaffield, W., van Deventer, S. J., Roberts, D. J., and Peppelenbosch, M. P. (2004). Indian Hedgehog is an antagonist of Wnt signaling in colonic epithelial cell differentiation. *Nat. Genet.* **36,** 277–282.

van der Flier, L. G., van Gijn, M. E., Hatzis, P., Kujala, P., Haegebarth, A., Stange, D. E., Begthel, H., van den Born, M., Guryev, V., Oving, I., van Es, J. H., Barker, N., *et al.* (2009). Transcription factor achaete scute-like 2 controls intestinal stem cell fate. *Cell* **136,** 903–912.

van Es, J. H., Jay, P., Gregorieff, A., van Gijn, M. E., Jonkheer, S., Hatzis, P., Thiele, A., van den Born, M., Begthel, H., Brabletz, T., Taketo, M. M., and Clevers, H. (2005). Wnt signalling induces maturation of Paneth cells in intestinal crypts. *Nat. Cell Biol.* **7,** 381–386.

Van Slambrouck, S., Grijelmo, C., De Wever, O., Bruyneel, E., Emami, S., Gespach, C., and Steelant, W. F. (2007). Activation of the FAK-src molecular scaffolds and p130Cas-JNK signaling cascades by α1-integrins during colon cancer cell invasion. *Int. J. Oncol.* **31,** 1501–1508.

Vanzomeren-Dohm, A., Sarro, J., Flannery, E., and Duman-Scheel, M. (2011). The Drosophila Netrin receptor Frazzled/DCC functions as an invasive tumor suppressor. *BMC Dev. Biol.* **11,** 41.

Varela-Echavarría, A., and Guthrie, S. (1997). Molecules making waves in axon guidance. *Genes Dev.* **11,** 545–557.

Varnat, F., Duquet, A., Malerba, M., Zbinden, M., Mas, C., Gervaz, P., and Ruiz i Altaba, A. (2009). Human colon cancer epithelial cells harbour active HEDGEHOG-GLI signaling that is essential for tumour growth, recurrence, metastasis and stem cell survival and expansion. *EMBO Mol. Med.* **1,** 338–351.

Varnat, F., Siegl-Cachedenier, I., Malerba, M., Gervaz, P., and Ruiz i Altaba, A. (2010). Loss of WNT-TCF addiction and enhancement of HH-GLI1 signaling define the metastatic transition of human colon carcinomas. *EMBO Mol. Med.* **2,** 440–457.

Vider, B. Z., Zimber, A., Chastre, E., Prevot, S., Gespach, C., Estlein, D., Wolloch, Y., Tronic, S. R., Gazit, A., and Yaniv, A. (1996). Evidence for the involvement of the Wnt 2 gene in human colorectal cancer. *Oncogene* **12,** 153–158.

Vielmetter, J., Chen, X. N., Miskevich, F., Lane, R. P., Yamakawa, K., Korenberg, J. R., and Dreyer, W. J. (1997). Molecular characterization of human neogenin, a DCC-related protein, and the mapping of its gene (NEO1) to chromosomal position 15q22.3-q23. *Genomics* **41,** 414–421.

Vogelstein, B., Fearon, E. R., Kern, S. E., Hamilton, S. R., Preisinger, A. C., Nakamura, Y., and White, R. (1989). Allelotype of colorectal carcinomas. *Science* **244,** 207–211.

von Kries, J. P., Winbeck, G., Asbrand, C., Schwarz-Romond, T., Sochnikova, N., Dell'Oro, A., Behrens, J., and Birchmeier, W. (2000). Hot spots in β-catenin for interactions with LEF-1, conductin and APC. *Nat. Struct. Biol.* **7,** 800–807.

Wang, W., Reeves, W. B., and Ramesh, G. (2008). Netrin-1 and kidney injury. I. Netrin-1 protects against ischemia-reperfusion injury of the kidney. *Am. J. Physiol. Renal Physiol.* **294,** F739–F747.

Wang, W., Reeves, W. B., and Ramesh, G. (2009a). Netrin-1 increases proliferation and migration of renal proximal tubular epithelial cells via the UNC5B receptor. *Am. J. Physiol. Renal Physiol.* **296,** F723–F729.

Wang, R., Wei, Z., Jin, H., Wu, H., Yu, C., Wen, W., Chan, L. N., Wen, Z., and Zhang, M. (2009b). Autoinhibition of UNC5b revealed by the cytoplasmic domain structure of the receptor. *Mol. Cell* **33**, 692–703.

Watt, F. M. (2004). Unexpected Hedgehog-Wnt interactions in epithelial differentiation. *Trends Mol. Med.* **10**, 577–580.

Wendt, M. K., Drury, L. J., Vongsa, R. A., and Dwinell, M. B. (2008). Constitutive CXCL12 expression induces anoikis in colorectal carcinoma cells. *Gastroenterology* **135**, 508–517.

Williams, M. E., Strickland, P., Watanabe, K., and Hinck, L. (2003). UNC5H1 induces apoptosis via its juxtamembrane region through an interaction with NRAGE. *J. Biol. Chem.* **278**, 17483–17490.

Wilson, N. H., and Key, B. (2007). Neogenin: one receptor, many functions. *Int. J. Biochem. Cell Biol.* **39**, 874–878.

Wilson, B. D., Ii, M., Park, K. W., Suli, A., Sorensen, L. K., Larrieu-Lahargue, F., Urness, L. D., Suh, W., Asai, J., Kock, G. A., Thorne, T., Silver, M., *et al.* (2006). Netrins promote developmental and therapeutic angiogenesis. *Science* **313**, 640–644.

Winder, T., and Lenz, H. J. (2010). Vascular endothelial growth factor and epidermal growth factor signaling pathways as therapeutic targets for colorectal cancer. *Gastroenterology* **138**, 2163–2176.

Wood, R. D., Mitchell, M., Sgouros, J., and Lindahl, T. (2001). Human DNA repair genes. *Science* **291**, 1284–1289.

Woods, C. G. (2004). Neuroscience. Crossing the midline. *Science* **304**, 1455–1456.

Wright, K. T., Seabright, R., Logan, A., Lilly, A. J., Khanim, F., Bunce, C. M., and Johnson, W. E. (2010). Extracellular Nm23H1 stimulates neurite outgrowth from dorsal root ganglia neurons in vitro independently of nerve growth factor supplementation or its nucleoside diphosphate kinase activity. *Biochem. Biophys. Res. Commun.* **398**, 79–85.

Wu, T. W., Li, W. W., and Li, H. (2008). Netrin-1 attenuates ischemic stroke-induced apoptosis. *Neuroscience* **156**, 475–482.

Xiang, R., Davalos, A. R., Hensel, C. H., Zhou, X. J., Tse, C., and Naylor, S. L. (2002). Semaphorin 3F gene from human 3p21.3 suppresses tumor formation in nude mice. *Cancer Res.* **62**, 2637–2643.

Xie, W., Jin, L., Mei, Y., and Wu, M. (2009). E2F1 represses beta-catenin/TCF activity by direct upregulation of Siah1. *J. Cell. Mol. Med.* **13**, 1719–1727.

Xie, J., Ponuwei, G. A., Moore, C. E., Willars, G. B., Tee, A. R., and Herbert, T. P. (2011). cAMP inhibits mammalian target of rapamycin complex-1 and -2 (mTORC1 and 2) by promoting complex dissociation and inhibiting mTOR kinase activity. *Cell. Signal.* **23**, 1927–1935.

Yam, P. T., Langlois, S. D., Morin, S., and Charron, F. (2009). Sonic hedgehog guides axons through a noncanonical, Src-family-kinase-dependent signaling pathway. *Neuron* **62**, 349–362.

Yamada, T., Mori, Y., Hayashi, R., Takada, M., Ino, Y., Naishiro, Y., Kondo, T., and Hirohashi, S. (2003). Suppression of intestinal polyposis in Mdr1-deficient ApcMin/+ mice. *Cancer Res.* **63**, 895–901.

Yamagishi, S., Hampel, F., Hata, K., Del Toro, D., Schwark, M., Kvachnina, E., Bastmeyer, M., Yamashita, T., Tarabykin, V., Klein, R., and Egea, J. (2011). FLRT2 and FLRT3 act as repulsive guidance cues for Unc5-positive neurons. *EMBO J.* **30**, 2920–2933.

Yamasaki, L., Jacks, T., Bronson, R., Goillot, E., Harlow, E., and Dyson, N. J. (1996). Tumor induction and tissue atrophy in mice lacking E2F-1. *Cell* **85**, 537–548.

Yamasaki, L., Bronson, R., Williams, B. O., Dyson, N. J., Harlow, E., and Jacks, T. (1998). Loss of E2F-1 reduces tumorigenesis and extends the lifespan of Rb1(+/−)mice. *Nat. Genet.* **18**, 360–364.

Yang, A. D., Fan, F., Camp, E. R., van Buren, G., Liu, W., Somcio, R., Gray, M. J., Cheng, H., Hoff, P. M., and Ellis, L. M. (2006). Chronic oxaliplatin resistance induces epithelial-to-mesenchymal transition in colorectal cancer cell lines. *Clin. Cancer Res.* **12**, 4147–4153.

Yang, Y., Zou, L., Wang, Y., Xu, K. S., Zhang, J. X., and Zhang, J. H. (2007). Axon guidance cue Netrin-1 has dual function in angiogenesis. *Cancer Biol. Ther.* **6**, 743–748.

Yang, L., Garbe, D. S., and Bashaw, G. J. (2009). A frazzled/DCC-dependent transcriptional switch regulates midline axon guidance. *Science* **324**, 944–947.

Yang, Y. H., Szabat, M., Bragagnini, C., Kott, K., Helgason, C. D., Hoffman, B. G., and Johnson, J. D. (2011). Paracrine signaling loops in adult human and mouse pancreatic islets: Netrins modulate β-cell apoptosis signalling *via* dependence receptors. *Diabetologia* **54**, 828–842.

Yebra, M., Montgomery, A. M., Diaferia, G. R., Kaido, T., Silletti, S., Perez, B., Just, M. L., Hildbrand, S., Hurford, R., Florkiewicz, E., Tessier-Lavigne, M., and Cirulli, V. (2003). Recognition of the neural chemoattractant Netrin-1 by integrins α6β4 and α3β1 regulates epithelial cell adhesion and migration. *Dev. Cell* **5**, 695–707.

Yebra, M., Diaferia, G. R., Montgomery, A. M., Kaido, T., Brunken, W. J., Koch, M., Hardiman, G., Crisa, L., and Cirulli, V. (2011). Endothelium-derived netrin-4 supports pancreatic epithelial cell adhesion and differentiation through integrins α2β1 and α3β1. *PLoS One* **6**, e22750.

Yee, C. S., Sybingco, S. S., Serdetchania, V., Kholkina, G., Bueno de Mesquita, M., Naqvi, Z., Park, S. H., Lam, K., and Killeen, M. T. (2011). ENU-3 is a novel motor axon outgrowth and guidance protein in *C. elegans*. *Dev. Biol.* **352**, 243–253.

Yin, Y., Sanes, J. R., and Miner, J. H. (2000). Identification and expression of mouse Netrin-4. *Mech. Dev.* **96**, 115–119.

Ying, L., Hofseth, A. B., Browning, D. D., Nagarkatti, M., Nagarkatti, P. S., and Hofseth, L. J. (2007). Nitric oxide inactivates the retinoblastoma pathway in chronic inflammation. *Cancer Res.* **67**, 9286–9293.

Yoo, A. S., Sun, A. X., Li, L., Shcheglovitov, A., Portmann, T., Li, Y., Lee-Messer, C., Dolmetsch, R. E., Tsien, R. W., and Crabtree, G. R. (2011). MicroRNA-mediated conversion of human fibroblasts to neurons. *Nature* **476**, 228–232.

Yoshikawa, K., Shimada, M., Miyamoto, H., Higashijima, J., Miyatani, T., Nishioka, M., Kurita, N., Iwata, T., and Uehara, H. (2009). Sonic hedgehog relates to colorectal carcinogenesis. *J. Gastroenterol.* **44**, 1113–1117.

Yu, F., Deng, H., Yao, H., Liu, Q., Su, F., and Song, E. (2010). Mir-30 reduction maintains self-renewal and inhibits apoptosis in breast tumor-initiating cells. *Oncogene* **29**, 4194–4204.

Yurchenco, P. D., and Wadsworth, W. G. (2004). Assembly and tissue functions of early embryonic laminins and etrins. *Curr. Opin. Cell Biol.* **16**, 572–579.

Zaynagetdinov, R., Ryzhov, S., Goldstein, A. E., Yin, H., Novitskiy, S. V., Goleniewska, K., Polosukhin, V. V., Newcomb, D. C., Mitchell, D., Morschl, E., Zhou, Y., Blackburn, M. R., *et al.* (2010). Attenuation of chronic pulmonary inflammation in A2B adenosine receptor knockout mice. *Am. J. Respir. Cell Mol. Biol.* **42**, 564–571.

Zhang, J., and Cai, H. (2010). Netrin-1 prevents ischemia/reperfusion-induced myocardial infarction via a DCC/ERK1/2/eNOS s1177/NO/DCC feed-forward mechanism. *J. Mol. Cell. Cardiol.* **48**, 1060–1070.

Zhang, C., Meng, F., Wang, C., Guo, H., Fan, M., Liu, S., Zhou, R., and He, F. (2004). Identification of a novel alternative splicing form of human Netrin-4 and analyzing the expression patterns in adult rat brain. *Brain Res. Mol. Brain Res.* **130**, 68–80.

Zhang, L., Yang, X., Yang, S., and Zhang, J. (2011a). The Wnt/β-catenin signaling pathway in the adult neurogenesis. *Eur. J. Neurosci.* **33**, 1–8.

Zhang, S., Huang, W. C., Li, P., Guo, H., Poh, S. B., Brady, S. W., Xiong, Y., Tseng, L. M., Li, S. H., Ding, Z., Sahin, A. A., Esteva, F. J., *et al.* (2011b). Combating trastuzumab resistance by targeting SRC, a common node downstream of multiple resistance pathways. *Nat. Med.* **17**, 461–469.

Zhou, B. P., Liao, Y., Xia, W., Zou, Y., Spohn, B., and Hung, M. C. (2001). HER-2/neu induces p53 ubiquitination via Akt-mediated MDM2 phosphorylation. *Nat. Cell Biol.* **3,** 973–982.
Zhou, W. J., Geng, Z. H., Chi, S., Zhang, W., Niu, X. F., Lan, S. J., Ma, L., Yang, X., Wang, L. J., Ding, Y. Q., and Geng, J. G. (2011). Slit-Robo signaling induces malignant transformation through Hakai-mediated E-cadherin degradation during colorectal epithelial cell carcinogenesis. *Cell Res.* **21,** 609–626.
Zhu, Y., Richardson, J. A., Parada, L. F., and Graff, J. M. (1998). Smad3 mutant mice develop metastatic colorectal cancer. *Cell* **94,** 703–714.
Ziel, J. W., and Sherwood, D. R. (2012). Roles for Netrin signaling outside of axon guidance: a view from the worm. *Dev. Dyn.* **239,** 1296–12305.
Ziel, J. W., Hagedorn, E. J., Audhya, A., and Sherwood, D. R. (2009). UNC-6 (Netrin) orients the invasive membrane of the anchor cell in *C. elegans*. *Nat. Cell Biol.* **11,** 183–189.
Zimber, A., and Gespach, C. (2008). Bile acids and derivatives, their nuclear receptors FXR, PXR and ligands. Role in health and disease and their therapeutic potential. *Anticancer Agents Med. Chem.* **8,** 540–563.
Zoncu, R., Efeyan, A., and Sabatini, D. M. (2011). mTOR: from growth signal integration to cancer, diabetes and ageing. *Nat. Rev. Mol. Cell Biol.* **12,** 21–35.

NOTE ADDED IN PROOF
Most Significant and Recent Papers in Guidance and Colorectal Cancer Fields Include:

Ahlquist, D. A., Zou, H., Domanico, M., Mahoney, D. W., Yab, T. C., Taylor, W. R., Butz, M. L., Thibodeau, S. N., Rabeneck, L. I., Paszat, L. F., Kinzler, K. W., Vogelstein, B., *et al.* (2012). Next-generation stool DNA test accurately detects colorectal cancer and large adenomas. *Gastroenterology* **142,** 248–256.
Albo, D., Akay, C. L., Marshall, C. L., Wilks, J. A., Verstovsek, G., Liu, H., Agarwal, N., Berger, D. H., and Ayala, G. E. (2011). Neurogenesis in colorectal cancer is a marker of aggressive tumor behavior and poor outcomes. *Cancer* **117,** 4834–4845.
Brasch, J., Harrison, O. J., Ahlsen, G., Liu, Q., and Shapiro, L. (2011). Crystal structure of the ligand binding domain of netrin G2. *J. Mol. Biol.* **414,** 723–734.
Carlsson, E., Ranki, A., Sipilä, L., Karenko, L., Abdel-Rahman, W. M., Ovaska, K., Siggberg, L., Aapola, U., Assämäki, R., Häyry, V., Niiranen, K., Helle, M., *et al.* (2012). Potential role of a navigator gene NAV3 in colorectal cancer. *Br. J. Cancer* **106,** 517–524.
Castets, M., Broutier, L., Molin, Y., Brevet, M., Chazot, G., Gadot, N., Paquet, A., Mazelin, L., Jarrosson-Wuilleme, L., Scoazec, J. Y., Bernet, A., and Mehlen, P. (2011). DCC constrains tumour progression via its dependence receptor activity. *Nature* **482,** 543–547.
Cotrufo, T., Pérez-Branguli, F., Muhaisen, A., Ros, O., Andrés, R., Baeriswyl, T., Fuschini, G., Tarrago, T., Pascual, M., Ureña, J., Blasi, J., Giralt, E., *et al.* (2011). A signaling mechanism coupling netrin-1/deleted in colorectal cancer chemoattraction to SNARE-mediated exocytosis in axonal growth cones. *J. Neurosci.* **31,** 14463–14480.
De Boeck, A., Pauwels, P., Hensen, K., Rummens, J. L., Westbroek, W., Hendrix, A., Maynard, D., Denys, H., Lambein, K., Braems, G., Gespach, C., Bracke, M., *et al.* (2012). Bone marrow-derived mesenchymal stem cells promote colorectal cancer progression through paracrine NRG-1/HER3 signaling. *Gut.*

Dorard, C., de Thonel, A., Collura, A., Marisa, L., Svrcek, M., Lagrange, A., Jego, G., Wanherdrick, K., Joly, A. L., Buhard, O., Gobbo, J., Penard-Lacronique, V., *et al.* (2011). Expression of a mutant HSP110 sensitizes colorectal cancer cells to chemotherapy and improves disease prognosis. *Nat. Med.* **17,** 1283–1289.

Hinkel, I., Duluc, I., Martin, E., Guenot, D., Freund, J. N., and Gross, I. (2011). Cdx2 controls expression of the protocadherin Mucdhl, an inhibitor of growth and β-catenin activity in colon cancer cells. *Gastroenterology* **142,** 875–885.

Hunter, K. W. (2012). Mouse models of cancer: does the strain matter? *Nat. Rev. Cancer* **12,** 144–149.

Jayakumar, C., Mohamed, R., Ranganathan, P. V., and Ramesh, G. (2011). Intracellular kinases mediate increased translation and secretion of netrin-1 from renal tubular epithelial cells. *PLoS One* **6,** e26776.

Jarjour, A. A., Durko, M., Luk, T. L., Marçal, N., Shekarabi, M., and Kennedy, T. E. (2011). Autocrine netrin function inhibits glioma cell motility and promotes focal adhesion formation. *PLoS One* **6,** e25408.

Kahlert, C., Lahes, S., Radhakrishnan, P., Dutta, S., Mogler, C., Herpel, E., Brand, K., Steinert, G., Schneider, M., Mollenhauer, M., Reissfelder, C., Klupp, F., *et al.* (2011). Overexpression of ZEB2 at the invasion front of colorectal cancer is an independent prognostic marker and regulates tumor invasion in vitro. *Clin. Cancer Res.* **17,** 7654–7663.

Klein, C., Mikutta, J., Krueger, J., Scholz, K., Brinkmann, J., Liu, D., Veerkamp, J., Siegel, D., Abdelilah-Seyfried, S., and le Noble, F. (2011). Neuron navigator 3a regulates liver organogenesis during zebrafish embryogenesis. *Development* **138,** 1935–1945.

Lao, V. V., and Grady, W. M. (2011). Epigenetics and colorectal cancer. *Nat. Rev. Gastroenterol. Hepatol.* **8,** 686–700.

Lee, C. C., Chen, W. S., Chen, C. C., Chen, L. L., Lin, Y. S., Fan, C. S., and Huang, T. S. (2012). TCF12 functions as a transcriptional repressor of E-cadherin and its overexpression is correlated with the metastasis of colorectal cancer. *J. Biol. Chem.* **287,** 2798–2809.

Matallanas, D., Romano, D., Al-Mulla, F., O'Neill, E., Al-Ali, W., Crespo, P., Doyle, B., Nixon, C., Sansom, O., Drosten, M., Barbacid, M., and Kolch, W. (2011). Mutant K-Ras activation of the proapoptotic MST2 pathway is antagonized by wild-type K-Ras. *Mol. Cell* **44,** 893–906.

Nagtegaal, I. D., Quirke, P., and Schmoll, H. J. (2012). Has the new TNM classification for colorectal cancer improved care? *Nat. Rev. Clin. Oncol.* **9,** 119–123.

Nguyen, L. V., Vanner, R., Dirks, P., and Eaves, C. J. (2012). Cancer stem cells: an evolving concept. *Nat. Rev. Cancer* **12,** 133–143.

Palmesino, E., Haddick, P. C., Tessier-Lavigne, M., and Kania, A. (2012). Genetic analysis of DSCAM's role as a netrin-1 receptor in vertebrates. *J. Neurosci.* **32,** 411–416.

Satow, R., Shitashige, M., Jigami, T., Fukami, K., Honda, K., Kitabayashi, I., and Yamada, T. (2012). β-Catenin inhibits promyelocytic leukemia protein tumor suppressor function in colorectal cancer cells. *Gastroenterology* **142,** 572–581.

Seetharaman, A., Selman, G., Puckrin, R., Barbier, L., Wong, E., D'Souza, S. A., and Roy, P. J. (2011). MADD-4 is a secreted cue required for midline-oriented guidance in Caenorhabditis elegans. *Dev. Cell* **21,** 669–680.

Sharma, V., Antonacopoulou, A. G., Tanaka, S., Panoutsopoulos, A. A., Bravou, V., Kalofonos, H. P., and Episkopou, V. (2011). Enhancement of TGF-β signaling responses by the E3 ubiquitin ligase Arkadia provides tumor suppression in colorectal cancer. *Cancer Res.* **71,** 6438–6449.

Smartt, H. J., Greenhough, A., Ordóñez-Morán, P., Talero, E., Cherry, C. A., Wallam, C. A., Parry, L., Al Kharusi, M., Roberts, H. R., Mariadason, J. M., Clarke, A. R., Huelsken, J., *et al.* (2012). β-Catenin represses expression of the tumour suppressor 15-prostaglandin dehydrogenase in the normal intestinal epithelium and colorectal tumour cells. *Gut* (Epub ahead of print).

van Gils, J. M., Derby, M. C., Fernandes, L. R., Ramkhelawon, B., Ray, T. D., Rayner, K. J., Parathath, S., Distel, E., Feig, J. L., Alvarez-Leite, J. I., Rayner, A. J., McDonald, T. O., *et al.* (2012). The neuroimmune guidance cue netrin-1 promotes atherosclerosis by inhibiting the emigration of macrophages from plaques. *Nat. Immunol.* **13**, 136–143.

Xiong, H., Hong, J., Du, W., Lin, Y. W., Ren, L. L., Wang, Y. C., Su, W. Y., Wang, J. L., Cui, Y., Wang, Z. H., and Fang, J. Y. (2012). Roles of STAT3 and ZEB1 in E-cadherin downregulation and human colorectal cancer epithelial-mesenchymal transition. *J. Biol. Chem.* **2787**, 5819–5832.

Zhu, W., Cai, M. Y., Tong, Z. T., Dong, S. S., Mai, S. J., Liao, Y. J., Bian, X. W., Lin, M. C., Kung, H. F., Zeng, Y. X., Guan, X. Y., and Xie, D. (2012). Overexpression of EIF5A2 promotes colorectal carcinoma cell aggressiveness by upregulating MTA1 through C-myc to induce epithelial-mesenchymal transition. *Gut* **61**, 562–575.

A Roundabout Way to Cancer

Mimmi S. Ballard and Lindsay Hinck

Department of Molecular, Cell and Developmental Biology, University of California, Santa Cruz, California, USA

The Slit family of secreted proteins and their transmembrane receptor, Robo, were originally identified in the nervous system where they function as axon guidance cues and branching factors during development. Since their discovery, a great number of additional roles have been attributed to Slit/Robo signaling, including regulating the critical processes of cell proliferation and cell motility in a variety of cell and tissue types. These processes are often deregulated during cancer progression, allowing tumor cells to bypass safeguarding mechanisms in the cell and the environment in order to grow and escape to new tissues. In the past decade, it has been shown that the expression of Slit and Robo is altered in a wide variety of cancer types, identifying them as potential therapeutic targets. Further, studies have demonstrated dual roles for Slits and Robos in cancer, acting as both oncogenes and tumor suppressors. This bifunctionality is also observed in their roles as axon guidance cues in the developing nervous system, where they both attract and repel neuronal migration. The fact that this signaling axis can have opposite functions depending on the cellular circumstance make its actions challenging to define. Here, we summarize our current understanding of the dual roles that Slit/Robo signaling play in development, epithelial tumor progression, and tumor angiogenesis.

Advances in CANCER RESEARCH, Volume 114

0065-230X/12 $35.00
DOI: 10.1016/B978-0-12-386503-8.00005-3

I. INTRODUCTION

The existence of axon guidance molecules was postulated by Ramon & Cajal in the late 1800s, but 100 years elapsed before their molecular identification. As their name suggests, these cues act to instruct the migration of neurons and their axons in the developing nervous system, establishing the initial pattern of axonal projections that is subsequently refined by activity-dependent mechanisms. Axon guidance cues are secreted by both target and non-target cells, and they are bifunctional, acting as both attractants and repellents when patterning the nervous system. Cell surface receptors are responsible for directing the response of a cell to these cues. These receptors translate directional information provided by the cues to the cytoskeleton, generating the movement and, in the case of axons, directional outgrowth required for pathfinding.

In addition to their role as axon guidance cues, these molecules also function outside the nervous system to regulate the development of other organs, including the immune and vascular systems, epithelial organs, and glands. In these contexts, "axon guidance" cues do not simply function as instructional signals, but rather act more broadly to control the growth, branching, adhesion, and position of cells in complex tissues. Like many molecules that play key roles in development, "axon guidance" cues are often deregulated in disease processes, especially cancer, with many of these cues acting to either promote or suppress tumor growth and progression.

Among the many families of "axon guidance" molecules, Slits, signaling through their Roundabout (Robo) receptors, constitute a relatively small group of factors. They were originally identified as chemorepellents that play a crucial role in preventing developing commissural neurons from inappropriately re-crossing the midline. Once believed to have only restricted functions in the developing nervous system as guidance cues and branching factors (Brose *et al.*, 1999; Wang *et al.*, 1999), members of this family are now identified as key regulators of many cellular processes in multiple tissue types, including the mammary gland, heart, lung, and kidney (Greenberg *et al.*, 2004; Hinck, 2004; Medioni *et al.*, 2010; Piper *et al.*, 2000). In addition, they have also been implicated in multiple human pathologies including cancer and inflammation (Legg *et al.*, 2008; London and Li, 2011; London *et al.*, 2010; Wu *et al.*, 2001).

As we learn more about the mechanisms of cancer progression, it is becoming clear that tumor cells hijack normal cellular processes to survive and metastasize to secondary tissues. Specifically, a cell's proliferative, adhesive, and migratory properties are often altered in the process of tumor cell transformation, allowing rapid proliferation and tumor growth, detachment from the surrounding tissue, and invasion into the vasculature leading to metastasis. Over the past decade, many studies have implicated Slit/Robo signaling in the regulation of cell proliferation, cell adhesion, and cell migration, raising the possibility that this pathway represents a key target for alteration in cells undergoing tumor cell transformation. In fact, Slit and Robo expression levels are altered in a majority of human cancers. However, recent work suggests that the role of Slit/Robo in tumor progression is anything but simple. Emerging evidence postulates that Slits and Robos function both as oncogenes and tumor suppressors, often in the same tissue. In this review, we summarize how Slit/Robo signaling confers both tumor suppressive and oncogenic effects on the progression of various types of cancers, focusing most of our discussion on vertebrate systems, though invertebrate studies will be touched upon when relevant.

II. THE SLIT/ROBO SIGNALING INTERACTION IS WELL CHARACTERIZED

A. Slit is a Large Secreted Factor

Slits are secreted extracellular matrix proteins expressed in many cell types and tissues. While flies and worms express only one Slit molecule, mice and humans express three (Slit1, Slit2, and Slit3) that share a high degree of structural conservation (Dickson and Gilestro, 2006). Slits are large multi-domain proteins with a unique tandem of four leucine-rich repeats (LRRs, D1–D4), each connected via disulfide bonds near the N-terminus. These LRRs are followed by seven to nine epidermal growth factor (EGF)-like domains and a laminin G-like domain capped by a C-terminal cysteine-rich module. Structural studies have revealed that the LRR domains each contain a conserved motif that creates a concave shape that might be important for modulating the interaction of Slits with their cognate receptors (Howitt *et al.*, 2004; Morlot *et al.*, 2007a,b). Further structural studies have shown that Slits undergo post-translational modifications (Fig. 1A). Slit proteins are proteolytically cleaved within the fifth EGF region to release an N-terminal fragment that binds Robo receptors and mediates all assayed cell guidance functions of Slit/Robo signaling (Nguyen Ba-Charvet *et al.*, 2001).

The most commonly studied member of the Slit family of proteins is Slit2. It is known to regulate many aspects of tissue morphogenesis and cell function, including cell migration, proliferation, adhesion, and death. While the function of Slit1 remains largely unknown, recent studies are beginning to ascribe functions to Slit3 in embryonic angiogenesis. Studies have shown that Slit3 may function as an angiogenic factor involved in regulating endothelial cell proliferation and motility, in addition to

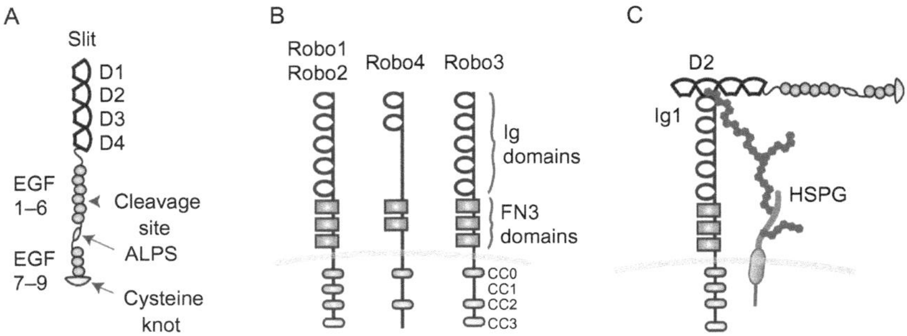

Fig. 1 Structural representation of Slits, Robos, and their interaction. (A) At their N-terminus, vertebrate and invertebrate Slits consist of four leucine-rich repeats (LRRs), termed D1–D4. These LRRs are followed by seven to nine epidermal growth factor (EGF)-like domains, a laminin G-like domain (ALPS), and a C-terminal cysteine-rich knot. Slits are proteolytically cleaved between two EGF-like domains. (B) Vertebrates have four Robos (Robo1–4); while fly, chick, and zebrafish have three (Robo1–3). Robo1, 2, and 3 contain five immunoglobulin (Ig) domains and three fibronectin type 3 (FN3) domains. Robo4 contains only two Ig domains and two FN3 domains. Zebrafish Robo4 is unique in that it contains three Ig domains instead of two. In their cytoplasmic tail, Robos contain between two and four conserved proline-rich domains (CC0–CC3). (C) The Slit/Robo signaling pair can be stabilized via heparan sulfate glycosaminoglycans (GAGs) that are either present in the extracellular matrix or attached to membrane-associated proteins such as the heparan sulfate proteoglycan (HSPG) syndecan. (See Page 6 in Color Section at the back of the book.)

regulating formation of vascular networks (Zhang *et al.*, 2009). Thus, while our current knowledge confirms that the Slit family of axon guidance molecules plays important roles in many aspects of development, it is also clear that there are likely many unknown functions yet to be discovered.

B. Robo is a Highly Conserved Transmembrane Receptor

The Robo family of receptors is highly conserved, though the number of Robo genes differs between invertebrates and vertebrates. Whereas worms have only one Robo (Sax-3) receptor; flies, chick, and zebrafish have three (Robo1–3); and mammals have four (Robo1/Dutt1, Robo2, Robo3/Rig-1, and Robo4/Magic Roundabout) (Fig. 1B; Challa *et al.*, 2001; Hohenester, 2008; Huminiecki *et al.*, 2002; Lee *et al.*, 2001). While Robo1–3 contain a high degree of structural and functional similarity, Robo4 appears to function distinctly from the other Robo family members and, therefore, will be discussed in a separate section.

Robo1–3 are single-pass transmembrane receptors that belong to the immunoglobulin (Ig) superfamily of cell adhesion molecules and are conserved between vertebrates and invertebrates (Hohenester, 2008). The ectodomain of mammalian Robo receptors contains five Ig domains followed by three fibronectin type 3 (FN3) repeats (Fig. 1B). The Robo intracellular domain has no inherent catalytic activity, but confers a downstream signal by recruiting various factors to conserved proline-rich domains, referred to as CC0–CC3, and a number of phosphorylatable sites present within the intracellular domain. It was recently demonstrated that the different functions of Robo receptors in neuronal development are due to both gene expression levels and specific characteristics of the different Robo receptors themselves (Spitzweck *et al.*, 2010). For example, by engineering the expression of each of the three Robo receptors from each *Robo* gene locus, the authors found that lateral positioning of longitudinal axon pathways depends on gene expression, not on the type of Robo expressed, whereas commissure formation depends on the expression of specific combinations of Robos. This suggests that the receptors have structural differences that confer specific signaling responses, such as prevention of midline crossing by Robo1 and promotion of midline crossing by Robo2 (Spitzweck *et al.*, 2010).

It has long been believed that the different signaling responses induced by each receptor in response to Slit binding are primarily due to variations in the number and combinations of CC motifs in their intracellular domains. However, recent studies reveal that both intracellular and extracellular domains dictate the resulting signaling response induced by Slit binding. In *Drosophila* neuronal development, Robo2 has both lateral positioning and midline crossing activities. In a 2010 study, Evans and Bashaw demonstrate that Ig1 and Ig3 are required for lateral positioning, whereas promidline crossing seems to be dictated by Ig2. The authors performed studies in which the cytoplasmic domains of Robo2 and Robo3 were replaced with that of Robo1 and found that stimulation with Slit confers wild-type activities, strongly suggesting that the specificity lies in the ectodomains. However, complete loss of the cytoplasmic domains ablates all receptor activity, indicating that these are also required for signaling. Further, the authors demonstrate that the differences in signaling response are not simply due to differences in Slit2 binding to the ectodomain, but rather due to differences in multimerization and receptor/ligand stoichiometry dictated by the Ig domains (Evans and Bashaw, 2010). These results suggest that while the cytoplasmic domain is important for downstream signaling, the specificity of the signaling response might be determined by the ectodomains. As Slits and Robos are known to regulate multiple types of signaling responses, including cell motility and cell proliferation, and to play critical roles in the development of many vital organs, including the kidney and breast, it will be interesting to learn if the Ig domains of Robo ectodomains play an equally important role in conferring Slit signaling in diverse systems.

C. Cleavage of Robo Reveals Additional Regulatory Complexities

Current efforts examining post-translational modifications of Robo suggest that regulation of this receptor, and consequently the signaling pathways it mediates, is more complex than initially proposed. This has been demonstrated using the Alexander hepatoma cell line, PLC/PRF/5, in which the intracellular domain of Robo is successively cleaved by metalloproteases and γ-secretases, yielding two distinct intracellular Robo1 fragments (Robo1-CTF1 and Robo1-CTF2; Seki *et al.*, 2010). The identification of several nuclear localization signals (NLSs) within these intracellular fragments of Robo1 suggests a potential transcriptional role for the receptor, at least in cancer cells. Biochemical fractionation of PLC/PRF/5 cells treated with the proteasome inhibitor MG-132 show Robo1-CTF2 exclusively located in the nucleus, whereas Robo1-CTF1 is found in each of the membrane, cytosolic, and nuclear fractions (Seki *et al.*, 2010). This suggests that the successive cleavage of the Robo1-CTF is critical for proper localization within the cell and may play a regulatory role. However, removal of all three potential NLSs does not abolish nuclear localization, suggesting that perhaps nuclear localization of Robo1-CTF relies on other currently unidentified effector molecules (Seki *et al.*, 2010). It is clear that further studies of NLS-containing Robo1-CTF binding partners are needed to elucidate the full mechanism of transcriptional regulation by Robo1-CTFs.

In addition to intracellular cleavage, a number of axon guidance receptors, including Robo, undergo extracellular cleavages, generating protein products that regulate a number of cellular functions, such as migration. Recently, studies in *Drosophila* aimed at elucidating the exact mechanism of Robo activation following Slit binding reveal a potential role for the metalloprotease–disintegrin Kuzbanian (ADAM10 in mammals) in generating a free ectodomain by extracellular cleavage. Although Kuz/ADAM10 is expressed in both neurons and midline glia, only neuronal expression is required for Slit/Robo repulsion, suggesting that the protease acts on Robo-expressing neurons and not the glia (Coleman *et al.*, 2010). Further, it appears that extracellular cleavage of Robo is required for receptor activation following Slit stimulation, as expression of an uncleavable form of Robo is unable to rescue a *Robo* mutant phenotype. It was also found that cleavage of Robo by Kuz/ADAM10 is necessary for the recruitment of Son of sevenless (Sos) and other factors required for Slit/Robo-mediated repulsion at the midline (Coleman *et al.*, 2010). Taken together, these data suggest that cleavage is an important mechanism that regulates the activation of Robo and its signal transduction, and it is likely that subsequent studies will reveal how cleavage regulates the many different functions of Slit/Robo signaling.

D. Structural Studies Have Defined the Interaction Between Slit and Robo

The interactions between Slit and Robo molecules are evolutionarily conserved, as evidenced by studies that show that human Slit2 is able to bind *Drosophila* Robo1 with similar affinity as its mammalian receptor, and vice versa, that *Drosophila* Slit successfully binds rat Robo1 and Robo2 (Brose *et al.*, 1999). Biochemical studies also show that the interaction between this receptor/ligand pair involves the highly conserved second LRR domain (D2) of Slit and the Ig1 domain of Robo, while Ig2–Ig5 and all FN3 domains of Robo1 appear to be dispensable for binding (Fig. 1C; Chen *et al.*, 2001; Fukuhara *et al.*, 2008; Howitt *et al.*, 2004; Liu *et al.*, 2004; Morlot *et al.*, 2007b). Thus, the binding between Slits and Robos is highly conserved and structurally well defined.

Recent insight into the structural requirements for binding between Slits and Robo1 has revealed that the complex can be stabilized by heparan sulfate glycosaminoglycans (GAGs), which are required for functional Slit/Robo signaling in both *Drosophila* and vertebrate neurological development (Fig. 1C; Fukuhara *et al.*, 2008; Hu, 2001; Hussain *et al.*, 2006; Inatani *et al.*, 2003; Ogata-Iwao *et al.*, 2011; Plump *et al.*, 2002; Schulz *et al.*, 2011; Smart *et al.*, 2011). There appears to be dual functions for these GAGs: first, they bind to Slit in the extracellular matrix and stabilize their interaction with Robo, and second, they act on target cells to mediate Slit/Robo signaling by serving as coreceptors.

There are numerous studies demonstrating the importance of GAGs in facilitating the functional interaction between Slits and Robos. Structural studies by Hussain and colleagues show that heparin, a highly sulfated variant of heparan sulfate, binds to Slit and forms a ternary complex with Robo, resulting in a 10-fold increase in the affinity between Slit and Robo (Hussain *et al.*, 2006). Mutational studies demonstrate a key role for the second LRR of Slit (termed D2) in binding heparin via a conserved basic patch, and binding Robo via the adjacent concave face. Further crystallographic studies reveal a contiguous HS/heparin binding surface, extending across the Slit–Robo interface and consistent with at least five HS disaccharide units, as required to support Slit/Robo signaling (Fukuhara *et al.*, 2008). These biochemical analyses are further supported by functional studies in this and others papers, which demonstrate that enzymatic removal of heparan sulfates from neurons using heparanases results in a loss of responsiveness to Slit (Hu, 2001; Hussain *et al.*, 2006; Piper *et al.*, 2006). In a different strategy, mutation of exostosin, an enzyme required for heparan sulfate synthesis, leads to patterning defects at the mouse optic chiasm that phenocopy those observed in the *Slit1* −/−; *Slit2* −/− knockout (Inatani *et al.*, 2003; Plump *et al.*, 2002). A similar study performed in zebrafish found that this loss of heparan sulfate synthesis phenocopies or even enhances the guidance defects observed in Robo2/astray mutants (Kastenhuber *et al.*, 2009; Lee *et al.*, 2004). Taken together, these studies show that heparan sulfates mediate the formation of stable Slit/Robo signaling complexes and are critical for their signaling function. However, it is unclear whether or not these GAGs constitute membrane bound proteoglycans. Genetic studies in *Drosophila* suggest that this may be the case, as the heparan sulfate proteoglycan, glypican, which is associated with the cell surface via a glycosylphosphatidylinositol (GPI) linkage, interacts with Slit and regulates its distribution (Liang *et al.*, 1999; Ronca *et al.*, 2001; Smart *et al.*, 2011; Zhang *et al.*, 2004). Thus, heparan sulfate GAGs, either cell associated or present as free sugars in the extracellular matrix, concentrate, and localize Slits, shaping the signaling environment by regulating their concentration and accessibility.

On the target cells, the heparan sulfate proteoglycan Syndecan (Sdc) plays a key role as a coreceptor for Robo (Johnson *et al.*, 2004; Rhiner *et al.*, 2005; Steigemann *et al.*, 2004). In the *Drosophila* embryo, Sdc is coexpressed with Robo on axons and is absent in Slit-secreting midline cells. Mutations in mammalian Sdc enhance the muscle and axonal patterning phenotypes observed in loss-of-function *Slit* and *Robo* animals (Johnson *et al.*, 2004; Steigemann *et al.*, 2004). Moreover, cell type-specific rescue experiments in *Sdc* mutants reveal that axon guidance defects of the *Sdc* mutant are entirely rescued by Sdc expression in neurons, while there is no rescuing activity in response to Sdc expression in midline cells (Johnson *et al.*, 2004; Steigemann *et al.*, 2004). These findings indicate that Sdc activity does not participate in the production and/or secretion of Slit, but rather is required for the reception and/or the transmission of Slit signals in Robo-expressing target cells. Further studies in neural development in *Caenorhabditis elegans* corroborate this important role for Sdc in regulating Slit/Robo-mediated axon guidance. Mutational analyses demonstrate that only the extracellular domain of Sdc is required for Slit/Robo signaling (Chanana *et al.*, 2009; Schulz *et al.*, 2011), and that the chondroitin sulfate modification of Sdc is necessary for its coreceptor function on target cells (Chanana *et al.*, 2009).

Together, functional and structural evidence supports a model in which heparan sulfate proteoglycans enhance the relatively low-affinity interaction between Slits and Robos by acting as secondary receptors. Further, studies in *Drosophila* suggest that there are different cellular requirements for proteoglycans, with Sdcs acting on target cells, and glypicans acting on the Slit-expressing cells. Glypicans are thought to sequester Slit and present it to the Sdc/Robo1 pair, thereby regulating the formation of the ternary signaling complex. Together, these collective interactions help to localize and fine-tune Slit/Robo signaling. While current data suggest that the interaction between

Slit, Robo, and GAGs represents a significant regulatory relationship, whether these heparan sulfate coreceptors are required for Slit/Robo signaling in all cell and tissues types in higher organisms, and whether they play a role in disease processes will require more in-depth studies in mammalian models.

III. SLIT/ROBO: ROLES IN EPITHELIAL TUMORIGENESIS

The role of axon guidance molecules in cancer progression has been studied for over a decade, yet their exact function remains elusive. By their nature, axon guidance molecules are bifunctional, acting as both attractant and repellents for migrating axons and cells. As such, Slits also have this dual role; in one example, this bifunctionality is displayed in a single trajectory of mesodermal cells in the *Drosophila* embryo. These cells move away from the ventral midline, repelled by Slit, and then migrate toward target muscles, attracted by Slit (Kramer *et al.*, 2001). Consistent with this dual role as both positive and negative cues, Slits are also capable of acting as both "friend and foe" in the progression of tumor cells; Slits have been shown to both promote and prevent tumor metastasis by suppressing or enhancing cellular attachments and migration depending on the cellular context (Tseng *et al.*, 2010; Zhou *et al.*, 2011). This duality of Slit function is also observed in the regulation of tumor cell proliferation and survival where they promote proliferation and angiogenesis in some contexts and prevent these same processes in others (Table I). Thus, current evidence renders it impossible to label Slits and Robos as either tumor suppressors or oncogenes. Nevertheless, it is clear that acting in either role they play important functions during tumor progression. This makes them attractive targets for cancer therapeutics and potential candidates for diagnostic purposes.

A. Slit and Robo Expressions are Altered in Cancer

The progressive transformation of normal cells into malignant progeny involves the accumulation of genetic changes, such as the loss or silencing of tumor suppressor genes and the induction of oncogenes. Studies show that Slit and Robo expression is altered in a long list of cancers (Tables II and III). In examining the literature, it is interesting to note that there are examples of both up- and downregulation of these genes, suggesting that the Slit/Robo pathway can function in both promoting and suppressing tumor cell survival, proliferation, and migration. Currently, it is unclear whether these genes

Table I The Function of Slit/Robo Signaling in Tumor and Nontumor Cell Proliferation Migration and Angiogenesis

Slit/Robo members	Action	Tissue/cell type	References
Promigration/metastasis/chemotaxis			
Slit2/Robo1	Induces malignant transformation promotes metastasis by regulating degradation of cadherins	HEK293 cells, colorectal epithelial carcinoma cells; colorectal carcinoma xenograft model	Zhou *et al.* (2011)
Slit/Robo	Mediates migration by downregulating N-cadherin, decreasing adhesion and increasing translocation of β-catenin to the nucleus	Axon growth cone migration, mouse fibroblast L-cells	Rhee *et al.* (2002, 2007)
Robo4	Mediates attractant signals in endothelial cells via Rho GTPases	Embryonic zebrafish vasculature	Kaur *et al.* (2006)
Slit2/Robo4:Robo1 heterodimer	Mediates the chemotactic response of endothelial cells	Zebrafish vasculature	Kaur *et al.* (2008)
Slit2/Robo1	Mediates the migration and vascular tube formation. Slit2 also acts as a lymphangiogenic factor	Mouse lymphatic endothelial cells	Yang *et al.* (2010)
Slit2/Robo1:Robo4 heterodimer	Promotes cell migration and angiogenesis by inducing formation of filopodia	Human umbilical vein endothelial cells (HUVECs)	Sheldon *et al.* (2009)
Slit/Robo	Enhances cell migration by decreasing e-cadherin function at the membrane	*Drosophila* heart tube formation	Santiago-Martinez *et al.* (2008)
Slit/Robo1	Promotes migration by positively regulating Rho GTPases	Src-transformed epithelial cells	Khusial *et al.* (2010)
Slit2/Robo1	Promotes eosinophil chemotaxis, possibly through activation of Cdc42	Ovalbumin airway inflammation	Yang *et al.* (2010), and Ye *et al.* (2010)
Slit2/Robo1	Promotes directed migration and metastasis of cancer cells	Breast cancer cell lines	Schmid *et al.* (2007)
Antimigration/metastasis/chemotaxis			
Slit2/Robo3	Prevents cell migration by stabilizing P-cadherin at the membrane	Oral mucosa, oral squamous cell carcinoma (OSCC)	Bauer *et al.* (2011)

Slit/Robo	Prevents metastasis by enhancing cell–cell adhesion via N-cadherin	Chick cranial trigeminal gangliogenesis	Shiau and Bronner-Fraser (2009)
Slit2/Robo1	Prevents epithelial cell migration by enhancing cell–cell adhesion via PI3K and β-catenin	Breast cancer cells	Prasad *et al.* (2008)
Slit2/Robo1	Prevents cell migration and promotes cell–cell adhesion by regulating E-cadherin expression	Human lung cancer cell lines and lung tumor samples	Tseng *et al.* (2010)
Slit2/Robo	Prevents VSMC migration by modulating cytoskeletal molecules	Vascular smooth muscle cells (VSMCs)	Liu *et al.* (2006)
Slit2/Robo	Inhibits migration of RASM cells by controlling WASP and Arp2/3 expressions	Rat airway smooth muscle (RASM) cells	Ning *et al.* (2011)
Slit2/Robo	Blocks angiocrine-induced tumor growth and migration	Endothelial cell lines, Invasive human ductal carcinoma samples	Brantley-Sieders *et al.* (2011)
Slit2/Robo1	Prevents cell migration/metastasis by attenuating Cdc42 activity	Epithelial cell lines, medulloblastoma, glioma cell lines and tumor samples, mouse glioma xenograft models	Stella *et al.* (2009), Werbowetski-Ogilvie *et al.* (2006), and Yiin *et al.* (2009)
Slit2/Robo	Inhibits migration of cells toward chemoattractant stimulus by attenuating Cdc42 and Rac2	Leukocytes, vascular smooth muscle cells (VSMCs), mouse model of chemical irritant peritonitis	Liu *et al.* (2006), Tole *et al.* (2009), and Wu *et al.* (2001)
Slit2/Robo1	Regulates eosinophil/neutrophil chemotaxis by modulating srGAP expression	Eosinophils, neutrophils in endotoxin-induced lung inflammation model	Ye *et al.* (2010)
Slit2/Robo1	Prevents CXCR4/CXCL12-mediated inhibiting downstream signaling	Breast cancer cell lines	Prasad *et al.* (2004)
Slit2/Robo1	Prevents metastasis of cancer cells by recruiting USP33 and redistributing Robo1 to the membrane	Breast cancer cells	Yuasa-Kawada *et al.* (2009)
Slit/Robo4	Inhibits endothelial cell migration and proliferation	Rat endothelial cells	Suchting *et al.* (2005)

(*continues*)

Table I *(continued)*

Slit/Robo members	Action	Tissue/cell type	References
Prosurvival/proliferation			
Slit/Robo	Increases proliferation by inducing the translocation of β-catenin to the nucleus	Axon growth cone migration, mouse fibroblast L-cells	Rhee *et al.* (2007) and Rhee *et al.* (2002)
Antigrowth, proliferation			
Slit2/Robo1	Restricts branch formation by inducing translocation of β-catenin to the membrane, decreasing basal cell proliferation	Mammary epithelial cells	Macias *et al.* (2011)
Slit2/Robo1	Prevents epithelial cell growth by restricting nuclear β-catenin localization	Breast cancer cells	Prasad *et al.* (2008)
Slit2/Robo1	Prevents tumor growth by restricting nuclear β-catenin levels via activation of the GSK-3β pathway	Human nonsmall cell lung cancer cell line	Tseng *et al.* (2010)
Slit2/Robo	Restricts tumor growth by regulating signaling downstream of CXCL12/CXCR4	Breast cancer xenograft models, mammary epithelial cells	Marlow *et al.* (2008)
Antiangiogenic			
Robo4	Restricts aortic sprouting, coordinates intersomitic vessel formation	Embryonic zebrafish vasculature	Bedell *et al.* (2005)
Slit/Robo4	Inhibits angiogenesis, vascular tube formation	Rat endothelial cells	Suchting *et al.* (2005)
Slit2/Robo1	Inhibits angiogenic remodeling in the presence of ephrin-A1	Mouse pulmonary microvascular endothelial cells	Dunaway *et al.* (2011)
Slit2/Robo1:Robo4 heterodimer	Inhibits pathological neovascular processes of the cornea by signaling through Robo1 and Robo4	Corneal neovascular rat model, primary HUVECs	Han and Zhang (2010)
Slit2/Robo4	Blocks angiogenesis by inhibiting signaling downstream of VEGF/VEGFR	Mammalian endothelial cell lines, rat and mouse retinal endothelium	Jones *et al.* (2008) and Jones *et al.* (2009)
Robo1	Restricts choroidal and retinal angiogenesis *in vitro*	Monkey choroidal retinal endothelial cells, retinopathy of prematurity	Huang *et al.* (2009a)

Slit/Robo4	Prevents angiogenesis by inhibiting signaling downstream of VEGF/VEGFR	Endothelial cells of the mammary gland stroma, blood vessel endothelial cells of the corneal stroma	Marlow *et al.* (2010) and Mulik *et al.* (2011)
Slit2	Decreases LPS-induced vascular permeability by increasing V/E-cadherin levels at the membrane	Blood vessel endothelium of the lung	London *et al.* (2010)
Proangiogenic			
Slit2/Robo1	Promotes tube formation and endothelial cell migration	Human malignant melanoma A375 cells	Wang *et al.* (2003)
Slit2/Robo1	Promotes tumor angiogenesis and growth *in vivo*	Oral carcinogenesis	Wang *et al.* (2008)
Slit2/Robo1	Promotes lymphangiogenesis and lymphatic metastasis	Lymphatic endothelial cells	Yang *et al.* (2010)
Slit2/Robo	Promotes angiogenesis in culture and *in vivo*	Mouse pulmonary microvascular endothelial cells	Dunaway *et al.* (2011)

Table II A List of Cancers in Which Expression of Slits and Robos are Decreased, and the Mode of Regulation, If Known

Elevated expression level		
Gene	Cancer type	References
Slits	Prostate cancer, nitrofen-hypoplastic lung cancer, and lobular breast carcinoma	Christgen *et al.* (2009), Doi *et al.* (2009), Latil *et al.* (2003), and Ma *et al.* (2004)
Slit2	Oral cheek mucosa with oral squamous cell carcinoma	Wang *et al.* (2008)
Robo1	Hepatocellular cancer, colorectal cancer, nonsmall cell lung cancer, Glioma cancer, monkey choroidal retinal endothelial cells, retinopathy of prematurity (ROP), and neovascularized cornea	Avci *et al.* (2008), Gorn *et al.* (2005), Grone *et al.* (2006), Han and Zhang (2010), Huang *et al.* (2009), Ito *et al.* (2006), Mertsch *et al.* (2008), and Xu *et al.* (2010)
Robo2	Hepatocellular cancer	Avci *et al.* (2008)
Robo4	Tumor endothelial cells, neovascularized cornea, monkey choroidal retinal endothelial cells, and colorectal cancer	Avci *et al.* (2008), Grone *et al.* (2006), Han and Zhang. (2010), Mura *et al.* (2011), and Seth *et al.* (2005)
Robo4	HSV-infected endothelial cells in the corneal stroma	Mulik *et al.* (2011)

are differentially regulated based on tumor type or stage, but mounting evidence suggests that changes in the expression of these genes play important roles in regulating tumor progression.

1. SLIT EXPRESSION IS ALTERED IN EPITHELIAL TUMOR PROGRESSION

The most frequently observed alteration of *Slit* expression is downregulation. This is evidenced by allelotyping studies of 44 breast carcinoma samples that show loss of heterozygosity at several regions on chromosome 4, one of which has been identified as the *Slit2* gene locus (4q25–26; Shivapurkar *et al.*, 1999a,b; Singh *et al.*, 2007). These studies report allelic deletion in 63% of breast carcinomas, 35% of cervical carcinoma, and >60% of small cell lung carcinoma and mesothelioma (Shivapurkar *et al.*, 1999a,b; Singh *et al.*, 2007). Thus, it seems that a common method for alteration of Slit/Robo signaling in cancer is via *Slit* gene silencing. In addition to gene loss by deletion, several other mechanisms of gene silencing occur at the *Slit* gene locus. Of these, the most commonly encountered mechanism is hypermethylation of the promoter region. Numerous studies

Table III A Summary of Cancers and Diseases in Which Expression of Slits and Robos are Elevated

Decreased expression level

Gene	Mode of silencing	Cancer type	References
Slit2	LOH, allelic deletion	63% of breast carcinoma, 35% of cervical carcinoma, and >60% of small lung carcinoma and mesothelioma	Shivapurkar *et al.* (1999a, 1999b) and Singh *et al.* (2007)
Slit2	Promoter hypermethylation	Breast carcinoma, nonsmall cell lung cancer, ovarian carcinoma, gliomas, hepatocellular carcinoma, colorectal carcinoma, and lymphocytic leukemia primary tumors	Dallol *et al.* (2005, 2003a, 2003b), Dunwell *et al.* (2009), Jin *et al.* (2009), Qiu *et al.* (2011), Sharma *et al.* (2007)
Slit2	Unknown	Corneal neovascularization	Han and Zhang (2010) and Wang *et al.* (2008)
Slit1, 3	Promoter hypermethylation	41% of breast, 33% of colorectal, and 29% of glioma tumor cell lines and primary tumors	Dallol *et al.* (2005) and Dickinson *et al.* (2004)
Slit3	Unknown	Hepatocellular carcinoma	Avci *et al.* (2008)
Slit2	Promoter hypermethylation catalyzed by polycomb group member EZH2	Human prostate cancer samples	Yu *et al.* (2010)
Robo1	Gene deletion	Small-cell lung cancer cell line (U2020)	Xian *et al.* (2001)
Robo1	Promoter hypermethylation	19% of primary invasive breast cancer, 18% of clear cell renal cell cancer, and 4% primary nonsmall cell lung cancer	Dallol *et al.* (2002b)
Robo1, 2	Promoter hypermethylation	Early dysplastic lesions of head and neck cancer	Ghosh *et al.* (2009)
Robo3	Promoter hypermethylation	Cervical cancer	Narayan *et al.* (2006)
Robo4	Unknown	Human breast tumor samples	Richardson *et al.* (2006)
Robo4	Unknown	Hepatocellular carcinoma	Avci *et al.* (2008)

have shown that regions frequently hypermethylated in cancers contain the genes for *Slit1*, *Slit2*, *Slit3*, *Robo1*, and *Robo3* (Dallol *et al.*, 2005). For example, *Slit2* is silenced through hypermethylation in the majority of samples from numerous tumor types including: breast, non-small cell lung, ovarian, gliomas, hepatocellular, colorectal cancers as well as lymphocytic leukemias (Dallol *et al.*, 2003a,b; Dunwell *et al.*, 2009; Jin *et al.*, 2009; Qiu *et al.*, 2011; Sharma *et al.*, 2007). In further support of a role for Slits in suppressing tumor growth and, consequently, being silenced during tumor progression, reexpression of Slit2 greatly inhibits the proliferation of transformed cell lines derived from many of these tumor types (Dallol *et al.*, 2002a, 2003b; Jin *et al.*, 2009; Qiu *et al.*, 2011). Although Slit2 is the most frequently studied of the Slit proteins expressed in mammals, similar expression studies reveal silencing of *Slit3* via hypermethylation in breast (41%), colorectal (33%), and glioma (29%) tumor cell lines, with similar frequencies of *Slit3* and *Slit1* promoter hypermethylation reported in these types of primary tumors (Dickinson *et al.*, 2004).

One additional consequence of epigenetic silencing of *Slit2* and *Slit3* is downregulation of microRNA (miR)-218-1 and miR-218-2, which are located within intron 15 of human *Slit2* and intron 14 of human *Slit3*, respectively (Angeloni *et al.*, 2006; Tie *et al.*, 2010). This miR negatively regulates *Robo1* expression in gastric, head, and nasopharyngeal cancers (Alajez *et al.*, 2011; Tie *et al.*, 2010), but its downregulation due to *Slit* methylation in many tumor types provides one explanation for the observation that *Robo1* appears to be infrequently silenced in the majority of tumor samples (Grone *et al.*, 2006; Ito *et al.*, 2006; Xu *et al.*, 2010). Interestingly, it is possible that loss of this negative feedback loop contributes to tumor progression because the expression of Robo1 in tumor cells could allow them to migrate in response to Slits that are provided by non-tumor cells in the surrounding environment (Alajez *et al.*, 2011), or by tumor cells that still secrete Slit because they have only partially silenced its expression (Tie *et al.*, 2010). Thus, hypermethylation of the *Slit* gene loci in solid tumors may contribute to tumor progression by switching Slit/Robo1 signaling from autocrine to paracrine, facilitating the metastasis of tumor cells that are responding to this deregulated pathway. Given the frequency of *Slit* hypermethylation in human tumors and its effect on miR-218 expression, this family of genes represents attractive candidates for therapeutic strategies that reverse epigenetic silencing or re-establish miR-218 expression.

Recently, a second mechanism of *Slit* gene silencing was observed in human cancer samples. A genome-wide location analysis of human prostate cancer samples identified *Slit2* as a target of epigenetic repression via the polycomb group (PcG) member EZH2 (Yu *et al.*, 2010). PcG proteins are transcriptional repressors that function through multimeric chromatin-associated polycomb repressive complexes to epigenetically silence gene expression by catalyzing the methylation of specific histone residues. In prostate cancer samples, low Slit2 expression correlates not only with high EZH2 expression level but also with the aggressiveness of the cancer and the degree of metastasis. Further, treatment with either methylation inhibitors or EZH2-suppressing compounds decreases metastasis and increases Slit2 expression (Yu *et al.*, 2010). Taken together, these studies introduce a novel mode of *Slit* silencing that had previously not been recognized in cancer samples.

In contrast to downregulation of Slits, which is well documented in the literature, there are relatively few papers that identify the upregulation of Slits in cancer as occurs, for example, in human ductal carcinoma samples, prostate, and nitrofen-hypoplastic lung cancers (Brantley-Sieders *et al.*, 2011; Doi *et al.*, 2009; Latil *et al.*, 2003). Data mining studies have also revealed their upregulation in lobular breast cancers, a type of breast carcinoma that has been ascribed only a few unique molecular characteristics (Christgen *et al.*, 2009; Ma *et al.*, 2004). These data suggest that at least some types of tumors are associated with Slit overexpression, but how it contributes to tumor development in these circumstances is currently unknown.

2. ROBO EXPRESSION IS ALTERED IN EPITHELIAL TUMOR PROGRESSION

Robo1 was discovered in *Drosophila* as a gene required for proper midline crossing of commissural axons during development (Kidd *et al.*, 1999). It was also found to be deleted in the small cell lung cancer cell line U2020, hence the name Deleted in U twenty twenty, or Dutt1 (Xian *et al.*, 2001). *Robo1* and *Dutt1* genes are derived from alternative promoters of the same gene and appear to have differential spatial and temporal patterns of transcriptional activity, with the *Dutt1* form expressed ubiquitously, and the *Robo1* form restricted primarily to embryogenesis (Clark *et al.*, 2002). In addition to being silenced by deletion, *Dutt1* is also hypermethylated in subsets of primary tumor samples, such as primary invasive breast cancer (19%), primary clear cell renal cell cancer (18%), and primary non-small cell lung cancer (4%). Of those tumors, 80% of breast and 75% of primary clear cell renal cell carcinomas also contain allelic losses in the genomic region containing *Dutt1*, an observation supporting a role for *Dutt1* as a tumor suppressor that obeys Knudson's two hit hypothesis (Dallol *et al.*, 2002b). More recently, hypermethylation at the *Robo1* and *Robo2* gene loci was reported in early dysplastic lesions of head and neck (Ghosh *et al.*, 2009), as well as in cervical cancer (Narayan *et al.*, 2006). While these data point to a tumor suppressor function for Robos, in fact the percentage of tumors displaying reduced or silenced *Robo* expression is much less than that seen for *Slit* genes. Indeed, the opposite is true, with *Robo1* expression elevated in numerous cancers, including human hepatocellular carcinoma, colorectal cancer, non-small cell lung cancer, and glioma samples (Gorn *et al.*, 2005; Grone *et al.*, 2006; Ito *et al.*, 2006; Mertsch *et al.*, 2008; Xu *et al.*, 2010). In one of these tumor samples, coordinate regulation of Slit and Robo was observed, with decreased Slit concomitant with increased Robo1 expression, as predicted by miR-218 regulation of the Slit/Robo signaling axis (Xu *et al.*, 2010). Thus, this finding of a negative regulatory loop that upregulates *Robo* when *Slit* is silenced generates a layer of complexity to the study of Slit/Robo1 in cancer cells. While it was previously simple to label the pathway as tumor suppressive when *Slits* were found to be silenced, this new finding introduces the possibility that, under these circumstances, Slit/Robo signaling can function oncogenically due to upregulation of Robo in the tumor setting.

In sum, since the first hint that Slit and Robo could play a role in tumor biology, there have been numerous studies documenting changes in their level of expression in tumor samples. Recent insights demonstrate the complexity of this regulation, as evidenced by the negative feedback loop for Robo expression under the control of non-coding RNAs encoded intronically in *Slit2* and *Slit3*. This regulatory relationship suggests that the level of both ligand and receptor must be assessed when drawing conclusions about the overall effect of Slit/Robo signaling on tumor progression. Moreover, because the Slit/Robo pathway regulates many common signaling pathways that are often deregulated in cancers, such as those mediated by the Rho family of small GTPases and β-catenin, it is becoming clear that changes in Slit and Robo expression have effects that extend beyond the roles originally identified for these proteins as instructive cues for cell migration. In fact, a growing body of literature shows that Slit/Robo signaling affects other aspects of tumor cell behavior, including their survival and growth.

B. Slits and Robos Mediate Tumor Cell Survival and Proliferation

Once the process of transformation has been initiated by driver mutations, the expansion and progression of premalignant cells to metastatic carcinomas depend on a multistep process involving the evasion of proapoptotic signals and the reception of prosurvival and proproliferative signals. Thus, the early stages of cancer development involve the response of nascent tumor cells to cues such as Slits in their surrounding environment. While there are only some suggestions in the literature that Slits may regulate cell survival, our understanding of its role in regulating cell proliferation is growing. Slits appear to control cell proliferation through β-catenin, a signaling target that has distinct functions at the plasma membrane, where it mediates cell–cell contact in association with the homotypic cell adhesion protein, E-cadherin, and in the nucleus, where it regulates cell proliferation in association with Lef/Tcf transcription factors. Consequently, through this one downstream target, Slits influence two aspects of tumor cell transformation: proliferation and adhesion.

1. SLITS AND ROBOS: INHIBITORS OF TUMOR CELL DEATH

A critical step of tumor cell transformation is achieving immortality. Cells employ several mechanisms to execute apoptosis in order to prevent the survival of rogue cells. Apoptosis can be triggered intrinsically, for example via p53, leading to the release of cytochrome c from the mitochondria and the activation of caspases via the apoptosome (Fulda and Debatin, 2006). Apoptosis can also be initiated extrinsically by signaling through death receptors, also culminating in the activation of caspases (Fulda and Debatin, 2006). An alternative extrinsic cell death pathway is mediated by so-called "dependence receptors" that require or "depend on" their ligand to prevent their own constitutive proapoptotic signaling, which occurs when their ligand falls below a critical concentration. While Robos have not been identified as dependence receptors, the receptors for the Netrin family of axon guidance cues, DCCs and UNC5s, do fit into this category of receptors. Consequently, in addition to mediating axon guidance through these receptors, Netrin also acts as a survival factor both for axons during normal development and for cancer cells during tumor progression (Delloye-Bourgeois *et al.*, 2009; Fitamant *et al.*, 2008; Furne *et al.*, 2008). Intriguingly, Slits have been found to regulate Netrins by binding and sequestering them (Stein and Tessier-Lavigne, 2001). Consequently, Slits may act as proapoptotic factors by allowing the concentration of Netrin to fall below threshold levels for survival. Although there is no direct evidence to support a link between the regulation of Netrin by Slit and the loss of apoptosis in cancer, this regulatory loop suggests a possible mechanism through which Slit exerts a tumor suppressive function by promoting the proapoptotic signaling of dependence receptors. The silencing of *Slits*, which occurs in many types of cancer, could, therefore, contribute to the immortality of tumor cells by disabling one of the pathways that culls rogue cells that have tumor forming potential.

2. SLITS AND ROBOS: REGULATORS OF TUMOR CELL PROLIFERATION

Following survival of a few transformed cells, cancer progression requires proliferation of these cells to generate tumor mass. Investigation into the role of Slit/Robo signaling in regulating cell proliferation has revealed both positive and negative effects, as predicted by its dual function as both attractant and repellent guidance cue. In support of a role for Slit in suppressing cell proliferation, numerous studies suggest that signaling through Robo1 regulates the subcellular localization of β-catenin, inhibiting its transcriptional function in the nucleus by promoting its localization at the membrane. In a recent study, we have shown that during murine breast (mammary gland) development, Slit/Robo signaling restricts the proliferation of the outer layer of basal cells by increasing the cytoplasmic and membrane pools of β-catenin at the expense of its nuclear pool (Macias *et al.*, 2011). This loss of growth control during early postnatal mammary gland development generates an overabundance of myoepithelial cells that produce an excess of growth factors, leading to an overall increase in cell proliferation and excessive branching morphogenesis. Eventually, these surplus myoepithelial cells invade the luminal population and disrupt cell adhesion (Strickland *et al.*, 2006), and, along with other changes that occur, such as upregulation of CXCR4 and SDF1, spur the development of hyperplastic lesions with basal characteristics (Marlow *et al.*, 2008).

The imbalance in growth control during early mammary gland development observed in *Robo1−/−* tissue provides a gratifying developmental correlate for the role of Slits in suppressing growth in models of breast and non-small cell lung cancer (Prasad *et al.*, 2008; Tseng *et al.*, 2010). In breast cancer cell lines, overexpression of Slits inhibits the transcriptional activity of β-catenin by activating glycogen synthase kinase-3beta (GSK-3β) through the phosphoinositol-3-kinase (PI3K)/Akt signaling pathway. These overexpressing cells display enhanced intercellular adhesion and greater colocalization of β-catenin with E-cadherin (Fig. 2A). Further, studies performed in xenograft models of breast cancer show that tumors generated from Slit-overexpressing cells are significantly smaller compared to control tumors (Marlow *et al.*, 2008; Prasad *et al.*, 2008). A comparable but converse experiment was performed in a cell line derived from non-small cell lung cancer in which knockdown of *Slit* increases the metastatic potential of the cells by inhibiting GSK-3β activity, again via the PI3K/Akt pathway (Tseng *et al.*, 2010). This, in turn, increases the levels of nuclear β-catenin and increases the expression of Snail, a crucial regulator of epithelial–mesenchymal transitions (EMT), resulting in decreased cadherin expression, reduced cell adhesion, and increased cell motility (Fig. 2A; Tseng *et al.*, 2010). Taken together, these studies show that, at least in breast and lung, Slits act as tumor suppressors promoting the adhesive role of β-catenin at the membrane at the expense of its proliferative role in the nucleus.

Reminiscent of the context-dependent roles that Slit plays as both attractant and repellent in cell migration (Kramer *et al.*, 2001), the opposite role for Slit as inducer of proliferation has also been documented. In cell lines derived from colorectal carcinoma, Slit/Robo1 signaling enhances tumor

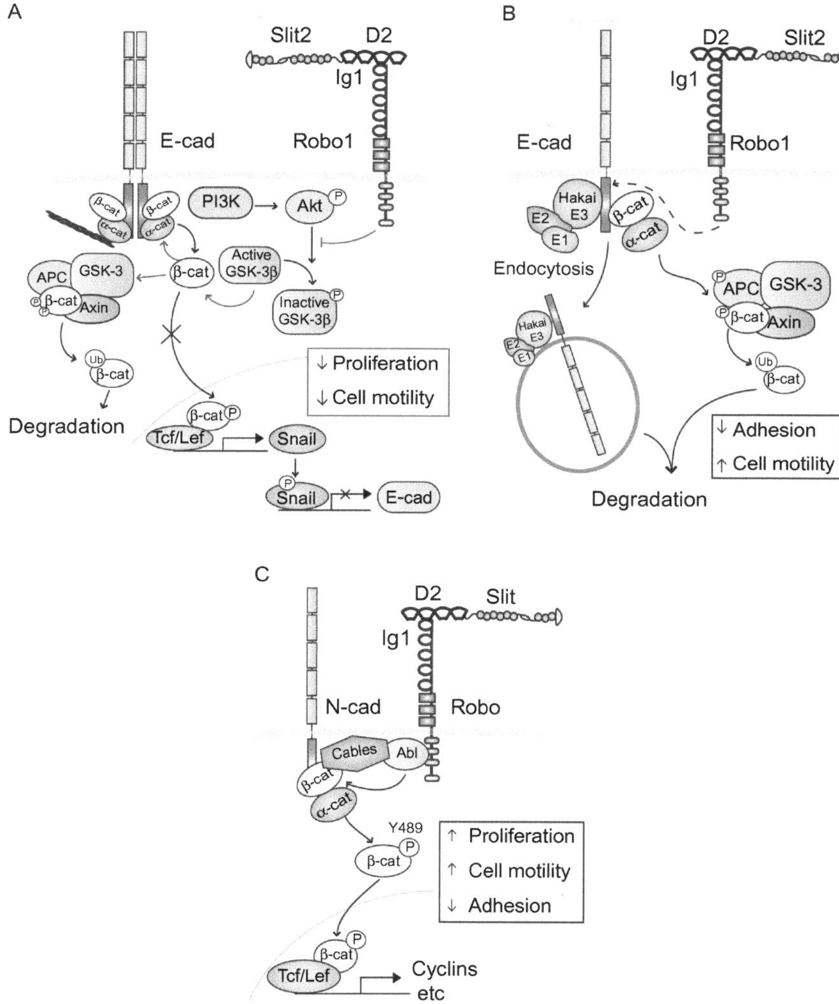

Fig. 2 Slit/Robo signaling regulates cell proliferation and cell–cell contacts by controlling the localization of β-catenin in the cell. (A) As illustrated by blue arrows, binding of Slit to Robo inhibits phosphatidylinositol kinase (PI3K)-induced Akt activity. Glycogen synthase kinase-3beta (GSK-3β) is consequently left in its non-phosphorylated, active form, and targets β-catenin for phosphorylation, excluding it from the nucleus and thus preventing its transcriptional activity. Cytoplasmic β-catenin either becomes ubiquitinated through the GSK-3β–adenomatous polyposis coli (APC)–Axin complex or transferred to the membrane where it interacts with E-cadherin, stabilizing cell–cell contacts, and preventing cell migration. A third effect of Slit2/Robo1 signaling is reduced nuclear β-catenin, which in turn inhibits Snail expression, thus relieving the repression on E-cadherin expression and enhancing cell–cell contacts. Slit/Robo signaling can also function to decrease cell–cell contacts and increase proliferation (B, C). (B) Slit/Robo signaling drives cell migration by recruiting the ubiquitin ligase Hakai to

growth and metastasis by regulating cadherin degradation and, thereby, increasing cell proliferation and migration (Zhou *et al.*, 2011). Overexpression of either Slit2 or Robo1 or recombinant Slit2 treatment of Robo1-expressing colorectal epithelial carcinoma cells results in recruitment of the ubiquitin ligase, Hakai, to E-cadherin and its subsequent ubiquitination and lysosomal degradation (Fig. 2B). Downregulation of E-cadherin in these cells is accompanied by an EMT, increased proliferation and increased migration, and in a xenograft model, this corresponds to increased tumor growth and metastasis. Clinical data corroborated these observations, showing an increase in Slit and Robo1 expression in metastatic, compared to nonmetastatic, human colorectal carcinoma samples. This increase inversely correlates with the overall survival of patients, supporting the idea that in some tumor contexts, Slit/Robo signaling can function oncogenically to promote cell growth and migration (Zhou *et al.*, 2011).

Studies on embryonic chick neural retinal cells have identified a second mechanism for downregulating cadherin through Slit/Robo. In this setting, Slit/Robo signaling induces the recruitment of Cables to the Abelson tyrosine kinase (Abl), which is bound by Robo (Rhee *et al.*, 2007). This causes Cables, in turn, to bind to β-catenin and form a complex with N-cadherin at the plasma membrane, which brings Abl into position to phosphorylate β-catenin on Y489. This triggers the dissociation of β-catenin from N-cadherin, compromising cell–cell adhesion, and allowing translocation of Y489-phosphorylated β-catenin to the nucleus where it activates Tcf/Lef-mediated transcription (Fig. 2C; Rhee *et al.*, 2007). Although a correlation between this Slit/Robo-mediated increase in nuclear β-catenin activity and enhanced cell proliferation is not reported in this study, an increase in proliferation in a different cell type from the retina, retinal pigment epithelial cells, has been observed in response to recombinant Slit2 treatment (Zhou *et al.*, 2011). Thus, several signaling pathways have been identified that support an oncogenic role for Slit in reducing cell adhesion and enhancing cell proliferation.

Taken together, these studies suggest that Slit/Robo1 signaling regulates cellular proliferation by targeting both cadherins and β-catenin in order to regulate the transcriptional activity of β-catenin. In events that suppress tumor growth, Slit/Robo1 directs the subcellular localization of β-catenin

E-cadherin and of β-catenin. This results in the proteasomal degradation of β-catenin and Hakai-mediated lysosomal degradation of E-cadherin, causing decreased cell–cell contacts and enhanced cell migration. (C) In another context, Slit/Robo signaling recruits the Abelson tyrosine kinase (Abl), which binds to the adaptor protein Cables. Cables links the Robo/Abl complex to the N-cadherin/β-catenin complex, thus enabling Abl to phosphorylate β-catenin on Y489, causing its translocation to the nucleus where it activates transcription of cell proliferation genes. (See Page 7 in Color Section at the back of the book.)

through the PI3K/Akt pathway, an effect that has been documented in both non-small cell lung and breast cancer models, as well as during normal breast development (Macias *et al.*, 2011; Prasad *et al.*, 2008; Tseng *et al.*, 2010). In contrast, two different mechanisms have been identified that achieve oncogenic outcomes downstream of Slit/Robo signaling. In both examples, the cadherin/β-catenin complex is disrupted, releasing β-catenin. However, in one mechanism this occurs through Slit/Robo-induced lysosomal degradation of cadherin and in the other, through targeted phosphorylation of β-catenin by an Abl/Robo complex. In conclusion, additional studies are required to determine the extent to which these proproliferative, promigratory mechanisms regulate Slit signaling in normal and disease settings.

C. Slits and Robos Mediate Tumor Cell Motility and Metastasis

Tumor cell metastasis requires multiple steps including: weakening associations between tumor and neighboring cells or between tumor cells and the environment, rearrangement of the actin cytoskeleton to drive actin protrusions and other structures necessary for cell motility, and sensitization of the cell to attractant signaling gradients. These changes occur while the cell is simultaneously desensitized to repellent signaling molecules in the environment, thus allowing cell migration. Slits and Robos have been implicated in each of these steps and, not surprisingly, they have been found to act as both oncogenes and tumor suppressors, enhancing and inhibiting tumor cell invasion, depending on the cellular context.

1. SLITS AND ROBOS: REGULATORS OF CELL–CELL ADHESIONS

Cadherins are expressed in all epithelial cells and play a key role in establishing contact between a cell and its environment. Cadherin expression is often misregulated in cancer cells, which leads to decreased cell attachment and a more metastatic phenotype (Blanco *et al.*, 2004). This allows tumor cells to migrate and invade the vasculature, leading to cancer metastasis. Slit/Robo signaling has been shown to regulate this first step toward metastasis by influencing cell adhesion through its action on cadherins and β-catenin. As discussed above, one consequence of this regulation is altered subcellular distribution of β-catenin, which increases proliferation with its translocation to the nucleus (Prasad *et al.*, 2008; Rhee *et al.*, 2007, 2002; Tseng *et al.*, 2010; Zhou *et al.*, 2011). There is additional evidence, mostly genetic and collected in developmental settings, that further demonstrates a role for Slit and Robo in regulating cadherin-mediated cell–cell adhesion. Again, unsurprisingly, given Slit's function as attractant and repellent in axon guidance, both increased and decreased cell adhesion have been attributed to Slit/Robo signaling, dependent on the biological context. For example, a positive role for Slit/Robo in enhancing cell–cell adhesion is observed during chick cranial trigeminal gangliogenesis when cells derived from neural crest and ectodermal placodes interact to generate ganglionic structures (Shiau and Bronner-Fraser, 2009). Trigeminal placode cells express N-cadherin and Robo1, while the intermingling neural crest cells express Slit1. Loss of either N-cadherin or Robo1 results in dispersed and disorganized placodal neurons within the trigeminal ganglion, suggesting that N-cadherin and Robo1 function in collaboration to mediate the proper coalescence of placode-derived neurons (Shiau and Bronner-Fraser, 2009). In concordant studies, overexpression of either Slit or Robo results in both the post-translational upregulation of N-cadherin and its redistribution to the placodal cell membrane, again leading to a model in which Slit/Robo signaling stabilizes sites of cell–cell contact by influencing the subcellular localization of cadherin (Shiau and Bronner-Fraser, 2009). It is important to note, however, that in this study no changes in the level or distribution of β-catenin were reported. In a second example of Slit mediating increased cell adhesion in collaboration with a cadherin, P-cadherin is shown to colocalize with Slit in the basal cell layers of normal oral mucosa, with this expression downregulated in oral squamous cell carcinoma (OSCC; Bauer *et al.*, 2011). In an OSCC cell line that overexpresses P-cadherin, a complex of P-cadherin and Robo3 is detected, and treatment of these cells with Slit results in a dose-dependent downregulation of cell migration that could be relieved using a small interfering RNA that reduces Robo3 expression (Bauer *et al.*, 2011). Taken together, these studies support a tumor suppressive role for the Slit/Robo signaling axis in maintaining cell–cell adhesion and, consequently, a non-invasive cellular state by enhancing cadherin function.

Conversely, Slit/Robo signaling has also been shown to inhibit cadherin function at the membrane during *Drosophila* heart tube formation, resulting in decreased cell–cell adhesion (Santiago-Martinez *et al.*, 2008). Again, genetic evidence suggests that Robo and E-cadherin/shotgun (shg) function together in modulating cardioblast adhesion, but in this biological context, their actions oppose one another. This is evidenced by the observation that *Robo* loss-of-function phenocopies *e-cadherin/shg* gain-of-function, generating embryos with no lumen due to enhanced cardioblast adhesion (Santiago-Martinez *et al.*, 2008). Similarly *Robo* gain-of-function phenocopies *e-cadherin/shg* loss-of-function, but in this circumstance lumen formation was blocked due to insufficient cardioblast adhesion. These studies support a role for Robo antagonizing E-cadherin/shg function, with Robo mediating a repulsive or antiadhesive signal that functions in opposition of the proadhesive actions of E-cadherin/shg.

By targeting cadherins and the cadherin/β-catenin complex, Slit/Robo signaling regulates two of the crucial steps in tumor progression: cell proliferation and cell adhesion. A challenge for researchers is to understand the circumstances that determine whether this signaling pathway acts positively to enable cell contacts, or negatively to deter them. For other guidance families, different complexes of receptors specify attraction versus repulsion. For example, attraction via Netrin is mediated by DCC in a complex with DSCAM, whereas repulsion requires an UNC5 receptor that acts either together with DCC or alone (Moore *et al.*, 2007). In contrast, no coreceptors have been identified that specifically regulate the attractant or repellent functions of Robo, although perhaps its interaction with cadherin, albeit indirect, serves this role. Regardless of these events at the plasma membrane, a central requirement for either the positive or the negative response of a cell to Slit is the interaction of Robo with the actin cytoskeleton, a topic that is discussed in the next section.

2. SLITS AND ROBOS: REGULATORS OF THE ACTIN CYTOSKELETON

Following detachment of cells from the surrounding tissue, tumor progression requires enhanced cell motility, which is accompanied by increased actin polymerization and the enhanced activity of proteins that optimize its turnover. Developmental studies show that Slit/Robo signaling affects cell motility by controlling the activity of several proteins involved in reorganizing the actin cytoskeleton, including the small GTPases comprising the Rho family (Rac, Cdc42, and RhoA) and other key regulators of the actin cytoskeleton, such as the nonreceptor tyrosine kinase, Abl, and Ena/Vasp proteins.

a. Rho GTPases

Many studies have shown that Rho GTPases play an important role in modulating the downstream action of Slit/Robo1 signaling. These proteins switch between active and inactive states, and are regulated by GEFs (guanine nucleotide exchange factors) and GAPs (GTPase-activating proteins): the former stimulate, and the latter inhibit GTPase function. During *Drosophila* neural development, Slit/Robo-mediated repulsion of commissural neurons at the midline requires the activation of rac (Fan *et al.*, 2003), and recruitment of both a rac GEF, called Sos, and a rac GAP, called vilse/crGAP (Hu *et al.*, 2005; Lundstrom *et al.*, 2004; Yang and Bashaw, 2006). Sos binds to Robo through the adaptor protein dock (Nck in mammals; Fan *et al.*, 2003; Yang and Bashaw, 2006), whereas vilse/crGAP interacts directly with Robo (Fig. 3A; Hu *et al.*, 2005; Lundstrom *et al.*, 2004). Studies show that the activities of both this GAP and GEF support Slit/Robo-mediated repulsion, as both Sos and vilse/crGAP mutants display mild defects in midline repulsion that can be significantly enhanced through loss of one copy of either *Slit* or *Robo* (Fig. 3A; Hu *et al.*, 2005; Yang and Bashaw, 2006). This raises the question, how do both the activation and the inhibition of Rac lead to axonal repulsion? One possibility is that these GAPs and GEFs function in distinct steps, with each required for different molecular actions that support repulsion. Alternatively, it could be that Rac cycling alone is sufficient for repulsion, which may not depend on the maintenance of a specific level of RacGTP. In any case, these studies demonstrate the importance of GAPs and GEFs in regulating Slit/Robo signaling by controlling the activity state of small GTPases.

With their central role in regulating the actin cytoskeleton, Rho GTPases are in the unique position to regulate cell motility in response to Slit.

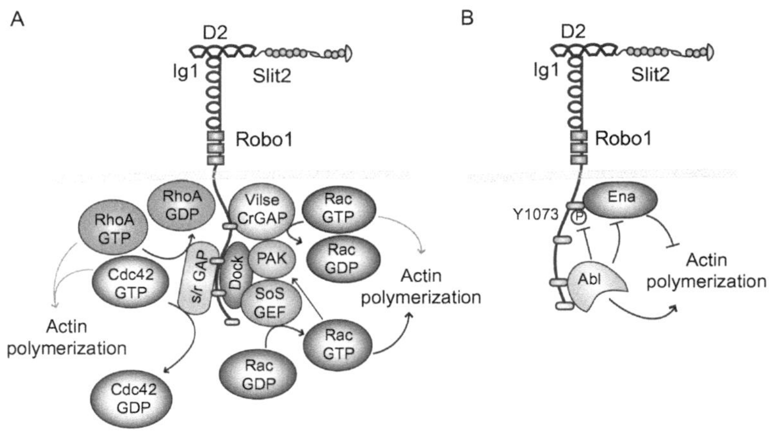

Fig. 3 Slit/Robo signaling regulates cell migration by controlling the activation state of actin cytoskeleton modulators. (A) Slit/Robo signaling regulates actin polymerization, and thus cell migration, by controlling the activity level of Rho GTPases. Slit/Robo signaling prevents cell migration by recruiting Slit/Robo (s/r)GAP to the CC1 and CC2 domains, which inactivates the small Rho GTPases RhoA and Cdc42, and vilse/crGAP to the CC0 domain, which exchanges RacGTP for RacGDP. In other contexts, Slit/Robo signaling drives actin polymerization by recruiting Dock, which in turn recruits son of sevenless (Sos) GEF and PAK p21-activated kinase. Sos GEF activates the small Rho GTPase Rac by exchanging GDP for GTP, leading to actin polymerization. RacGTP in turn activates PAK, which also drives actin polymerization. (B) The antimigratory function of Slit/Robo signaling is regulated by the Abl kinase, which attenuates Robo signaling via phosphorylation of Robo at Y1073 near CC0, possibly preventing substrate binding, and by directly targeting the Robo effector protein Enabled (Ena). In the absence of Abl, Ena binds to CC0 and functions to inhibit cell migration by preventing actin polymerization. (See Page 8 in Color Section at the back of the book.)

Accordingly, a number of studies demonstrate the importance of Cdc42 in mediating the aggressive spread of tumor cells and the role of Slit/Robo1 signaling in inhibiting this invasion by attenuating Cdc42 activation. For example, a study on medulloblastoma reveals expression of Slit and Robo in a variety of tumor samples and cell lines, with no evidence that *Slit* is silenced by methylation (Werbowetski-Ogilvie *et al.*, 2006). Treatment of cells derived from these tumors with Slit2 inhibits their invasiveness, without affecting the direction of their migration or their proliferation. The authors suggest that these effects are due to a reduction in the activity of Cdc42 (Werbowetski-Ogilvie *et al.*, 2006). This result is also observed in two studies on glioma cell lines and tumor samples which report reduced expression of Slit in primary human glioma specimens and invasive glioma cells, compared to normal brain cells and astrocytes (Parsons *et al.*, 2008; Yiin *et al.*, 2009). Further, treatment of these glioma cells with recombinant Slit2 or its overexpression in these cell lines inhibits cell migration and reduces invasion by decreasing Cdc42 activity, an effect that is prevented by siRNA-mediated reduction of either Slit or Robo1 expression. This inhibition of cell invasion was further confirmed in xenograft studies that demonstrate decreased infiltration of Slit2-expressing glioma cells implanted into the brains of mice. Interestingly, there is no evidence that these effects are mediated through the regulation of β-catenin by Slit, as neither the level of β-catenin, its phosphorylation status, nor its association with N-cadherin is altered in Slit-expressing glioma cells (Yiin *et al.*, 2009). Taken together these studies on two different types of neural tumors demonstrate that the small GTPase, Cdc42, is subject to negative regulation by Slit/Robo signaling.

Slit/Robo (s/r) GAP is a strong candidate for the GAP that inhibits Cdc42 activity downstream of Slit and Robo. s/rGAP has been shown to reduce the activation of both Cdc42 and Rho, but not Rac (Wong *et al.*, 2001). Studies demonstrate that a dominant negative form of this GAP blocks the inactivation of Cdc42 by Slit and also the migration of cells in response to Slit (Wong *et al.*, 2001). Further, a study in tumor and non-transformed cultured epithelial cells has shown that Slit2/Robo1 counteracts hepatocyte growth factor (HGF)-induced migration by directly targeting and inhibiting Cdc42 and, as a consequence, actin-based protrusive forces (Fig. 3A; Stella *et al.*, 2009). Taken together, these data show that Slits inhibit the motility of tumor cells by negatively regulating Cdc42 Rho GTPase, and that one of the consequences of losing Slit expression during tumor progression is inappropriate cellular migration due to this deregulation.

b. Abl and Ena

In addition to the Rho family of small GTPases, Slit binding to Robo also leads to the recruitment of at least one kinase that regulates both actin cytoskeletal rearrangements and the activity of Robo itself. Structural and genetic studies show that the Abl kinase and its substrate Enabled (Mena in mammals) interact directly with the cytoplasmic domains of Robo (Fig. 3B; Bashaw *et al.*, 2000). Genetic studies in *Drosophila* demonstrate opposing roles for Ena and Abl in Robo-mediated axonal repulsion, whereby Abl antagonizes repulsive Robo signaling and Ena enhances it (Bashaw *et al.*, 2000). In a series of studies, it has been shown that reducing the level of Abl suppresses *Robo* loss-of-function phenotypes, while its overexpression inhibits Robo function. The opposite is true for Ena, as reducing its levels enhances *Robo* loss-of-function phenotypes and suppresses *Robo* gain-of-function. Further, Abl phosphorylates Robo to inhibit its function because a Y–F mutation in a conserved tyrosine that is targeted by Abl generates a hyperactive Robo receptor. In contrast, deleting the cytoplasmic domain of Robo that binds Ena reduces the ability of this Robo mutant to rescue *Robo* loss-of-function phenotypes (Bashaw *et al.*, 2000).

While these studies demonstrate the consequences of Abl and Ena interactions with Robo, the molecular mechanism by which these proteins mediate their effect on the directional outgrowth of an axon through Robo is still poorly understood. For Abl, one possibility is that it binds to and phosphorylates unliganded Robo, inhibiting the ability of signaling proteins to interact with their docking sites on the Robo cytoplasmic domain, until Slit binds Robo and relieves the inhibition. Abl is known to phosphorylate Ena, but the significance of this phosphorylation is poorly understood. The Ena proteins (Mena, Vasp, and EVL in vertebrates) contain N- and C-terminal Ena/Vasp homology domains that flank a proline-rich central region. They are generally thought of as positive regulators of actin assembly that function in promoting the growth of long, sparsely branched actin filament networks. Consequently, it is still unclear how Ena, which enhances actin polymerization and filopodial/lamellipodial protrusion, plays a role in repulsive axon guidance downstream of Robo (Bear and Gertler, 2009), except that it may direct the growth of the cell away from Slit by promoting assembly at sites distal to high ligand concentration. Moreover, few studies have been published on how Ena and Abl contribute to the migration of tumor cells in response to Slit. It has been shown that transformation of cells with the Src oncogene results in Abl activation, which stabilizes Robo1 at the plasma membrane, leading to the activation of both Cdc42 and Rac, as well as Slit-independent cell migration (Khusial *et al.*, 2010). Thus, under these circumstances, the activation of Robo1 positively regulates downstream Rho GTPases and induces migration, whereas during central nervous system development and in other tumor types, the opposite effect is observed. Taken together, these studies demonstrate that Slit/Robo signaling communicates with the actin cytoskeleton to regulate cell motility, although the nature of the response depends upon the developmental and disease context. It is clear from many studies that Rho GTPases, Abl, and Ena all play important roles in promoting tumor cell metastasis (Allington and Schiemann, 2011; Gertler and Condeelis, 2011; Hall, 2009). However, additional research is required to determine how Slits regulate these cytoskeletal effectors in tumor cells and whether Slits could be potential therapeutic targets for hindering tumor cell motility by interfering with these signaling routines.

3. SLITS AND ROBOS: REGULATORS OF CELL CHEMOTAXIS

Cells must decipher and integrate a complex set of signals in order to migrate toward targets. Indeed, even the metastatic migration of tumor cells is not a random walk and many types of cancers preferentially target specific organs. While Slit is one of the cues cells respond to in the extracellular environment, there are many others, such as chemokines. Compared to large, extracellular matrix-associated Slits, chemokines are small (8–10 kDa) soluble factors, first identified in the immune system, but now with documented roles in regulating the migration of many cell types, including tumor cells. A number of studies examine how Slits affect the motility of immune and tumor cells in response to other extracellular factors, notably chemokines, but also growth factors such as HGF and platelet-derived growth factor (PDGF). Although the preponderance of data supports a role for Slits in inhibiting the migration of cells responding to stimulant, there are two examples of Slit increasing cell migration in response to chemokines (Schmid *et al.*, 2007; Ye *et al.*, 2010).

The first of these studies was published over a decade ago. Standard Transwell assays were used to evaluate the effects of Slit on leukocyte migration from the upper to lower chamber in response the chemokine CXCL12 (SDF1) or bacterial chemotactic factor, *N*-formyl peptide f-Met-Leu-Phe. It was found that Slit reduces the chemotactic migration of leukocytes when added to either upper, lower, or both chambers, indicating that Slit reduces the overall motility of cells, rather than acting as a repulsive cue to guide their migration (Wu *et al.*, 2001). Further studies have refined our understanding of the underlying molecular mechanisms. Videomicroscopic live cell tracking demonstrates that Slit2 selectively impairs chemotaxis, defined as the directional migration of cells, but not chemokinesis, which is the random movement of neutrophils in response to stimulant (Tole *et al.*, 2009). Slit2 achieves this effect by suppressing the activation of Cdc42 and Rac2 that would normally occur in response to stimulation, with consequent disruption of actin free barbed end formation. A similar inhibition of migration and downregulation of Rac activity is observed in vascular smooth muscle cells in response to PDGF (Liu *et al.*, 2006).

These observations were translated *in vivo* using a mouse model of chemical irritant peritonitis. Pre-administration of Slit2 either intraperitoneally or by tail vein injection significantly reduces the recruitment of neutrophils to the site of inflammation (Tole *et al.*, 2009). These data suggest that localized or systemic delivery of Slit2 reduces leukocyte recruitment and, consequently, the tissue damage associated with inflammation. This finding is in accordance with observations from other inflammation models, including glomerulonephritis-associated kidney injury, global cerebral ischemia, and skin sensitization to allergin (Altay *et al.*, 2007; Guan *et al.*, 2003; Kanellis *et al.*, 2004), in which Slit functions similarly in an anti-inflammatory manner. However, a recently published study using two different models of allergic airway inflammation suggests that Slits have a more complex role in the immune system. In the first model of ovalbumin (OVA) airway inflammation, Slits enhance eosinophil chemotaxis, while in the second model of endotoxin-induced lung inflammation, Slits suppress neutrophil chemotaxis (Ye *et al.*, 2010). Eosinophils and neutrophils both express Robo1, while Clara cells in the bronchial epithelium secrete Slit2. Aerosol challenge of wild-type mice with OVA triggers leukocytes, primarily eosinophils, to infiltrate into lung. This infiltration is significantly enhanced in Slit2 transgenic (Slit2-Tg) mice, which overexpress Slit2 under the control of the cPMV promoter (Yang *et al.*, 2010). These data suggest that Slit augments eosinophil recruitment. Similarly aerosol challenge of wild-type mice with endotoxin again triggers leukocyte infiltration into lung, but in this case primarily neutrophils are mobilized. With endotoxin challenge, however, significantly fewer neutrophils are observed in the lungs of Slit2-Tg mice, an effect that is reversed by the application of a function-blocking antibody directed against the extracellular domain of Robo1. These data suggest that Slit inhibits neutrophil recruitment. Together with *in vitro* studies that demonstrate enhanced chemokine-induced eosinophil migration in response to Slit, but reduced neutrophil migration (Yang *et al.*, 2010), these data suggest that, depending on the cellular circumstance, Slits can have differential effects on leukocytes. The molecular basis for these distinct responses can be traced to levels of s/rGAP expression, with eosinophils containing significantly lower levels of this Slit/Robo effector compared to neutrophils. This results in the activation of Cdc42 in eosinophils, rather than inhibition, which occurs in neutrophils when s/rGAP is present (Yang *et al.*, 2010). Thus, even though almost a decade of work has pointed to a single role for Slit as an inhibitory factor in the immune system, with this recent finding, it appears that this is not the case and that, once again, depending on the cellular context, Slit has a dual role as activator and inhibitor of cellular response.

The role of Slit as an inhibitor of inflammation has potentially far-reaching implications in terms of its role in cancer biology as a tumor suppressor. In normal tissue during wound healing, removing the irritant, or completing the repair limits the inflammatory response. In contrast, tumors become essentially unhealed wounds, characterized by chronic inflammation, which promotes rather than suppresses tumor growth by releasing growth and survival factors, creating genomic instability, promoting angiogenesis, and remodeling the extracellular matrix to facilitate invasion. That Slits inhibit the infiltration of not only leukocytes, but also dendritic cells (Guan *et al.*, 2003), T lymphocytes, and monocytes (Prasad *et al.*, 2007), could be harnessed therapeutically to normalize the inflammatory network and restrict infiltrating cells with tumor-promoting properties, while attracting those cells with tumor-suppressing properties.

Another way that Slit could function as a therapeutic agent in the war on cancer is by inhibiting metastasizing cells. This has been evidenced by a number of studies using breast cancer models that demonstrate the ability of Slit/Robo signaling to counter the promigratory, prometastatic consequences of the CXCL12/CXCR4 chemokine axis. A study by Muller and colleagues a decade ago demonstrates that the pattern of breast cancer metastases is governed, at least in part, by this chemokine axis (Muller *et al.*, 2001). CXCR4 is upregulated in breast cancer cells (Salvucci *et al.*, 2006), and, upon metastasis, guides these cells to organ sites with high CXCL12 levels such as the lung, liver, and bone. The involvement of CXCR4 in metastasis is not confined to breast cancer, as it is also expressed in other tumor cell lines that respond to CXCL12, such as astrogliomas, prostate carcinomas, B-cell lymphomas, and chronic lymphocytic leukemias (Moore, 2001). Using Transwell filters, two studies have demonstrated that Slit2 has the capacity to counteract CXCL12-induced chemotaxis of breast cancer cell lines that express both Robo1 and CXCR4 (Prasad *et al.*, 2004; Schmid *et al.*, 2007). By signaling through Robo1, Slit2 inhibits a number of downstream effectors that are activated by CXCR4, such as the focal adhesion components RAFTK/Pyk2, focal adhesion kinase, paxillin, PI3K, p44/42 MAP kinase, and metalloproteases 2 and 9 (Prasad *et al.*, 2004). In the absence of Slit2, which is downregulated in over 50% of sampled breast tumors, the expression of both CXCL12 and CXCR4 is upregulated (Marlow *et al.*, 2010), contributing to the development of hyperplastic lesions in *Slit2* and *Robo1* knockout mammary glands. Another study on the role of Slit/Robo signaling in inhibiting breast cancer cell migration in response to CXCL12 implicates ubiquitin-specific protease 33 (USP33), a deubiquitinating enzyme. The authors provide evidence that Slit stalls the chemotaxis of breast cancer cells by inducing the redistribution of Robo to the plasma membrane, a process that is dependent on USP33 (Yuasa-Kawada *et al.*, 2009). Taken together, these data once again raise the possibility that Slits could function therapeutically, in this case to combat tumor metastasis by inhibiting tumor cell migration in response to CXCL12.

IV. ROBO4 IS AN UNCONVENTIONAL ROBO RECEPTOR

Robo4 was a late addition to the family of Robo receptors due to the lack of structural homology between it and the other Robos. Nevertheless, studies over the past decade have shown that Robo4 is a key member of the Slit/Robo signaling axis, especially in the vasculature, where it is expressed on the surface of endothelial cells and functions in regulating angiogenesis (Huminiecki *et al.*, 2002; Park *et al.*, 2003). Robo4 was considered an endothelial-specific member of the Robo family until very recently with the publication of a study showing its expression in the developing brain, where it appears to regulate the radial migration of newborn neurons (Zheng *et al.*, 2011). These new data raise the possibility that Robo4 has, as yet undiscovered, roles in different organs.

Robo4 is structurally unique compared to the other receptors in the Robo family. In contrast to the five Ig and three FN3 domains in the extracellular region of Robo1–3, Robo4 contains only two extracellular Ig and FN3 domains, and only two intracellular CC domains, CC0 and CC2 (Huminiecki and Bicknell, 2000; Huminiecki *et al.*, 2002). Moreover, while strong evidence for an interaction between the extracellular domain of Robo1–3 and the D2 domain of Slits exists, current data do not support a direct interaction between Slits and Robo4. Biochemical and structural studies carried out by Morlot and colleagues have identified the critical amino acids in the Robo ectodomain required for Slit binding, and Robo4 lacks these amino acids (Morlot *et al.*, 2007b). BiaCore analysis on recombinant proteins also fails to provide evidence for a Slit/Robo4 interaction (Suchting *et al.*, 2005), even though Slit2 and Robo4 can be coimmunoprecipitated from cell lysates and positive immunostaining is observed on cells incubated with recombinant Slits (Park *et al.*, 2003; Zhang *et al.*, 2009). One explanation for the apparent association between Slits and Robo4 when they are in a cellular context, but not in purified forms, is that they exist in a protein complex on the cell and that a coreceptor is present to transmit Slit binding into Robo4 activation. Candidates for this coreceptor are transmembrane heparin sulfate proteoglycans such as Sdc (Hu, 2001; Steigemann *et al.*, 2004). Indeed, *in vitro* studies by Hussain and colleagues found that heparan sulfate is required for functional Slit–Robo4 signaling (Hohenester *et al.*, 2006). Additionally, a non-exclusive alternative is that Robo1 fulfills this coreceptor function as Robo1/Robo4 complexes have been documented, as well as a requirement for this interaction in inhibiting endothelial chemotaxis in response to Slit2 (Kaur *et al.*, 2008; Sheldon *et al.*, 2009). More recently, UNC5B, a chemorepellent receptor for Netrin1 (Leonardo *et al.*, 1997), was identified in a protein–protein interaction screen as an alternative binding partner for Robo4 in the vasculature

(Koch *et al.*, 2011). While additional studies are required to investigate this novel interaction, it reveals a new layer of complexity in Robo4 function. The role of Robo4 as a regulator of angiogenesis will be described in more detail below, but suffice it to say that, while it is unlikely that Robo4 interacts directly with Slits, there is ample evidence that it transduces Slit signaling, possibly via coreceptors such as heparan sulfate proteoglycans or Robo1, or even via reverse signaling through UNC5B.

V. SLIT/ROBO: ROLES IN TUMOR ANGIOGENESIS

Angiogenesis describes the process of new vessel growth from mature, pre-existing vessels. This process occurs naturally throughout development and later in the adult during both wound healing and pregnancy, both times of increased tissue remodeling. Angiogenesis involves several steps: first, endothelial cells migrate toward a proangiogenic stimulus, such as vascular endothelial growth factor (VEGF), released extracellularly into the environment; and second, endothelial cells become migratory and congregate at the source of the angiogenic stimulus to form loops, and later vessels, as more and more cells arrive (Potente *et al.*, 2011). Tumor cells often hijack several aspects of this process to enable additional tumor growth and metastasis, both of which require an increased supply of oxygen and other nutrients. Without increased angiogenesis, tumors are limited in their growth potential and become necrotic. Thus, tumor cells must induce angiogenesis to transition from a small group of cells to a large, malignant tumor, and to ultimately metastasize to other tissues. These processes require intimate communication between tumor cells and endothelial cells, and studies from the past decade provide strong support for the Slit/Robo pathway in mediating this cross-talk. However, how Slit/Robo signaling affects tumor angiogenesis remains unclear, as reports demonstrate both pro- and antiangiogenic functions in pathological systems (Table I). What is clear, however, is that Slits and Robos play key roles in regulating the process of tumor angiogenesis.

A. Vascular Expression of Slits and Robos

Of the three Slit proteins expressed in vertebrates, multiple studies have reported Slit2 and Slit3 in the vasculature of normal tissues, expressed by vasculature smooth muscle cells/pericytes that encircle blood vessels (Jones *et al.*, 2009; Liu *et al.*, 2006; Marlow *et al.*, 2010), and also by endothelial cells (Brantley-Sieders *et al.*, 2011; Zhang *et al.*, 2009). In contrast, only one study has reported detection of Slit1 in the vasculature (Abdollahi *et al.*,

2007), suggesting that it may not play a key role in developmental and tumor angiogenesis, although future studies may change this view. As for Slit receptors, Robo4 appears to be expressed by all endothelial cells, whereas Robo1 may only be expressed on some types of blood vessels (Huminiecki *et al.*, 2002; Legg *et al.*, 2008; Mura *et al.*, 2011; Park *et al.*, 2003; Sheldon *et al.*, 2009; Verissimo *et al.*, 2009).

B. Function of Slits and Robos in the Vasculature

Robo4 expression was initially identified by bioinformatic data mining, with the accompanying analysis of its expression revealing its presence exclusively at sites of active angiogenesis, notably tumor vessels (Huminiecki *et al.*, 2002). Robo4 was also independently identified in a study aimed at identifying genes whose expression is perturbed in an Activin receptor-like kinase (Alk) mutant mouse model (Park *et al.*, 2003). Alk is a member of the TGF-β superfamily of receptors and is involved in the normal development of the vasculature (Johnson *et al.*, 1996). Mice lacking Alk expression develop abnormal connections between arterial and venous vascular beds and die at midgestation (Urness *et al.*, 2000). Such a developmental role for Robo4 is supported by studies in zebrafish in which either morpholino knockdown of Robo4 or its overexpression results in stunted or absent intersomitic vessels, although normal patterning of axial vessels are seen, suggesting both primary and redundant roles for *Robo4* in this system (Bedell *et al.*, 2005). Further, numerous studies show that Slits affect the migration of endothelial cells, acting as both a chemoattractant (Howitt *et al.*, 2004; Kaur *et al.*, 2006; Wang *et al.*, 2003, 2008) and chemorepellent (Marlow *et al.*, 2010; Park *et al.*, 2003; Seth *et al.*, 2005; Zhang *et al.*, 2009). More recently, a study aimed at identifying novel tumor endothelial markers that can be used as targets by antiangiogenic therapeutics found that Robo4 is highly expressed on tumor vessels compared to normal tissue vessels, which display little or no immunostaining (Mura *et al.*, 2011). Taken together, these studies support a role for Robo4 in both normal development of the vasculature and tumor angiogenesis. It was therefore surprising when it was discovered that *Robo4 −/−* mice display normal vessel patterning in a variety of contexts: intersomitic and cephalic vessels during early embryogenesis, stereotypical nerve–artery alignment in embryonic limb skin, and normal vascularization of the mammary gland during postnatal organogenesis (Jones *et al.*, 2008; Marlow *et al.*, 2010). These data raise the question as to the role of Robo4 in mammalian endothelium.

Although a comprehensive understanding of the function of Robo4 in the vasculature has been hindered by contradictory findings, accumulating evidence suggests that Slit/Robo4 signaling functions to downregulate VEGF signaling in the mature vasculature, thus restraining angiogenesis during pathological neovascular processes (Han and Zhang, 2010; Huang *et al.*, 2009a; Jones *et al.*, 2008, 2009; Koch *et al.*, 2011; London and Li, 2011; Mulik *et al.*, 2011), as well as during normal periods of robust sprouting angiogenesis such as occurs during pregnancy (Marlow *et al.*, 2010). The analysis of *Robo4* −/− mice provided the first insight into this function by showing that in wild type, but not knockout animals, Slit2 suppresses VEGF-induced hyperpermeability of retinal endothelium (Jones *et al.*, 2009). Further experiments employed two models of vascular disease, oxygen-induced retinopathy and laser-induced choroidal neovascularization, both of which result in pathological angiogenesis (Jones *et al.*, 2009). In both cases, intravitreal administration of Slit2 reduces angiogenesis in wild type, but not *Robo4* −/−, mice. Moreover, the opposite effect of elevated angiogenesis was observed in concordant experiments looking at breast development and cancer. In these contexts, loss of Slit/Robo4 signaling results in excessive angiogenesis only when there is increased VEGF expression in the gland, which occurs during pregnancy and preneoplasia (Marlow *et al.*, 2010). There is no *Robo4* −/− mammary gland phenotype in the absence of proangiogenic stimulation (Marlow *et al.*, 2010). In all these examples, Slit/Robo4 regulates angiogenesis by inhibiting signaling downstream of VEGF/VEGFR, including the activation of Src, Rac, and FAK (Jones *et al.*, 2008; Marlow *et al.*, 2010). Further analysis of the pathway has shown that Robo4 interacts directly with paxillin and the ArfGAP (ADP-ribosylation factor-directed GTPase-activating proteins), GIT1, to block the activation of Arf6 in response to VEGF and fibronectin (Fig. 4A; Jones *et al.*, 2009).

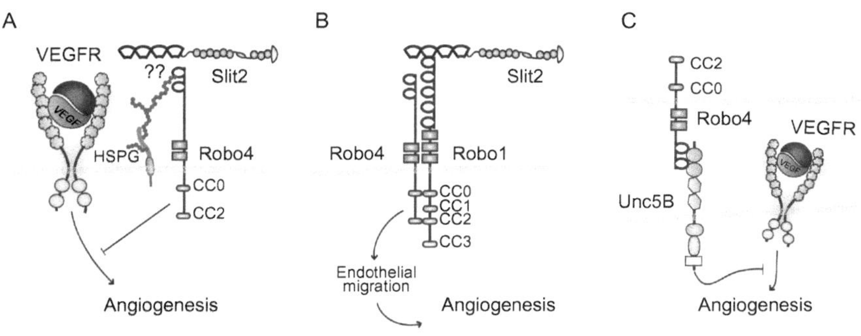

Fig. 4 Slit/Robo signaling regulates the process of tumor angiogenesis. (A) Slit2 binds to Robo4/proteoglycan complex and signals to block proangiogenic signaling downstream of VEGF/VEGFR. (B) Slit promotes angiogenesis by binding to a Robo1/Robo4 heterodimer and driving endothelial migration. (C) Robo4 binds UNC5B, and signals to block proangiogenic signaling downstream of VEGF/VEGFR. (See Page 8 in Color Section at the back of the book.)

Additional studies on disease models support the role of Slit/Robo4 signaling in inhibiting angiogenesis and enhancing vascular stability. One study used infection with herpes simplex virus (HSV) to generate chronic inflammatory lesions, called stromal keratitis, in the cornea. This pathological condition is associated with enhanced angiogenesis that is driven by VEGF (Mulik *et al.*, 2011; Suryawanshi *et al.*, 2011). Following HSV infection, *Robo4* is upregulated in endothelial cells of corneal stroma. However, a corresponding increase in *Slit2* is not observed, suggesting that the Slit/Robo4 signaling axis is unable to control angiogenesis in response to this infection because Slit production is limited. To investigate, Slit2 was subconjunctivally administered. This treatment resulted in reduced neovascularization, a result that is not observed when Slit expression is knocked down using shRNA. The researchers also examined the activity status of Arf6 and Rac, and found that they were reduced after Slit2 treatment, supporting previous studies that Slit/Robo4 signaling opposes VEGF/VEGFR signaling by modulating downstream signaling pathways. In other studies, an inflammatory reaction was triggered by the administration of lipopolysaccharide (LPS), which induces endothelial hyperpermeability (London *et al.*, 2010). In this context, Slit2 treatment reduces LPS-induced vascular permeability by increasing the localization of V/E-cadherin to cell/cell contacts and stabilizing its interaction with p120-catenin. Taken together, these studies have paved the way to a new view on the function of Robo4, in which it functions as a guardian of blood vessel stability by countering proangiogenic and proinflammatory signals.

There are, nevertheless, several aspects to this model of Robo4 function that remain unresolved. First, is the question concerning the role of Robo1, which is (1) expressed on at least some endothelial cells; (2) heterodimerizes with Robo4 (Kaur *et al.*, 2006, 2008; Sheldon *et al.*, 2009); and (3) is required for some (Sheldon *et al.*, 2009), but not all (Marlow *et al.*, 2010), of Robo4 functions. The papers summarized above describing the antiangiogenic and antiinflammatory roles of Robo4 do not address the function of Robo1 in these processes. However, there is some evidence that Robo1 binds Robo4 and may be required to transduce the Slit signal. Because Robo4 lacks many of the amino acids critical for Slit2 binding (Morlot *et al.*, 2007b), it is considered unlikely that it binds Slit directly. Instead, Robo1 may serve as a coreceptor for Robo4, and in this circumstance, the heterodimer may function to promote endothelial cell migration and angiogenesis (Fig. 4B; Kaur *et al.*, 2006, 2008; Sheldon *et al.*, 2009). The notion that Robo1 plays a functional role in angiogenesis, either alone or in complex with Robo4, is supported by a number of studies, including one on tumor angiogenesis (Wang *et al.*, 2003). In addition, the analysis of gene expression levels during retinal development reveals fluctuations in Robo1 levels that coincide with times of active retinal vascular development (Huang *et al.*, 2009b).

Further, these researchers show that loss of *Robo1* in monkey choroidal retinal endothelial cells perturbs tube formation, in addition to lowering cell proliferation and migration (Huang *et al.*, 2009b). Elevated levels of Robo1 are seen in several pathological animal models, such as retinopathy of prematurity and neovascularized corneas (Han and Zhang, 2010; Huang *et al.*, 2009a). Robo1 was also identified as a putative proangiogenic gene in assays to elucidate human genes whose expression correlates with either increased or decreased angiogenesis (Abdollahi *et al.*, 2007). Taken together these data suggest that Robo1 functions to restrict angiogenesis, either on its own or in a complex with Robo4, but additional studies are required to fully elucidate its role during developmental and pathological angiogenesis.

A second issue concerning Robo4 and its role in angiogenesis is the uncertain status of Slit as ligand. The discovery by Koch and colleagues that Robo4 binds UNC5B suggests that there may be an alternative mechanism for Robo4 signaling in the vasculature (Koch *et al.*, 2011). Further characterization of this interaction revealed the surprising finding that Robo4 acts as the ligand, not the receptor, in this relationship. Soluble Robo4 protein rescues vessel hyperpermeability in the *Robo4−/−* mice and reduces VEGF-induced hyperpermeability in wild-type mice, but not in mice treated with anti-UNC5B, supporting a model whereby Robo4 maintains vessel integrity by binding and signaling through UNC5B. In these studies, Robo4/UNC5B signaling was shown to counter the activation of Src kinase by VEGF/VEGFR (Koch *et al.*, 2011), the same mechanism that was previously identified for Slit/Robo4 signaling (Fig. 4C; Jones *et al.*, 2009). There is, however, no evidence that the vascular phenotypes of *Robo4−/−* and *Unc5b−/−* mice are similar, a result that may be expected for proteins in a ligand/receptor relationship. Altogether, these data support a role for Robo4 in blocking signaling pathways downstream of VEGF/VEGFR and, consequently, in inhibiting VEGF-induced changes in blood vessels, but it is currently unclear whether this requires Slit, UNC5B, or both (Fig. 4A).

C. Slit and Robo Expression is Altered in Tumor Angiogenesis

Like other members of the Robo family of receptors, there are documented changes in Robo4 expression in samples from tumors and other diseases. Increased Robo4 expression on endothelial cells has been reported in a number of pathologies, including tumor samples that display increased angiogenesis (Table II; Grone *et al.*, 2006; Han and Zhang, 2010; Huang *et al.*, 2009c; Mura *et al.*, 2011; Seth *et al.*, 2005). Conversely, decreased

Robo4 expression has also been observed in hepatocellular carcinoma samples analyzed by quantitative RT-PCR (Avci *et al.*, 2008), and in datasets from microarray analyses on human breast tumor samples (Table III; Richardson *et al.*, 2006). Thus, while it is possible that Robo4 is both up- and downregulated in disease models, at this point the data suggest that it is most often increased, suggesting a link between Slit/Robo4 signaling and new vessel formation, at least in a tumorigenic setting. There is some documentation of Robo1 expression on endothelial cells in tumors and other tissues undergoing neovascularization (Han and Zhang, 2010; Wang *et al.*, 2003), although whether Robo1 is expressed and how it functions in blood vessels remains controversial. The regulation of Slits in tissues surrounding blood vessels has already been discussed, with examples of both up- and downregulation in tumors and other tissues undergoing neovascularization (Han and Zhang, 2010; Wang *et al.*, 2008; Yang *et al.*, 2010). There is also some data suggesting that Slit expression is regulated by EphA2 tyrosine kinase, and that loss of EphA2 expression in a tumor setting elevates Slit expression, which acts in a proangiogenic manner (Brantley-Sieders *et al.*, 2011).

D. Slits and Robos: Regulators of Tumor Angiogenesis

Tumor growth and metastasis require increased supplies of oxygen and nutrients, and this requirement can be met by enhanced angiogenesis in the surrounding vasculature (Potente *et al.*, 2011). In order to achieve increased angiogenesis, tumor cells must downregulate antiangiogenic signals and then secrete proangiogenic cues to increase blood vessel growth (Kerbel, 2008). Currently there are no studies that directly examine the functional role of Robo4 in tumor angiogenesis, although in a preneoplastic setting in breast, loss of *Robo4* results in enhanced angiogenesis in response to epithelial-derived VEGF and CXCL12 (Marlow *et al.*, 2010). These data are in line with the documented antiangiogenic role of Robo4 in stabilizing the vasculature during pathological angiogenesis as described above.

In contrast, Wang and colleagues have shown that Robo1 is expressed on tumor endothelial cells and plays a role in new vessel formation in tumors that upregulate Slit expression (Wang *et al.*, 2003, 2008). In these studies, the extracellular domain of Robo1, or a function-blocking antibody called R5, are shown to inhibit both the chemoattractive migration of endothelial cells and their tube formation in response to Slit2, as well as tumor angiogenesis in xenograft models, suggesting that Slit2 functions proangiogenically in tumors that overexpress it, and further, that its signal is transduced via Robo1, and not Robo4 (Wang *et al.*, 2003, 2008). Further, overexpression of Slit2 in a nonmetastatic pancreatic islet cell model of carcinogenesis

(RIP1-Tag2) is reported by the same group to promote tumor lymphangiogenesis and lymphatic metastasis by signaling through Robo1, expressed on the lymphatic endothelial cells. Again, administration of the R5 function-blocking antibody reverses the enhanced angiogenesis and decreases tumor formation (Yang *et al.*, 2010).

A possible explanation for the apparent pro- versus antiangiogenic functions of Slit2/Robo signaling comes from a study by Dunaway and colleagues in which Slit2 promotes angiogenesis on its own, but inhibits it in the presence of ephrin-A1 (Dunaway *et al.*, 2011). Ephrin-A1 is the primary GPI-linked ligand for the EphA2 receptor, whose expression was shown by these researchers in a previous paper to inversely correlate with Slit expression (Brantley-Sieders *et al.*, 2011). The present study, however, shows that Slit2 stimulates angiogenesis through the activation of Akt and Rac GTPase, an effect that is inhibited in the presence of ephrin-A1. Thus, this study echoes the common theme that, because Slit/Robo signaling targets common signaling pathways, the context in which it signals determines the outcome. Clearly, additional studies are required to understand the complex and intertwined signaling that regulates neovascularization during disease processes. The current data are incomplete and conflicting, but there is one take home message from all these studies: Slit plays a central role in endothelial cell biology. As such, it holds promise as a therapeutic agent that could be used to treat pathologies and cancers that are made worse by enhanced angiogenesis.

VI. CONCLUSION

Since their early discovery as key regulators of axon migration in the developing nervous system, the Slit/Robo signaling pair has been implicated in a wide variety of developmental and pathological processes. More specifically, Slit/Robo signaling has been found to have a significant impact on a cell's behavior, from regulating cell migration to controlling cell growth, both processes critical in tumor cell progression. When activated aberrantly, these signaling events can promote tumor cell growth and migration, contributing to tumor metastasis, and poor patient prognosis. As such, Slits and Robos are promising candidates for anticancer therapeutics, but care must be exercised because the bifunctionality of this signaling axis could result in both tumor suppressive and oncogenic outcomes. For example, in its promigratory role, Slit-induced Robo signaling causes decreased cell–cell attachments and initiates promigratory pathways. In this context, treatment with RoboN, the ectodomain of Robo1, causes sequestration of Slit in the environment, thus stabilizing cell–cell attachments and preventing metastasis. However, in other contexts, Slit/Robo signaling has the opposite effect, enhancing cell–cell contacts by increasing E-cadherin stability and down-regulating transcriptional programs that promote proliferation. In this case, sequestering Slit by RoboN treatment would potentially weaken cell–cell contacts, enabling, rather than inhibiting, promigratory signals and tumor cell metastasis. Thus, since the downstream effects of Slit/Robo signaling vary greatly depending on both the extracellular and the intracellular milieu, development of effective therapeutics will require a better understanding of their disparate signaling consequences. Further, it will be necessary to design therapeutics with elaborate delivery mechanisms that ensure delivery of the drug to specific tissues in order to prevent unintended effects caused by changing normal Slit/Robo signaling in nearby tissues, as loss of Slit/Robo signaling in normal tissues and cells could cause deleterious effects. Despite these cautions, it is increasingly clear that Slits and Robos are key regulators of a wide variety of developmental and adult processes, from the epithelium to the vasculature, and that they hold promise as therapeutic targets in the fight against cancer and other diseases.

ACKNOWLEDGMENTS

This work was supported by funds from the NIH RO1 (GM-098897, L.H.), Santa Cruz Cancer Benefit Group (L.H.), and the California Institute of Regenerative Medicine predoctoral fellowship program (TG2-01157, M.B.)

REFERENCES

Abdollahi, A., Schwager, C., Kleeff, J., Esposito, I., Domhan, S., Peschke, P., Hauser, K., Hahnfeldt, P., Hlatky, L., Debus, J., Peters, J. M., Friess, H., *et al.* (2007). Transcriptional network governing the angiogenic switch in human pancreatic cancer. *Proc. Natl. Acad. Sci. USA* **104**, 12890–12895.

Alajez, N. M., Lenarduzzi, M., Ito, E., Hui, A. B., Shi, W., Bruce, J., Yue, S., Huang, S. H., Xu, W., Waldron, J., O'Sullivan, B., and Liu, F. F. (2011). MiR-218 suppresses nasopharyngeal cancer progression through downregulation of survivin and the SLIT2-ROBO1 pathway. *Cancer Res.* **71**, 2381–2391.

Allington, T. M., and Schiemann, W. P. (2011). The Cain and Abl of epithelial-mesenchymal transition and transforming growth factor-beta in mammary epithelial cells. *Cells Tissues Organs* **193**, 98–113.

Altay, T., McLaughlin, B., Wu, J. Y., Park, T. S., and Gidday, J. M. (2007). Slit modulates cerebrovascular inflammation and mediates neuroprotection against global cerebral ischemia. *Exp. Neurol.* **207**, 186–194.

Angeloni, D., ter Elst, A., Wei, M. H., van der Veen, A. Y., Braga, E. A., Klimov, E. A., Timmer, T., Korobeinikova, L., Lerman, M. I., and Buys, C. H. (2006). Analysis of a new homozygous deletion in the tumor suppressor region at 3p12.3 reveals two novel intronic noncoding RNA genes. *Genes Chromosomes Cancer* **45**, 676–691.

Avci, M. E., Konu, O., and Yagci, T. (2008). Quantification of SLIT-ROBO transcripts in hepatocellular carcinoma reveals two groups of genes with coordinate expression. *BMC Cancer* **8,** 392.

Bashaw, G. J., Kidd, T., Murray, D., Pawson, T., and Goodman, C. S. (2000). Repulsive axon guidance: Abelson and Enabled play opposing roles downstream of the roundabout receptor. *Cell* **101,** 703–715.

Bauer, K., Dowejko, A., Bosserhoff, A. K., Reichert, T. E., and Bauer, R. (2011). Slit-2 facilitates interaction of P-cadherin with Robo-3 and inhibits cell migration in an oral squamous cell carcinoma cell line. *Carcinogenesis* **32,** 935–943.

Bear, J. E., and Gertler, F. B. (2009). Ena/VASP: towards resolving a pointed controversy at the barbed end. *J. Cell Sci.* **122,** 1947–1953.

Bedell, V. M., Yeo, S. Y., Park, K. W., Chung, J., Seth, P., Shivalingappa, V., Zhao, J., Obara, T., Sukhatme, V. P., Drummond, I. A., Li, D. Y., and Ramchandran, R. (2005). Roundabout4 is essential for angiogenesis *in vivo*. *Proc. Natl. Acad. Sci. USA* **102,** 6373–6378.

Blanco, D., Vicent, S., Elizegi, E., Pino, I., Fraga, M. F., Esteller, M., Saffiotti, U., Lecanda, F., and Montuenga, L. M. (2004). Altered expression of adhesion molecules and epithelial-mesenchymal transition in silica-induced rat lung carcinogenesis. *Lab. Invest.* **84,** 999–1012.

Brantley-Sieders, D. M., Dunaway, C. M., Rao, M., Short, S., Hwang, Y., Gao, Y., Li, D., Jiang, A., Shyr, Y., Wu, J. Y., and Chen, J. (2011). Angiocrine factors modulate tumor proliferation and motility through EphA2 repression of Slit2 tumor suppressor function in endothelium. *Cancer Res.* **71,** 976–987.

Brose, K., Bland, K. S., Wang, K. H., Arnott, D., Henzel, W., Goodman, C. S., Tessier-Lavigne, M., and Kidd, T. (1999). Slit proteins bind Robo receptors and have an evolutionarily conserved role in repulsive axon guidance. *Cell* **96,** 795–806.

Challa, A. K., Beattie, C. E., and Seeger, M. A. (2001). Identification and characterization of roundabout orthologs in zebrafish. *Mech. Dev.* **101,** 249–253.

Chanana, B., Steigemann, P., Jackle, H., and Vorbruggen, G. (2009). Reception of Slit requires only the chondroitin-sulphate-modified extracellular domain of Syndecan at the target cell surface. *Proc. Natl. Acad. Sci. USA* **106,** 11984–11988.

Chen, J. H., Wen, L., Dupuis, S., Wu, J. Y., and Rao, Y. (2001). The N-terminal leucine-rich regions in Slit are sufficient to repel olfactory bulb axons and subventricular zone neurons. *J. Neurosci.* **21,** 1548–1556.

Christgen, M., Bruchhardt, H., Hadamitzky, C., Rudolph, C., Steinemann, D., Gadzicki, D., Hasemeier, B., Romermann, D., Focken, T., Krech, T., Ballmaier, M., Schlegelberger, B., *et al.* (2009). Comprehensive genetic and functional characterization of IPH-926: a novel CDH1-null tumour cell line from human lobular breast cancer. *J. Pathol.* **217,** 620–632.

Clark, K., Hammond, E., and Rabbitts, P. (2002). Temporal and spatial expression of two isoforms of the Dutt1/Robo1 gene in mouse development. *FEBS Lett.* **523,** 12–16.

Coleman, H. A., Labrador, J. P., Chance, R. K., and Bashaw, G. J. (2010). The Adam family metalloprotease Kuzbanian regulates the cleavage of the roundabout receptor to control axon repulsion at the midline. *Development* **137,** 2417–2426.

Dallol, A., Da Silva, N. F., Viacava, P., Minna, J. D., Bieche, I., Maher, E. R., and Latif, F. (2002a). SLIT2, a human homologue of the Drosophila Slit2 gene, has tumor suppressor activity and is frequently inactivated in lung and breast cancers. *Cancer Res.* **62,** 5874–5880.

Dallol, A., Forgacs, E., Martinez, A., Sekido, Y., Walker, R., Kishida, T., Rabbitts, P., Maher, E. R., Minna, J. D., and Latif, F. (2002b). Tumour specific promoter region methylation of the human homologue of the Drosophila Roundabout gene DUTT1 (ROBO1) in human cancers. *Oncogene* **21,** 3020–3028.

Dallol, A., Krex, D., Hesson, L., Eng, C., Maher, E. R., and Latif, F. (2003a). Frequent epigenetic inactivation of the SLIT2 gene in gliomas. *Oncogene* **22,** 4611–4616.

Dallol, A., Morton, D., Maher, E. R., and Latif, F. (2003b). SLIT2 axon guidance molecule is frequently inactivated in colorectal cancer and suppresses growth of colorectal carcinoma cells. *Cancer Res.* **63,** 1054–1058.

Dallol, A., Dickinson, R. E., and Latif, F. (2005). Epigenetic Disruption of the SLIT-ROBO Interactions in Human Cancer. Springer, New York, NY.

Delloye-Bourgeois, C., Fitamant, J., Paradisi, A., Cappellen, D., Douc-Rasy, S., Raquin, M. A., Stupack, D., Nakagawara, A., Rousseau, R., Combaret, V., Puisieux, A., Valteau-Couanet, D., *et al.* (2009). Netrin-1 acts as a survival factor for aggressive neuroblastoma. *J. Exp. Med.* **206,** 833–847.

Dickinson, R. E., Dallol, A., Bieche, I., Krex, D., Morton, D., Maher, E. R., and Latif, F. (2004). Epigenetic inactivation of SLIT3 and SLIT1 genes in human cancers. *Br. J. Cancer* **91,** 2071–2078.

Dickson, B. J., and Gilestro, G. F. (2006). Regulation of commissural axon pathfinding by slit and its Robo receptors. *Annu. Rev. Cell Dev. Biol.* **22,** 651–675.

Doi, T., Hajduk, P., and Puri, P. (2009). Upregulation of Slit-2 and Slit-3 gene expressions in the nitrofen-induced hypoplastic lung. *J. Pediatr. Surg.* **44,** 2092–2095.

Dunaway, C. M., Hwang, Y., Lindsley, C. W., Cook, R. S., Wu, J. Y., Boothby, M., Chen, J., and Brantley-Sieders, D. M. (2011). Cooperative signaling between Slit2 and Ephrin-A1 regulates a balance between angiogenesis and angiostasis. *Mol. Cell. Biol.* **31,** 404–416.

Dunwell, T. L., Dickinson, R. E., Stankovic, T., Dallol, A., Weston, V., Austen, B., Catchpoole, D., Maher, E. R., and Latif, F. (2009). Frequent epigenetic inactivation of the SLIT2 gene in chronic and acute lymphocytic leukemia. *Epigenetics* **4,** 265–269.

Evans, T. A., and Bashaw, G. J. (2010). Functional diversity of Robo receptor immunoglobulin domains promotes distinct axon guidance decisions. *Curr. Biol.* **20,** 567–572.

Fan, X., Labrador, J. P., Hing, H., and Bashaw, G. J. (2003). Slit stimulation recruits Dock and Pak to the roundabout receptor and increases Rac activity to regulate axon repulsion at the CNS midline. *Neuron* **40,** 113–127.

Fitamant, J., Guenebeaud, C., Coissieux, M. M., Guix, C., Treilleux, I., Scoazec, J. Y., Bachelot, T., Bernet, A., and Mehlen, P. (2008). Netrin-1 expression confers a selective advantage for tumor cell survival in metastatic breast cancer. *Proc. Natl. Acad. Sci. USA* **105,** 4850–4855.

Fukuhara, N., Howitt, J. A., Hussain, S. A., and Hohenester, E. (2008). Structural and functional analysis of slit and heparin binding to immunoglobulin-like domains 1 and 2 of Drosophila Robo. *J. Biol. Chem.* **283,** 16226–16234.

Fulda, S., and Debatin, K. M. (2006). Extrinsic versus intrinsic apoptosis pathways in anticancer chemotherapy. *Oncogene* **25,** 4798–4811.

Furne, C., Rama, N., Corset, V., Chedotal, A., and Mehlen, P. (2008). Netrin-1 is a survival factor during commissural neuron navigation. *Proc. Natl. Acad. Sci. USA* **105,** 14465–14470.

Gertler, F., and Condeelis, J. (2011). Metastasis: tumor cells becoming MENAcing. *Trends Cell Biol.* **21,** 81–90.

Ghosh, S., Ghosh, A., Maiti, G. P., Alam, N., Roy, A., Roychoudhury, S., and Panda, C. K. (2009). Alterations of ROBO1/DUTT1 and ROBO2 loci in early dysplastic lesions of head and neck: clinical and prognostic implications. *Hum. Genet.* **125,** 189–198.

Gorn, M., Anige, M., Burkholder, I., Muller, B., Scheffler, A., Edler, L., Boeters, I., Panse, J., Schuch, G., Hossfeld, D. K., and Laack, E. (2005). Serum levels of Magic Roundabout protein in patients with advanced non-small cell lung cancer (NSCLC). *Lung Cancer* **49,** 71–76.

Greenberg, J. M., Thompson, F. Y., Brooks, S. K., Shannon, J. M., and Akeson, A. L. (2004). Slit and robo expression in the developing mouse lung. *Dev. Dyn.* **230,** 350–360.

Grone, J., Doebler, O., Loddenkemper, C., Hotz, B., Buhr, H. J., and Bhargava, S. (2006). Robol/Robo4: differential expression of angiogenic markers in colorectal cancer. *Oncol. Rep.* **15**, 1437–1443.

Guan, H., Zu, G., Xie, Y., Tang, H., Johnson, M., Xu, X., Kevil, C., Xiong, W. C., Elmets, C., Rao, Y., Wu, J. Y., and Xu, H. (2003). Neuronal repellent Slit2 inhibits dendritic cell migration and the development of immune responses. *J. Immunol.* **171**, 6519–6526.

Hall, A. (2009). The cytoskeleton and cancer. *Cancer Metastasis Rev.* **28**, 5–14.

Han, X., and Zhang, M. C. (2010). Potential anti-angiogenic role of Slit2 in corneal neovascularization. *Exp. Eye Res.* **90**, 742–749.

Hinck, L. (2004). The versatile roles of "axon guidance" cues in tissue morphogenesis. *Dev. Cell* **7**, 783–793.

Hohenester, E. (2008). Structural insight into Slit-Robo signalling. *Biochem. Soc. Trans.* **36**, 251–256.

Hohenester, E., Hussain, S., and Howitt, J. A. (2006). Interaction of the guidance molecule Slit with cellular receptors. *Biochem. Soc. Trans.* **34**, 418–421.

Howitt, J. A., Clout, N. J., and Hohenester, E. (2004). Binding site for Robo receptors revealed by dissection of the leucine-rich repeat region of Slit. *EMBO J.* **23**, 4406–4412.

Hu, H. (2001). Cell-surface heparan sulfate is involved in the repulsive guidance activities of Slit2 protein. *Nat. Neurosci.* **4**, 695–701.

Hu, H., Li, M., Labrador, J. P., McEwen, J., Lai, E. C., Goodman, C. S., and Bashaw, G. J. (2005). Cross GTPase-activating protein (CrossGAP)/Vilse links the Roundabout receptor to Rac to regulate midline repulsion. *Proc. Natl. Acad. Sci. USA* **102**, 4613–4618.

Huang, L., Xu, Y., Yu, W., Li, X., Liqun, C., He, X., and Peiying, H. (2009a). Robo1: a potential role in ocular angiogenesis. *Curr. Eye Res.* **34**, 1019–1029.

Huang, L., Yu, W., Li, X., Niu, L., Li, K., and Li, J. (2009b). Robo1/robo4: different expression patterns in retinal development. *Exp. Eye Res.* **88**, 583–588.

Huang, L., Yu, W., Li, X., Xu, Y., Niu, L., He, X., Dong, J., and Yan, Z. (2009c). Expression of Robo4 in the fibrovascular membranes from patients with proliferative diabetic retinopathy and its role in RF/6A and RPE cells. *Mol. Vis.* **15**, 1057–1069.

Huminiecki, L., and Bicknell, R. (2000). In silico cloning of novel endothelial-specific genes. *Genome Res.* **10**, 1796–1806.

Huminiecki, L., Gorn, M., Suchting, S., Poulsom, R., and Bicknell, R. (2002). Magic roundabout is a new member of the roundabout receptor family that is endothelial specific and expressed at sites of active angiogenesis. *Genomics* **79**, 547–552.

Hussain, S. A., Piper, M., Fukuhara, N., Strochlic, L., Cho, G., Howitt, J. A., Ahmed, Y., Powell, A. K., Turnbull, J. E., Holt, C. E., and Hohenester, E. (2006). A molecular mechanism for the heparan sulfate dependence of slit-robo signaling. *J. Biol. Chem.* **281**, 39693–39698.

Inatani, M., Irie, F., Plump, A. S., Tessier-Lavigne, M., and Yamaguchi, Y. (2003). Mammalian brain morphogenesis and midline axon guidance require heparan sulfate. *Science* **302**, 1044–1046.

Ito, H., Funahashi, S., Yamauchi, N., Shibahara, J., Midorikawa, Y., Kawai, S., Kinoshita, Y., Watanabe, A., Hippo, Y., Ohtomo, T., Iwanari, H., Nakajima, A., *et al.* (2006). Identification of ROBO1 as a novel hepatocellular carcinoma antigen and a potential therapeutic and diagnostic target. *Clin. Cancer Res.* **12**, 3257–3264.

Jin, J., You, H., Yu, B., Deng, Y., Tang, N., Yao, G., Shu, H., Yang, S., and Qin, W. (2009). Epigenetic inactivation of SLIT2 in human hepatocellular carcinomas. *Biochem. Biophys. Res. Commun.* **379**, 86–91.

Johnson, D. W., Berg, J. N., Baldwin, M. A., Gallione, C. J., Marondel, I., Yoon, S. J., Stenzel, T. T., Speer, M., Pericak-Vance, M. A., Diamond, A., Guttmacher, A. E., Jackson, C. E., *et al.* (1996). Mutations in the activin receptor-like kinase 1 gene in hereditary haemorrhagic telangiectasia type 2. *Nat. Genet.* **13**, 189–195.

Johnson, K. G., Ghose, A., Epstein, E., Lincecum, J., O'Connor, M. B., and Van Vactor, D. (2004). Axonal heparan sulfate proteoglycans regulate the distribution and efficiency of the repellent slit during midline axon guidance. *Curr. Biol.* **14,** 499–504.

Jones, C. A., London, N. R., Chen, H., Park, K. W., Sauvaget, D., Stockton, R. A., Wythe, J. D., Suh, W., Larrieu-Lahargue, F., Mukouyama, Y. S., Lindblom, P., Seth, P., *et al.* (2008). Robo4 stabilizes the vascular network by inhibiting pathologic angiogenesis and endothelial hyperpermeability. *Nat. Med.* **14,** 448–453.

Jones, C. A., Nishiya, N., London, N. R., Zhu, W., Sorensen, L. K., Chan, A. C., Lim, C. J., Chen, H., Zhang, Q., Schultz, P. G., Hayallah, A. M., Thomas, K. R., *et al.* (2009). Slit2-Robo4 signalling promotes vascular stability by blocking Arf6 activity. *Nat. Cell Biol.* **11,** 1325–1331.

Kanellis, J., Garcia, G. E., Li, P., Parra, G., Wilson, C. B., Rao, Y., Han, S., Smith, C. W., Johnson, R. J., Wu, J. Y., and Feng, L. (2004). Modulation of inflammation by slit protein *in vivo* in experimental crescentic glomerulonephritis. *Am. J. Pathol.* **165,** 341–352.

Kastenhuber, E., Kern, U., Bonkowsky, J. L., Chien, C. B., Driever, W., and Schweitzer, J. (2009). Netrin-DCC, Robo-Slit, and heparan sulfate proteoglycans coordinate lateral positioning of longitudinal dopaminergic diencephalospinal axons. *J. Neurosci.* **29,** 8914–8926.

Kaur, S., Castellone, M. D., Bedell, V. M., Konar, M., Gutkind, J. S., and Ramchandran, R. (2006). Robo4 signaling in endothelial cells implies attraction guidance mechanisms. *J. Biol. Chem.* **281,** 11347–11356.

Kaur, S., Samant, G. V., Pramanik, K., Loscombe, P. W., Pendrak, M. L., Roberts, D. D., and Ramchandran, R. (2008). Silencing of directional migration in roundabout4 knockdown endothelial cells. *BMC Cell Biol.* **9,** 61.

Kerbel, R. S. (2008). Tumor angiogenesis. *N. Engl. J. Med.* **358,** 2039–2049.

Khusial, P. R., Vadla, B., Krishnan, H., Ramlall, T. F., Shen, Y., Ichikawa, H., Geng, J. G., and Goldberg, G. S. (2010). Src activates Abl to augment Robo1 expression in order to promote tumor cell migration. *Oncotarget* **1,** 198–209.

Kidd, T., Bland, K. S., and Goodman, C. S. (1999). Slit is the midline repellent for the robo receptor in Drosophila. *Cell* **96,** 785–794.

Koch, A. W., Mathivet, T., Larrivee, B., Tong, R. K., Kowalski, J., Pibouin-Fragner, L., Bouvree, K., Stawicki, S., Nicholes, K., Rathore, N., Scales, S. J., Luis, E., *et al.* (2011). Robo4 maintains vessel integrity and inhibits angiogenesis by interacting with UNC5B. *Dev. Cell* **20,** 33–46.

Kramer, S. G., Kidd, T., Simpson, J. H., and Goodman, C. S. (2001). Switching repulsion to attraction: changing responses to slit during transition in mesoderm migration. *Science* **292,** 737–740.

Latil, A., Bieche, I., Chene, L., Laurendeau, I., Berthon, P., Cussenot, O., and Vidaud, M. (2003). Gene expression profiling in clinically localized prostate cancer: a four-gene expression model predicts clinical behavior. *Clin. Cancer Res.* **9,** 5477–5485.

Lee, J. S., Ray, R., and Chien, C. B. (2001). Cloning and expression of three zebrafish roundabout homologs suggest roles in axon guidance and cell migration. *Dev. Dyn.* **221,** 216–230.

Lee, J. S., von der Hardt, S., Rusch, M. A., Stringer, S. E., Stickney, H. L., Talbot, W. S., Geisler, R., Nusslein-Volhard, C., Selleck, S. B., Chien, C. B., and Roehl, H. (2004). Axon sorting in the optic tract requires HSPG synthesis by ext2 (dackel) and extl3 (boxer). *Neuron* **44,** 947–960.

Legg, J. A., Herbert, J. M., Clissold, P., and Bicknell, R. (2008). Slits and Roundabouts in cancer, tumour angiogenesis and endothelial cell migration. *Angiogenesis* **11,** 13–21.

Leonardo, E. D., Hinck, L., Masu, M., Keino-Masu, K., Ackerman, S. L., and Tessier-Lavigne, M. (1997). Vertebrate homologues of C. elegans UNC-5 are candidate netrin receptors. *Nature* **386,** 833–838.

Liang, Y., Annan, R. S., Carr, S. A., Popp, S., Mevissen, M., Margolis, R. K., and Margolis, R. U. (1999). Mammalian homologues of the Drosophila slit protein are ligands of the heparan sulfate proteoglycan glypican-1 in brain. *J. Biol. Chem.* **274,** 17885–17892.

Liu, Z., Patel, K., Schmidt, H., Andrews, W., Pini, A., and Sundaresan, V. (2004). Extracellular Ig domains 1 and 2 of Robo are important for ligand (Slit) binding. *Mol. Cell. Neurosci.* **26,** 232–240.

Liu, D., Hou, J., Hu, X., Wang, X., Xiao, Y., Mou, Y., and De Leon, H. (2006). Neuronal chemorepellent Slit2 inhibits vascular smooth muscle cell migration by suppressing small GTPase Rac1 activation. *Circ. Res.* **98,** 480–489.

London, N. R., and Li, D. Y. (2011). Robo4-dependent Slit signaling stabilizes the vasculature during pathologic angiogenesis and cytokine storm. *Curr. Opin. Hematol.* **18,** 186–190.

London, N. R., Zhu, W., Bozza, F. A., Smith, M. C., Greif, D. M., Sorensen, L. K., Chen, L., Kaminoh, Y., Chan, A. C., Passi, S. F., Day, C. W., Barnard, D. L., *et al.* (2010). Targeting Robo4-dependent Slit signaling to survive the cytokine storm in sepsis and influenza. *Sci. Transl. Med.* **2,** 23ra19.

Lundstrom, A., Gallio, M., Englund, C., Steneberg, P., Hemphala, J., Aspenstrom, P., Keleman, K., Falileeva, L., Dickson, B. J., and Samakovlis, C. (2004). Vilse, a conserved Rac/Cdc42 GAP mediating Robo repulsion in tracheal cells and axons. *Genes Dev.* **18,** 2161–2171.

Ma, X. J., Wang, Z., Ryan, P. D., Isakoff, S. J., Barmettler, A., Fuller, A., Muir, B., Mohapatra, G., Salunga, R., Tuggle, J. T., Tran, Y., Tran, D., *et al.* (2004). A two-gene expression ratio predicts clinical outcome in breast cancer patients treated with tamoxifen. *Cancer Cell* **5,** 607–616.

Macias, H., Moran, A., Samara, Y., Moreno, M., Compton, J. E., Harburg, G., Strickland, P., and Hinck, L. (2011). SLIT/ROBO1 signaling suppresses mammary branching morphogenesis by limiting basal cell number. *Dev. Cell* **20,** 827–840.

Marlow, R., Strickland, P., Lee, J. S., Wu, X., Pebenito, M., Binnewies, M., Le, E. K., Moran, A., Macias, H., Cardiff, R. D., Sukumar, S., and Hinck, L. (2008). SLITs suppress tumor growth *in vivo* by silencing Sdf1/Cxcr4 within breast epithelium. *Cancer Res.* **68,** 7819–7827.

Marlow, R., Binnewies, M., Sorensen, L. K., Monica, S. D., Strickland, P., Forsberg, E. C., Li, D. Y., and Hinck, L. (2010). Vascular Robo4 restricts proangiogenic VEGF signaling in breast. *Proc. Natl. Acad. Sci. USA* **107,** 10520–10525.

Medioni, C., Bertrand, N., Mesbah, K., Hudry, B., Dupays, L., Wolstein, O., Washkowitz, A. J., Papaioannou, V. E., Mohun, T. J., Harvey, R. P., and Zaffran, S. (2010). Expression of Slit and Robo genes in the developing mouse heart. *Dev. Dyn.* **239,** 3303–3311.

Mertsch, S., Schmitz, N., Jeibmann, A., Geng, J. G., Paulus, W., and Senner, V. (2008). Slit2 involvement in glioma cell migration is mediated by Robo1 receptor. *J. Neurooncol* **87,** 1–7.

Moore, M. A. (2001). The role of chemoattraction in cancer metastases. *Bioessays* **23,** 674–676.

Moore, S. W., Tessier-Lavigne, M., and Kennedy, T. E. (2007). Netrins and their receptors. *Adv. Exp. Med. Biol.* **621,** 17–31.

Morlot, C., Hemrika, W., Romijn, R. A., Gros, P., Cusack, S., and McCarthy, A. A. (2007a). Production of Slit2 LRR domains in mammalian cells for structural studies and the structure of human Slit2 domain 3. *Acta Crystallogr. D Biol. Crystallogr.* **63,** 961–968.

Morlot, C., Thielens, N. M., Ravelli, R. B., Hemrika, W., Romijn, R. A., Gros, P., Cusack, S., and McCarthy, A. A. (2007b). Structural insights into the Slit-Robo complex. *Proc. Natl. Acad. Sci. USA* **104,** 14923–14928.

Mulik, S., Sharma, S., Suryawanshi, A., Veiga-Parga, T., Reddy, P. B., Rajasagi, N. K., and Rouse, B. T. (2011). Activation of endothelial roundabout receptor 4 reduces the severity of virus-induced keratitis. *J. Immunol.* **186,** 7195–7204.

Muller, A., Homey, B., Soto, H., Ge, N., Catron, D., Buchanan, M. E., McClanahan, T., Murphy, E., Yuan, W., Wagner, S. N., Barrera, J. L., Mohar, A., *et al.* (2001). Involvement of chemokine receptors in breast cancer metastasis. *Nature* **410,** 50–56.

Mura, M., Swain, R. K., Zhuang, X., Vorschmitt, H., Reynolds, G., Durant, S., Beesley, J. F., Herbert, J. M., Sheldon, H., Andre, M., Sanderson, S., Glen, K., *et al.* (2011). Identification and angiogenic role of the novel tumor endothelial marker CLEC14A. *Oncogene* **31,** 293–305.

Narayan, G., Goparaju, C., Arias-Pulido, H., Kaufmann, A. M., Schneider, A., Durst, M., Mansukhani, M., Pothuri, B., and Murty, V. V. (2006). Promoter hypermethylation-mediated inactivation of multiple Slit-Robo pathway genes in cervical cancer progression. *Mol. Cancer* **5,** 16.

Nguyen Ba-Charvet, K. T., Brose, K., Ma, L., Wang, K. H., Marillat, V., Sotelo, C., Tessier-Lavigne, M., and Chedotal, A. (2001). Diversity and specificity of actions of Slit2 proteolytic fragments in axon guidance. *J. Neurosci.* **21,** 4281–4289.

Ning, Y., Sun, Q., Dong, Y., Xu, W., Zhang, W., Huang, H., and Li, Q. (2011). Slit2-N inhibits PDGF-induced migration in rat airway smooth muscle cells: WASP and Arp2/3 involved. *Toxicology* **283,** 32–40.

Ogata-Iwao, M., Inatani, M., Iwao, K., Takihara, Y., Nakaishi-Fukuchi, Y., Irie, F., Sato, S., Furukawa, T., Yamaguchi, Y., and Tanihara, H. (2011). Heparan sulfate regulates intra-retinal axon pathfinding by retinal ganglion cells. *Invest. Ophthalmol. Vis. Sci.* **52,** 6671–6679.

Park, K. W., Morrison, C. M., Sorensen, L. K., Jones, C. A., Rao, Y., Chien, C. B., Wu, J. Y., Urness, L. D., and Li, D. Y. (2003). Robo4 is a vascular-specific receptor that inhibits endothelial migration. *Dev. Biol.* **261,** 251–267.

Parsons, D. W., Jones, S., Zhang, X., Lin, J. C., Leary, R. J., Angenendt, P., Mankoo, P., Carter, H., Siu, I. M., Gallia, G. L., Olivi, A., McLendon, R., *et al.* (2008). An integrated genomic analysis of human glioblastoma multiforme. *Science* **321,** 1807–1812.

Piper, M., Georgas, K., Yamada, T., and Little, M. (2000). Expression of the vertebrate Slit gene family and their putative receptors, the Robo genes, in the developing murine kidney. *Mech. Dev.* **94,** 213–217.

Piper, M., Anderson, R., Dwivedy, A., Weinl, C., van Horck, F., Leung, K. M., Cogill, E., and Holt, C. (2006). Signaling mechanisms underlying Slit2-induced collapse of Xenopus retinal growth cones. *Neuron* **49,** 215–228.

Plump, A. S., Erskine, L., Sabatier, C., Brose, K., Epstein, C. J., Goodman, C. S., Mason, C. A., and Tessier-Lavigne, M. (2002). Slit1 and Slit2 cooperate to prevent premature midline crossing of retinal axons in the mouse visual system. *Neuron* **33,** 219–232.

Potente, M., Gerhardt, H., and Carmeliet, P. (2011). Basic and therapeutic aspects of angiogenesis. *Cell* **146,** 873–887.

Prasad, A., Fernandis, A. Z., Rao, Y., and Ganju, R. K. (2004). Slit protein-mediated inhibition of CXCR4-induced chemotactic and chemoinvasive signaling pathways in breast cancer cells. *J. Biol. Chem.* **279,** 9115–9124.

Prasad, A., Qamri, Z., Wu, J., and Ganju, R. K. (2007). Slit-2/Robo-1 modulates the CXCL12/CXCR4-induced chemotaxis of T cells. *J. Leukoc. Biol.* **82,** 465–476.

Prasad, A., Paruchuri, V., Preet, A., Latif, F., and Ganju, R. K. (2008). Slit-2 induces a tumor-suppressive effect by regulating beta-catenin in breast cancer cells. *J. Biol. Chem.* **283,** 26624–26633.

Qiu, H., Zhu, J., Yu, J., Pu, H., and Dong, R. (2011). SLIT2 is epigenetically silenced in ovarian cancers and suppresses growth when activated. *Asian Pac. J. Cancer Prev.* **12,** 791–795.

Rhee, J., Mahfooz, N. S., Arregui, C., Lilien, J., Balsamo, J., and VanBerkum, M. F. (2002). Activation of the repulsive receptor Roundabout inhibits N-cadherin-mediated cell adhesion. *Nat. Cell Biol.* **4,** 798–805.

Rhee, J., Buchan, T., Zukerberg, L., Lilien, J., and Balsamo, J. (2007). Cables links Robo-bound Abl kinase to N-cadherin-bound beta-catenin to mediate Slit-induced modulation of adhesion and transcription. *Nat. Cell Biol.* **9**, 883–892.

Rhiner, C., Gysi, S., Frohli, E., Hengartner, M. O., and Hajnal, A. (2005). Syndecan regulates cell migration and axon guidance in C. elegans. *Development* **132**, 4621–4633.

Richardson, A. L., Wang, Z. C., De Nicolo, A., Lu, X., Brown, M., Miron, A., Liao, X., Iglehart, J. D., Livingston, D. M., and Ganesan, S. (2006). X chromosomal abnormalities in basal-like human breast cancer. *Cancer Cell* **9**, 121–132.

Ronca, F., Andersen, J. S., Paech, V., and Margolis, R. U. (2001). Characterization of Slit protein interactions with glypican-1. *J. Biol. Chem.* **276**, 29141–29147.

Salvucci, O., Bouchard, A., Baccarelli, A., Deschenes, J., Sauter, G., Simon, R., Bianchi, R., and Basik, M. (2006). The role of CXCR4 receptor expression in breast cancer: a large tissue microarray study. *Breast Cancer Res. Treat.* **97**, 275–283.

Santiago-Martinez, E., Soplop, N. H., Patel, R., and Kramer, S. G. (2008). Repulsion by Slit and Roundabout prevents Shotgun/E-cadherin-mediated cell adhesion during Drosophila heart tube lumen formation. *J. Cell Biol.* **182**, 241–248.

Schmid, B. C., Rezniczek, G. A., Fabjani, G., Yoneda, T., Leodolter, S., and Zeillinger, R. (2007). The neuronal guidance cue Slit2 induces targeted migration and may play a role in brain metastasis of breast cancer cells. *Breast Cancer Res. Treat.* **106**, 333–342.

Schulz, J. G., Ceulemans, H., Caussinus, E., Baietti, M. F., Affolter, M., Hassan, B. A., and David, G. (2011). Drosophila syndecan regulates tracheal cell migration by stabilizing Robo levels. *EMBO Rep.* **12**, 1039–1046.

Seki, M., Watanabe, A., Enomoto, S., Kawamura, T., Ito, H., Kodama, T., Hamakubo, T., and Aburatani, H. (2010). Human ROBO1 is cleaved by metalloproteinases and gamma-secretase and migrates to the nucleus in cancer cells. *FEBS Lett.* **584**, 2909–2915.

Seth, P., Lin, Y., Hanai, J., Shivalingappa, V., Duyao, M. P., and Sukhatme, V. P. (2005). Magic roundabout, a tumor endothelial marker: expression and signaling. *Biochem. Biophys. Res. Commun.* **332**, 533–541.

Sharma, G., Mirza, S., Prasad, C. P., Srivastava, A., Gupta, S. D., and Ralhan, R. (2007). Promoter hypermethylation of p16INK4A, p14ARF, CyclinD2 and Slit2 in serum and tumor DNA from breast cancer patients. *Life Sci.* **80**, 1873–1881.

Sheldon, H., Andre, M., Legg, J. A., Heal, P., Herbert, J. M., Sainson, R., Sharma, A. S., Kitajewski, J. K., Heath, V. L., and Bicknell, R. (2009). Active involvement of Robo1 and Robo4 in filopodia formation and endothelial cell motility mediated via WASP and other actin nucleation-promoting factors. *FASEB J.* **23**, 513–522.

Shiau, C. E., and Bronner-Fraser, M. (2009). N-cadherin acts in concert with Slit1-Robo2 signaling in regulating aggregation of placode-derived cranial sensory neurons. *Development* **136**, 4155–4164.

Shivapurkar, N., Sood, S., Wistuba, I. I., Virmani, A. K., Maitra, A., Milchgrub, S., Minna, J. D., and Gazdar, A. F. (1999a). Multiple regions of chromosome 4 demonstrating allelic losses in breast carcinomas. *Cancer Res.* **59**, 3576–3580.

Shivapurkar, N., Virmani, A. K., Wistuba, I. I., Milchgrub, S., Mackay, B., Minna, J. D., and Gazdar, A. F. (1999b). Deletions of chromosome 4 at multiple sites are frequent in malignant mesothelioma and small cell lung carcinoma. *Clin. Cancer Res.* **5**, 17–23.

Singh, R. K., Indra, D., Mitra, S., Mondal, R. K., Basu, P. S., Roy, A., Roychowdhury, S., and Panda, C. K. (2007). Deletions in chromosome 4 differentially associated with the development of cervical cancer: evidence of slit2 as a candidate tumor suppressor gene. *Hum. Genet.* **122**, 71–81.

Smart, A. D., Course, M. M., Rawson, J., Selleck, S., Van Vactor, D., and Johnson, K. G. (2011). Heparan sulfate proteoglycan specificity during axon pathway formation in the Drosophila embryo. *Dev. Neurobiol.* **71**, 608–618.

Spitzweck, B., Brankatschk, M., and Dickson, B. J. (2010). Distinct protein domains and expression patterns confer divergent axon guidance functions for Drosophila Robo receptors. *Cell* **140,** 409–420.

Steigemann, P., Molitor, A., Fellert, S., Jackle, H., and Vorbruggen, G. (2004). Heparan sulfate proteoglycan syndecan promotes axonal and myotube guidance by slit/robo signaling. *Curr. Biol.* **14,** 225–230.

Stein, E., and Tessier-Lavigne, M. (2001). Hierarchical organization of guidance receptors: silencing of netrin attraction by slit through a Robo/DCC receptor complex. *Science* **291,** 1928–1938.

Stella, M. C., Trusolino, L., and Comoglio, P. M. (2009). The Slit/Robo system suppresses hepatocyte growth factor-dependent invasion and morphogenesis. *Mol. Biol. Cell* **20,** 642–657.

Strickland, P., Shin, G. C., Plump, A., Tessier-Lavigne, M., and Hinck, L. (2006). Slit2 and netrin 1 act synergistically as adhesive cues to generate tubular bi-layers during ductal morphogenesis. *Development* **133,** 823–832.

Suchting, S., Heal, P., Tahtis, K., Stewart, L. M., and Bicknell, R. (2005). Soluble Robo4 receptor inhibits *in vivo* angiogenesis and endothelial cell migration. *FASEB J.* **19,** 121–123.

Suryawanshi, A., Mulik, S., Sharma, S., Reddy, P. B., Sehrawat, S., and Rouse, B. T. (2011). Ocular neovascularization caused by herpes simplex virus type 1 infection results from breakdown of binding between vascular endothelial growth factor A and its soluble receptor. *J. Immunol.* **186,** 3653–3665.

Tie, J., Pan, Y., Zhao, L., Wu, K., Liu, J., Sun, S., Guo, X., Wang, B., Gang, Y., Zhang, Y., Li, Q., Qiao, T., *et al.* (2010). MiR-218 inhibits invasion and metastasis of gastric cancer by targeting the Robo1 receptor. *PLoS Genet.* **6,** e1000879.

Tole, S., Mukovozov, I. M., Huang, Y. W., Magalhaes, M. A., Yan, M., Crow, M. R., Liu, G. Y., Sun, C. X., Durocher, Y., Glogauer, M., and Robinson, L. A. (2009). The axonal repellent, Slit2, inhibits directional migration of circulating neutrophils. *J. Leukoc. Biol.* **86,** 1403–1415.

Tseng, R. C., Lee, S. H., Hsu, H. S., Chen, B. H., Tsai, W. C., Tzao, C., and Wang, Y. C. (2010). SLIT2 attenuation during lung cancer progression deregulates beta-catenin and E-cadherin and associates with poor prognosis. *Cancer Res.* **70,** 543–551.

Urness, L. D., Sorensen, L. K., and Li, D. Y. (2000). Arteriovenous malformations in mice lacking activin receptor-like kinase-1. *Nat. Genet.* **26,** 328–331.

Verissimo, A. R., Herbert, J. M., Heath, V. L., Legg, J. A., Sheldon, H., Andre, M., Swain, R. K., and Bicknell, R. (2009). Functionally defining the endothelial transcriptome, from Robo4 to ECSCR. *Biochem. Soc. Trans.* **37,** 1214–1217.

Wang, K. H., Brose, K., Arnott, D., Kidd, T., Goodman, C. S., Henzel, W., and Tessier-Lavigne, M. (1999). Biochemical purification of a mammalian slit protein as a positive regulator of sensory axon elongation and branching. *Cell* **96,** 771–784.

Wang, B., Xiao, Y., Ding, B. B., Zhang, N., Yuan, X., Gui, L., Qian, K. X., Duan, S., Chen, Z., Rao, Y., and Geng, J. G. (2003). Induction of tumor angiogenesis by Slit-Robo signaling and inhibition of cancer growth by blocking Robo activity. *Cancer Cell* **4,** 19–29.

Wang, L. J., Zhao, Y., Han, B., Ma, Y. G., Zhang, J., Yang, D. M., Mao, J. W., Tang, F. T., Li, W. D., Yang, Y., Wang, R., and Geng, J. G. (2008). Targeting Slit-Roundabout signaling inhibits tumor angiogenesis in chemical-induced squamous cell carcinogenesis. *Cancer Sci.* **99,** 510–517.

Werbowetski-Ogilvie, T. E., Seyed Sadr, M., Jabado, N., Angers-Loustau, A., Agar, N. Y., Wu, J., Bjerkvig, R., Antel, J. P., Faury, D., Rao, Y., and Del Maestro, R. F. (2006). Inhibition of medulloblastoma cell invasion by Slit. *Oncogene* **25,** 5103–5112.

Wong, K., Ren, X. R., Huang, Y. Z., Xie, Y., Liu, G., Saito, H., Tang, H., Wen, L., Brady-Kalnay, S. M., Mei, L., Wu, J. Y., Xiong, W. C., *et al.* (2001). Signal transduction in neuronal migration: roles of GTPase activating proteins and the small GTPase Cdc42 in the Slit-Robo pathway. *Cell* **107,** 209–221.

Wu, J. Y., Feng, L., Park, H. T., Havlioglu, N., Wen, L., Tang, H., Bacon, K. B., Jiang, Z., Zhang, X., and Rao, Y. (2001). The neuronal repellent Slit inhibits leukocyte chemotaxis induced by chemotactic factors. *Nature* **410,** 948–952.

Xian, J., Clark, K. J., Fordham, R., Pannell, R., Rabbitts, T. H., and Rabbitts, P. H. (2001). Inadequate lung development and bronchial hyperplasia in mice with a targeted deletion in the Dutt1/Robo1 gene. *Proc. Natl. Acad. Sci. USA* **98,** 15062–15066.

Xu, Y., Li, W. L., Fu, L., Gu, F., and Ma, Y. J. (2010). Slit2/Robo1 signaling in glioma migration and invasion. *Neurosci. Bull.* **26,** 474–478.

Yang, L., and Bashaw, G. J. (2006). Son of sevenless directly links the Robo receptor to rac activation to control axon repulsion at the midline. *Neuron* **52,** 595–607.

Yang, X. M., Han, H. X., Sui, F., Dai, Y. M., Chen, M., and Geng, J. G. (2010). Slit-Robo signaling mediates lymphangiogenesis and promotes tumor lymphatic metastasis. *Biochem. Biophys. Res. Commun.* **396,** 571–577.

Ye, B. Q., Geng, Z. H., Ma, L., and Geng, J. G. (2010). Slit2 regulates attractive eosinophil and repulsive neutrophil chemotaxis through differential srGAP1 expression during lung inflammation. *J. Immunol.* **185,** 6294–6305.

Yiin, J. J., Hu, B., Jarzynka, M. J., Feng, H., Liu, K. W., Wu, J. Y., Ma, H. I., and Cheng, S. Y. (2009). Slit2 inhibits glioma cell invasion in the brain by suppression of Cdc42 activity. *Neuro Oncol.* **11,** 779–789.

Yu, J., Cao, Q., Wu, L., Dallol, A., Li, J., Chen, G., Grasso, C., Cao, X., Lonigro, R. J., Varambally, S., Mehra, R., Palanisamy, N., *et al.* (2010). The neuronal repellent SLIT2 is a target for repression by EZH2 in prostate cancer. *Oncogene* **29,** 5370–5380.

Yuasa-Kawada, J., Kinoshita-Kawada, M., Rao, Y., and Wu, J. Y. (2009). Deubiquitinating enzyme USP33/VDU1 is required for Slit signaling in inhibiting breast cancer cell migration. *Proc. Natl. Acad. Sci. USA* **106,** 14530–14535.

Zhang, F., Ronca, F., Linhardt, R. J., and Margolis, R. U. (2004). Structural determinants of heparan sulfate interactions with Slit proteins. *Biochem. Biophys. Res. Commun.* **317,** 352–357.

Zhang, B., Dietrich, U. M., Geng, J. G., Bicknell, R., Esko, J. D., and Wang, L. (2009). Repulsive axon guidance molecule Slit3 is a novel angiogenic factor. *Blood* **114,** 4300–4309.

Zheng, W., Geng, A. Q., Li, P. F., Wang, Y., and Yuan, X. B. (2011). Robo4 regulates the radial migration of newborn neurons in developing neocortex. *Cereb. Cortex* [Epub ahead of print].

Zhou, W. J., Geng, Z. H., Chi, S., Zhang, W., Niu, X. F., Lan, S. J., Ma, L., Yang, X., Wang, L. J., Ding, Y. Q., and Geng, J. G. (2011). Slit-Robo signaling induces malignant transformation through Hakai-mediated E-cadherin degradation during colorectal epithelial cell carcinogenesis. *Cell Res.* **21,** 609–626.

Role of the VEGF/VEGFR Axis in Cancer Biology and Therapy

Annamaria Rapisarda* and Giovanni Melillo†

**SAIC-Frederick, Inc., Frederick National Laboratory for Cancer Research, Frederick, Maryland, USA*
†Discovery Medicine-Oncology, Bristol-Myers Squibb, Princeton, New Jersey, USA

New vessel formation (angiogenesis) is an essential physiological process for embryologic development, normal growth, and tissue repair. Angiogenesis is tightly regulated at the molecular level; however, this process is dysregulated in several pathological conditions such as cancer. The imbalance between pro- and antiangiogenic signaling molecules within tumors creates an abnormal vascular network that is characterized by dilated, tortuous, and leaky vessels. The pathoph ysiological consequences of these vascular abnormalities include temporal and spatial heterogeneity in tumor blood flow, oxygenation, and increased tumor interstitial fluid pressure. The resultant microenvironment deeply impacts on tumor progression, and also leads to a reduction in therapy efficacy. The discovery of vascular endothelial growth factor (VEGF) as a major driver of tumor angiogenesis has led to efforts to develop novel therapeutics aimed at inhibiting its

Advances in CANCER RESEARCH, Volume 114

0065-230X/12 $35.00
DOI: 10.1016/B978-0-12-386503-8.00006-5

activity. Anti-VEGF therapy has become an important option for the management of several human malignancies; however, a significant number of patients do not respond to anti-VEGF therapy when used either as single agent or in combination with chemotherapy. In addition, the benefit of antiangiogenic therapy is relatively short lived and the majority of patients relapse and progress. An increasing amount of reports suggest several potential mechanisms of resistance to antiangiogenic therapy including, but not limited to, tumor hypoxia. This chapter discusses the role of the VEGF axis in tumor biology and highlights the clinical application of anti-VEGF therapies elaborating on pitfalls and strategies to improve clinical outcome.

I. VASCULAR ENDOTHELIAL GROWTH FACTORS AND THEIR RECEPTORS IN CANCER BIOLOGY

A. Vascular Endothelial Growth Factors

There are five structurally related Vascular Endothelial Growth Factors (VEGF) ligands (VEGFA, VEGFB, VEGFC, VEGFD, and placenta growth factor (PIGF)). VEGFs are disulfide-bonded homodimers, although VEGFA and PIGF heterodimers have also been described (DiSalvo *et al.*, 1995). Each VEGF ligand is expressed as several different variants due to alternative splicing or posttranslational processing. Each variant binds differently to VEGF receptors (VEGFRs) and their coreceptors and therefore induces different biological responses, such as angiogenesis, lymphangiogenesis, permeability, inflammatory cell recruitment, and fatty acid uptake (see Table I). VEGFs are produced by many different cell types and act in an autocrine and paracrine manner. Knockout mice lacking expression of different VEGF ligands have demonstrated the critical role of VEGFs in vessel formation and function. The most striking effects are seen for VEGFA, where even one deleted allele is lethal (Carmeliet *et al.*, 1996; Ferrara *et al.*, 1996). VEGFA is critical for development of endothelial cells during embryogenesis and for organization of the vasculature, as well as for their survival.

B. VEGF Receptors

VEGFs bind to three structurally related receptor tyrosine kinases (RTKs), VEGFR1, VEGFR2, and VEGFR3. In addition, a number of coreceptors (such as neuropilins, NRPs) that lack intrinsic catalytic activity bind VEGF and modulate the effect of the VEGFRs. VEGFRs have a high degree of homology within the kinase domain; however, their signaling properties greatly differ.

Table I Functions, Binding Properties, and Biological Implications of VEGFs

VEGF isoform	Receptor	Coreceptor	Biological function
VEGFA165	VEGFR1, VEGFR2	NRP1, NRP2	Angiogenesis (permeability, survival, migration of EC)
VEGFA121	VEGFR1, VEGFR2	NRP1[a]	Angiogenic/antiangiogenic properties[b]
VEGFA145	VEGFR1, VEGFR2	NRP2	Angiogenesis
VEGFA189	VEGFR1, VEGFR2	NRP1	Angiogenesis
VEGFA(xxx)b	VEGFR1, VEGFR2	No	Antiangiogenic properties
VEGFB	VEGFR1	NRP1	Fatty acid uptake in EC of the heart
VEGFC[c]	VEGFR3 (VEGFR2)	NRP2	Lymphangiogenesis
VEGFD[c]	VEGFR3 (VEGFR2)	NRP2	Lymphangiogenesis
PIGF	VEGFR1	NRP1, NRP2	Inflammatory cell recruitment

Abbreviations: EC, endothelial cells.
[a]VEGFA121 binds NRP1 but does not bridge to VEGFRs (Pan *et al.*, 2007).
[b]VEGFA121 has been described as antiangiogenic (Nowak *et al.*, 2008).
[c]Processed.

1. VEGFR1

VEGFR1 (alternatively denoted as Fms-like tyrosine kinase 1, Flt1, in the mouse) is a single-transmembrane glycoprotein. Interestingly, VEGFR1 binds VEGFA with at least 10-fold higher affinity than VEGFR2, yet it is poorly activated (Ferrara and Davis-Smyth, 1997). A study by Gille *et al.* (2000) of chimeric VEGFR1 and VEGFR2 revealed that the juxtamembrane domain of VEGFR1 plays an inhibitory role in VEGFR1 signaling pathways, although the precise mechanism requires further investigation. Accumulating evidence indicates that a soluble form of VEGFR1 (sVEGFR1) has a negative regulatory role in human physiology, presumably by trapping VEGFA (Kendall *et al.*, 1994). Moreover, sVEGFR1 levels are elevated in patients with breast cancer, pancreatic cancer, leukemia, and colorectal cancer, where it is associated with a favorable prognosis (Scheufler *et al.*, 2003; Toi *et al.*, 2002). VEGFR1 is expressed not only in vascular endothelial cells but also in other cell types (monocytes and macrophages, human trophoblasts, renal mesangial cells, vascular smooth muscle cells, dendritic cells, and various types of cancer cells) (Shibuya and Claesson-Welsh, 2006). The fact that VEGFR1 is usually expressed at low levels has limited the progress in elucidating its signal transduction pathways (Fig. 1).

Notably, VEGFR1 plays a role in tumor progression and dissemination. Indeed, the rate of tumor growth of melanoma and glioma tumor models is considerably reduced in VEGFR1 $TK^{-/-}$ mice (Kerbel, 2008; Muramatsu *et al.*, 2010). In addition, VEGFR1 activity has been shown to play a role in

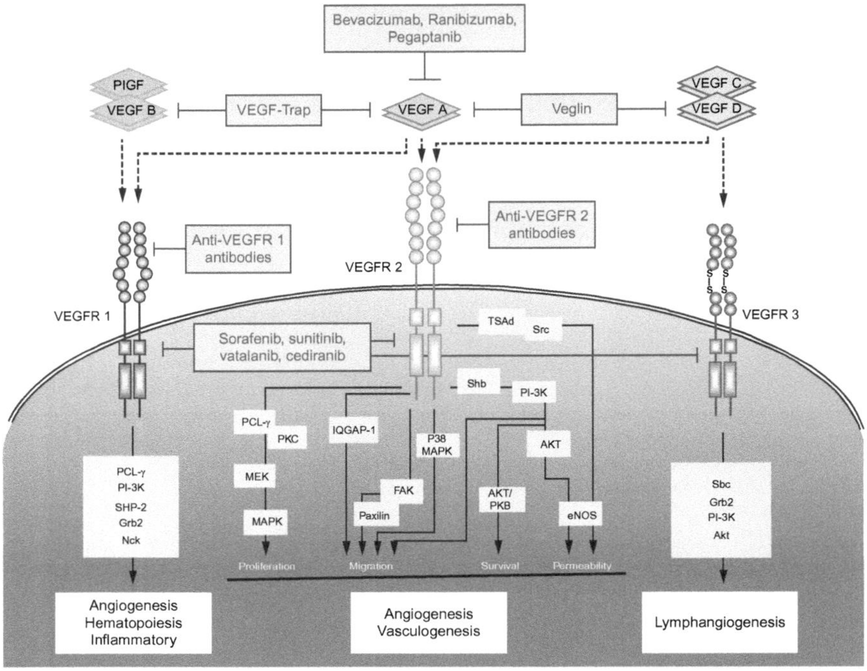

Fig. 1 Signaling and biological processes mediated by the VEGF/VEGFRs axis and therapeutic agents. (For color version of this figure, the reader is referred to the Web version of this chapter.)

metastatic dissemination, and expression of VEGFR1 in tumor cells seems to increase tumor invasiveness (Mylona *et al.*, 2007; Seto *et al.*, 2006). Furthermore, VEGFR1 has been shown to activate extracellular signal-regulated kinase 1/2, stress-activated protein kinase/c-Jun NH_2-terminal kinase (Fan *et al.*, 2005), and Src family kinases (Lesslie *et al.*, 2006) to mediate growth and migration of human colorectal carcinoma cells. Finally, activation of VEGFR1 in breast cancer cells supports their growth and survival (Wu *et al.*, 2006a), strongly arguing in favor of the importance of VEGFR1-mediated signaling in these models.

Regulation of inflammatory cell recruitment by VEGFR1 appears to be exerted mainly through PIGF. Notably, the expression of PIGF is very low under physiological conditions, but it may be strongly upregulated in various cell types by different pathological stimuli such as hypoxia, inflammatory cytokines, or oncogenes (Failla *et al.*, 2000; Green *et al.*, 2001; Larcher *et al.*, 2003). PIGF has been regarded as an attractive candidate for anti-angiogenic therapy. Indeed, it has been shown that PIGF plays a key role in promoting pathological angiogenesis associated with tumor progression (Carmeliet *et al.*, 2001), and overexpression of PIGF in a mouse melanoma model resulted in increased tumor growth and metastasis (Li *et al.*, 2006).

2. VEGFR2

There is much evidence that VEGFR2 (KDR) is the major mediator of VEGFA-driven responses in endothelial cells and it is considered to be a crucial signal transducer in both physiologic and pathologic angiogenesis (Hicklin and Ellis, 2005). In addition, VEGFR2 binds proteolytically processed VEGFC and VEGFD (McColl *et al.*, 2003). The signaling pathways triggered by engagement of VEGFR2 are relatively well understood (see Fig. 1).

VEGFR2 is expressed in most if not all adult vascular endothelial cells, as well as in circulating endothelial progenitor cells, pancreatic duct cells, retinal progenitor cells, megakaryocytes, and hematopoietic cells (Hicklin and Ellis, 2005). VEGFR2, often in combination with VEGFR3, is significantly upregulated in the tumor vascular endothelium in most common human solid tumor types (Smith *et al.*, 2010). Tumor cells may also express VEGFR2, although epithelial and mesenchymal tumor cells typically express VEGFR1 rather than VEGFR2 (Hicklin and Ellis, 2005; Podar and Anderson, 2005). Nevertheless, increased expression of VEGFR2 on tumor cells has been noted for melanoma and hematological malignancies (Youssoufian *et al.*, 2007). It has been shown that VEGFR2-mediated signaling led to survival of cancer cells under chronic hypoxic conditions and might contribute to a more aggressive phenotype (Calvani *et al.*, 2006). Growing evidence supports an important link between chronic inflammation and tumor development. Induction of VEGFR2 expression in tumor cells, and also in intestinal epithelium during colitis, is mediated by the proinflammatory cytokine interleukin 6, which is a strong promoter of tumor growth in experimental colitis-associated colon cancer (Waldner *et al.*, 2010). sVEGFR2 has been described and may have important biological roles. sVEGFR2 binds VEGFC and thus prevents activation of VEGFR3, consequently inhibiting lymphatic endothelial cell proliferation (Albuquerque *et al.*, 2009). Notably, it has been recently shown that downregulation of sVEGFR2 in advanced metastatic neuroblastoma may promote lymphogenic spread of metastases (Becker *et al.*, 2010).

3. VEGFR3

VEGFR3 (alternatively denoted Fms-like tyrosine kinase 4, Flt4, in the mouse) is activated by the binding of VEGFC or VEGFD. VEGFR3 and its ligands are key players in the regulation of normal and tumor lymphangiogenesis (Shibuya and Claesson-Welsh, 2006). Indeed, gene inactivation to eliminate expression of VEGFC alone, or combined deletion of VEGFC and VEGFD, unexpectedly resulted in defects mainly in lymphatic vessels, while blood vessels remained unaffected in mouse models (Haiko *et al.*, 2008). In adult tissues, VEGFR3 has an essential role in lymphatic endothelial cells, but its expression is also induced in endothelial cells engaged in active angiogenesis (Carmeliet *et al.*, 2009), such as in tumor vessels (Laakkonen *et al.*, 2007). The expression of VEGFR3 in tumor cells is controversial (Petrova *et al.*, 2008); however, it has been clearly demonstrated that inhibition of VEGFR3 activity arrests tumor vascularization, leading to decreased vascular density in several tumor models (Laakkonen *et al.*, 2007). The axis VEGFC/VEGFR3 plays a fundamental role in the tumor microenvironment by promoting the formation of new lymphatic vessels from preexisting ones (He *et al.*, 2004). VEGFC, produced by tumor cells, induces lymphatic endothelial destabilization, resulting in endothelial sprouting as well as leakage and enlargement of the vessels. These changes facilitate entry of tumor cells into the lymphatics and further dissemination of metastasis to sentinel lymph nodes (Achen and Stacker, 2008; He *et al.*, 2005).

4. NEUROPILINS

There are two NRP homologues, NRP1 and NRP2. The NRPs were first identified as receptors for class 3 semaphorins, a family of soluble molecules with neuronal guidance functions, and are now implicated in the development of the nervous and vascular systems (Hicklin and Ellis, 2005). Importantly, NRPs are also coreceptors for VEGF ligands and are being investigated as possible therapeutic targets to arrest angiogenesis as well as lymphangiogenesis in cancer. Interestingly, increased NRP expression in human leukemia and lymphoma (Karjalainen *et al.*, 2011) and in many solid tumors is associated with increased metastasis (Cohen *et al.*, 2002; Hansel *et al.*, 2004; Kawakami *et al.*, 2002; Lantuejoul *et al.*, 2003; Latil *et al.*, 2000; Stephenson *et al.*, 2002; Vanveldhuizen *et al.*, 2003). However, it is still controversial whether, and to which extent, cancer cells express NRPs.

C. VEGF/VEGFR Axis and the Tumor Microenvironment

The fine balance between the supply of oxygen and nutrients by blood vessels and the proliferation of cancer cells determines the onset of intratumor hypoxia and contributes to the angiogenic switch. Tumors that fail to activate the angiogenic pathway remain dormant and do not progress. The key regulator of hypoxia-induced angiogenesis is the transcription factor hypoxia-inducible factor (HIF)-1. Multiple HIF-1 target genes are involved in different steps of angiogenesis: arterial destabilization (VEGFA, PIGF,

VEGFR1), increased vascular permeability (VEGFA, VEGFR1, angiopoietin 2, Tie-2), extracellular matrix remodeling (MMPs, collagen prolyl-4-hydroxylase, uPAR), migration and proliferation of endothelial cells (VEGFA, PIGF, FGF2, angiopoietin 1, MCP-1, PDGF, SDF-1, CXCR4), endothelial cells sprouting (angiopoietin 2, Tie-2), endothelial tube formation and cell-to-cell interaction (VEGFA, PIGF, angiopoietin 1, integrins), and recruitment of and interaction with pericytes (PDGF, PAI-1, angiopoietin 1, Tie-2) (Hirota and Semenza, 2006). VEGFA exerts multiple effects within the tumor microenvironment, which aggravates tumor growth and metastatic spread and reduces treatment efficacy. Antibodies that bind VEGF and thereby prevent its binding to VEGFRs inhibit angiogenesis and have been exploited clinically for cancer therapy (Ferrara, 2005).

II. TARGETING VEGF/VEGFR FOR CANCER THERAPY

Despite the existence of many pathways that contribute to the angiogenic process, the VEGF/VEGFRs pathway is considered a key regulator of angiogenesis and this realization has led to considerable interest and efforts to exploit this pathway for cancer therapy. It is, therefore, not surprising that most of the antiangiogenic agents currently in preclinical and clinical development focus on inhibition of the VEGF pathway (Fig. 1). Several anti-VEGF strategies have been developed, including neutralizing antibodies to VEGF or VEGFRs, soluble VEGFR/VEGFR hybrids, and tyrosine kinase inhibitors of VEGFRs (Ellis *et al.*, 1996; Gerber *et al.*, 2000; Kim *et al.*, 1993; Klohs and Hamby, 1999; Prewett *et al.*, 1999). Table II summarizes some of the principal antiangiogenic molecules that are currently being used in clinical trials to target VEGF signaling.

A. Antibodies and Decoy Receptor-Based Therapies

1. BEVACIZUMAB

One of the earliest strategies used to inhibit VEGF activity has involved neutralizing antibodies to VEGF. In preclinical studies, a murine anti-VEGF monoclonal antibody inhibited angiogenesis and growth of human tumor xenografts (Gerber *et al.*, 2000; Kim *et al.*, 1993; Prewett *et al.*, 1999). Notably, the anti-VEGF antibody bevacizumab (Avastin®; Genentech Inc.) has been the first antiangiogenic agent to be approved for cancer therapy, in combination with chemotherapy, by the U.S. Food and Drug Administration. It was initially approved for the treatment of metastatic colorectal

Table II Antiangiogenic Agents in Advanced Clinical Development

Therapeutic agent	Type	Target	Clinical development	References
Bevacizumab/Avastin	mAb	VEGFA	Approved in 2004 (CC), 2006 (NSCLC), 2008 (BC), 2009 (RCC, glioblastoma)	Van Meter and Kim (2010)
Ramucirumab/IMC-1121B	mAb	VEGFR2	Phase II/III	Spratlin (2011)
MF-1/IMC-18F1	mAb	VEGFR1	Phase I	Wu *et al.* (2006b)
CDP791	PEG di-Fab conjugate	VEGFR2	Phase II	Youssoufian *et al.* (2007)
VEGF-Trap/aflibercept	Fusion protein	VEGFA, PIGF	Phase II/III	Teng *et al.* (2010)
VEGFAS/Veglin	Oligonucleotide	VEGFA, VEGFC, VEGFD	Phase I	Levine *et al.* (2006)
SU11248/sunitinib (Sutent)	RTKI	VEGFR1–3, PDGFR, c-kit, Flt3	Approved in 2006 (GIST and RCC)	Sulkes (2010)
Sorafenib (Nexavar)	RTKI	VEGFR2–3, PDGFR, Raf-1, Flt-3, c-kit	Approved in 2005 (RCC), 2008 (HCC)	Sulkes (2010)
Pazopanib (Votrient)	RTKI	VEGFR1–3, PDGFR, Flt-3, c-kit	Approved in 2009 (RCC)	Sternberg *et al.* (2010)
AG013736/axitinib	RTKI	VEGFR1–3, PDGFR, c-kit	Phase II/III	Kelly *et al.* (2010)
AZD6474/vandetanib (Zactima)	RTKI	VEGFR1–3, EGFR, RET	Phase II/III	Morabito *et al.* (2009)
AZD2171/cediranib (Resentin)	RTKI	VEGFR1–3, c-kit	Phase II/III	Lindsay *et al.* (2009)
Brivanib alanitate	RTKI	VEGFR2, FGFR1	Phase II/III	Diaz-Padilla and Siu (2011)
AV-951/tivozanib	RTKI	VEGFR1–3, PDGFR	Phase II/III	De Luca and Normanno (2010)
PTK787/vatalanib	RTKI	VEGFR1–3, PDGFR, c-kit	Phase II	Scott *et al.* (2007)
AE941/Neovastat	Shark cartilage component	VEGF/VEGFR binding, MMP2, MMP9	Phase II/III	White (2010)

Abbreviations: BC, breast cancer; CC, colorectal carcinoma; HCC, hepatocellular carcinoma; mAb, monoclonal antibody; NSCLC, nonsmall cell lung carcinoma; RCC, renal cell cancer; RTKI, receptor tyrosine kinase inhibitor.

cancer in combination with intravenous 5-fluorouracil-based chemotherapy (Hurwitz *et al.*, 2004). Subsequently, bevacizumab has been approved for various indications in nonsquamous cell lung carcinoma (NSCLC), metastatic renal cell carcinoma, and glioblastoma multiforme (Escudier *et al.*, 2010; Friedman *et al.*, 2009; Kreisl *et al.*, 2009; Rini *et al.*, 2008; Sandler *et al.*, 2006; Van Meter and Kim, 2010). The antitumor activity of bevacizumab is primarily manifested in combination with chemotherapy, except for renal cell carcinoma, where it has shown efficacy as a single agent (Yang *et al.*, 2003). Presently, bevacizumab is being used in nearly 1000 clinical trials, and despite promising results, its effects in many types of cancer are modest or even irrelevant (Van Meter and Kim, 2010). Furthermore, recent studies have raised the possibility that treatment with bevacizumab is associated with a more aggressive invasive tumor phenotype, particularly in glioblastoma (Keunen *et al.*, 2011). Although the clinical impact of these results is far from clear, it is obvious that antiangiogenic therapy will have to be closely evaluated depending on disease stage and molecular profile.

2. VEGFR2-TARGETING ANTIBODIES

Preclinical data with anti-VEGFR2 antibodies have demonstrated a reduction in VEGF-induced signaling as well as angiogenesis and primary or metastatic growth in a variety of different tumor models (Bruns *et al.*, 2002; Prewett *et al.*, 1999; Shaheen *et al.*, 2001; Zhu *et al.*, 1999); therefore, the specific, antibody-based blockade of VEGFR2 has also received special attention in clinical trials. Ramucirumab (IMC-1121B; Imclone Systems) is currently being tested in several clinical trials, including breast cancer, gastric cancer, and HCC (Spratlin, 2011). Based on preliminary results, this antibody has shown activity in patients previously treated with other antiangiogenic agents, suggesting a more efficient antitumor response with direct targeting of VEGFR2.

3. VEGF–PIGF DECOY RECEPTOR

After showing a remarkable inhibitory activity in different experimental models, VEGF-Trap (aflibercept; Sanofi-Aventi, Regeneron), the soluble decoy receptor with very high affinity for VEGFA and PIGF, entered clinical trials. Phase 3 trials with aflibercept in metastatic colon cancer and prostate cancer are still underway; however, studies in patients with NSCLC failed to reach the primary endpoint of improvement in overall survival (OS).

B. RTKs Small Molecule Inhibitors

Small molecule inhibitors of VEGFR tyrosine kinase activity represent another major approach to blocking VEGF-mediated angiogenesis. Several tyrosine kinase inhibitors have been developed to selectively inhibit VEGFR2, but they have also activity on other VEGFRs and tyrosine kinase receptors, including basic fibroblast growth factor (FGF) receptor, EGFR family members, PDGFR-α, PDGFR-β, c-kit, and Flt3.

1. SUNITINIB AND SORAFENIB

Sunitinib was approved in 2006 for its clinical use in imatinib-resistant gastrointestinal stromal tumors and advanced metastatic renal cell carcinoma (Demetri *et al.*, 2006; Motzer *et al.*, 2007), whereas sorafenib received FDA approval for the treatment of metastatic renal cell carcinoma (Escudier *et al.*, 2007) and HCC (Llovet *et al.*, 2008). Notably, sunitinib and sorafenib have shown clinical efficacy as single agents, possibly due to their ability to inhibit multiple RTKs and in particular those regulating tumor angiogenesis. Additional clinical trials aimed to evaluate combinations of sorafenib and sunitinib with different chemotherapeutic agents and other antiangiogenic agents are ongoing.

It is important to point out that preclinical studies have challenged the classic schedule of administration currently used for sunitinib in clinical trials. Indeed, short-term treatment with sunitinib was associated with an accelerated metastatic tumor growth and invasiveness in different tumor models (Ebos *et al.*, 2009), stressing the importance of fully understanding the potential responses to antiangiogenic therapies and optimizing dose and schedule in clinical trials. Interestingly, continuous daily administration of sunitinib in patients with advanced pancreatic neuroendocrine tumors showed clear improvement in both progression free and OS in a phase 3 trial (Raymond *et al.*, 2011), which led to FDA approval of sunitinib for the treatment of pancreatic neuroendocrine tumors.

2. PAZOPANIB

Pazopanib (Votrient), a pan-VEGFR inhibitor developed by GlaxoSmithKline, is currently being tested in a broad clinical program across multiple tumor types. It received approval by the FDA for use in advanced renal cell carcinoma (Sternberg *et al.*, 2010). A phase 3 clinical is being conducted to compare pazopanib with sunitinib for treatment of metastatic renal cell carcinoma based on the potential better toxicity profile associated with administration of pazopanib.

III. CHALLENGES OF VEGF/VEGFR TARGETED THERAPY: LIMITED THERAPEUTIC RESPONSE AND DEVELOPMENT OF RESISTANCE

Antiangiogenic therapy has become an important option for the treatment of cancer. However, its systematic application remains problematic because of poor understanding of mechanisms of action and occurrence of resistance (Jain *et al.*, 2006). Indeed, a significant fraction of patients do not respond to antiangiogenic therapy (Burris III and Rocha-Lima, 2008), whereas those who respond have relatively modest benefits, mostly in progression-free survival rather than in OS. In addition, a number of significant toxicities have been observed in patients treated with antiangiogenic agents, emphasizing that a careful assessment of the risk-benefit ratio needs to be conducted in individual patients. Despite disease stabilization and an increase in the proportion of patients with progression-free survival, tumors eventually become resistant to antiangiogenic agents and relapse (Bergers and Hanahan, 2008; Ellis and Hicklin, 2008a; Kerbel, 2008; Shojaei and Ferrara, 2008b). Ultimately, which patients may potentially benefit from the addition of an antiangiogenic agent to the therapeutic regimen remains poorly understood.

Multiple mechanisms may account for the activity of anti-VEGF agents in cancer patients including, but not limited to, their effect on tumor vasculature (Ellis and Hicklin, 2008b). Evidence has been provided supporting both a vascular regression, which is presumably associated with increased intratumor hypoxia (Kerbel and Folkman, 2002), and a so-called normalization of tumor vasculature, with a consequent decrease in interstitial pressure and better delivery of chemotherapy (Jain, 2005b). These conflicting and still largely controversial observations emphasize how important it is to better understand the effects of antiangiogenic agents on the tumor microenvironment to eventually better characterize the mechanisms that mediate resistance.

A. Intrinsic Resistance to VEGF-Targeted Therapies

A substantial fraction of patients treated with antiangiogenic agents, including bevacizumab, sorafenib, or sunitinib, fail to show even a transient clinical benefit (Batchelor *et al.*, 2007; Burris III and Rocha-Lima, 2008). This lack of clinical benefit could be interpreted as a rapid adaptation to and escape from the effects of antiangiogenic agents. Alternatively, in some cases, there may be preexisting resistance. It is conceivable that a number of pathways may be activated in human cancers that eventually confer intrinsic resistance to antiangiogenic therapy, such as redundancy of angiogenic factors (FGFs, PDGFs, PIGF) (Fischer *et al.*, 2007; Relf *et al.*, 1997), increased metastatic and invasive potential without an angiogenic switch (Casanovas *et al.*, 2005), high levels of infiltrating inflammatory cells that produce a number of proangiogenic factors (Shojaei and Ferrara, 2008b) or hypovascularity, such as in pancreatic ductal adenocarcinoma (Saif, 2007).

B. Acquired Resistance to Antiangiogenic Agents

Considering the results of both preclinical and clinical research showing modest effects of antiangiogenic therapy in patients with solid tumors, it is now widely recognized that tumors rapidly adapt to the effects of anti-VEGF agents to resume growth. Apart from instances of intrinsic resistance, most tumors acquire resistance to antiangiogenic therapies by upregulating pathways that sustain tumor growth and progression. Acquired resistance to antiangiogenic agents has been attributed to a number of potential mechanisms, including upregulation of alternative proangiogenic signals, increased production of proangiogenic factors by stromal cells, recruitment of bone marrow-derived proangiogenic cells, increased vascular pericyte coverage, and activation of an invasive phenotype. In addition, hypoxia-dependent responses may also play a role in several of these adaptive mechanisms. For instance, elevated CA9 (a HIF-1 target gene) and HIF-2α levels are inversely correlated with response to bevacizumab and irinotecan in malignant astrocytoma (Sathornsumetee *et al.*, 2008), suggesting that intra-tumor hypoxia may be an important factor in mediating resistance to antiangiogenic agents.

1. UPREGULATION OF COMPENSATORY PROANGIOGENIC PATHWAYS

A compensatory increase of FGFs was one of the first mechanisms of resistance identified in preclinical models (Casanovas *et al.*, 2005). The potential relevance of these findings is supported by clinical data that reported the induction of FGF2 in serum of patients that progressed on anti-VEGF therapy (Batchelor *et al.*, 2007). In addition, both in preclinical and clinical studies, PIGF was shown to be upregulated following anti-VEGF therapy (Batchelor *et al.*, 2007), while blockade of PIGF using monoclonal antibodies reduced tumor angiogenesis and metastasis in mouse models, regardless of whether tumors were sensitive or resistant to anti-VEGF therapy (Fischer *et al.*, 2007). Anti-PIGF therapies might play a complementary role to anti-VEGF therapy; however, clinical development of VEGF-Trap (that binds both VEGF and PIGF) has not shown any additional benefit compared to bevacizumab.

Recent data emphasize the role of the cell membrane-bound Notch ligand/receptor system in the development of resistance to antiangiogenic therapy (Li *et al.*, 2011). Moreover, tumors that have an intrinsic resistance to anti-VEGF agents appear to be sensitive to inhibition of Dll4 (Delta-like ligand 4; Yan and Plowman, 2007).

2. PRODUCTION OF PROANGIOGENIC FACTORS BY STROMAL CELLS

Reduced efficacy of antiangiogenic therapy may be due to the involvement of the stromal compartment in tumor angiogenesis. In particular, tumor-associated fibroblasts (TAFs) are thought to play a major role in tumor growth and possibly in resistance to antiangiogenic therapy (Liang *et al.*, 2006). Notably, it has been shown that TAFs from tumors resistant to anti-VEGF therapy can support tumor growth and angiogenesis by producing PDGF-C, proposing yet another potential mechanism of resistance (Crawford *et al.*, 2009). Indeed, these observations emphasize the role that the tumor microenvironment plays in drug resistance in general and to antiangiogenic agents in particular, strongly suggesting that the stromal cellular component needs to be understood in order to improve efficacy of anticancer therapies.

3. RECRUITMENT OF BONE MARROW-DERIVED PROANGIOGENIC CELLS

Induction of intratumor hypoxia during therapy with antiangiogenic agents may lead not only to an increase in the production of proangiogenic factors by tumor and stromal cells but also to recruitment of bone marrow-derived cells (BMDCs) that have the capacity to elicit angiogenesis and tumor growth. Proangiogenic BMDCs consist of vascular progenitors (such as endothelial and pericytes progenitors) and vascular modulators (such as tumor-associated macrophages, immature monocytic cells, myeloid cells) (Kerbel, 2008). Indeed, a marked mobilization of circulating BMDCs occurs rapidly after treatment of tumor-bearing mice with vascular disrupting agents, along with massive induction of tumor hypoxia (Shaked *et al.*, 2006). Moreover, circulating endothelial cells (CECs) have been shown to contribute to the rapid regrowth of tumors. Of interest, an increase in FGF2, SDF-1, and viable CECs was observed when tumors progressed following treatment with the VEGF RTK inhibitor AZD2171 in glioblastoma patients (Batchelor *et al.*, 2007).

More recently, it has been suggested that a specific myeloid cell population migrates to tumors and mediates tumor angiogenesis and resistance to anti-VEGF agents (Shojaei *et al.*, 2007). Interestingly, tumor and stromal cell production of G-CSF, IL6, and SDF-1 mediates the mobilization of $CD11b^{+}Gr1^{+}$ myeloid cells to the tumor, where they elicit angiogenesis and confer resistance to anti-VEGF therapy (Shojaei and Ferrara, 2008a; Shojaei *et al.*, 2007).

4. INCREASED PERICYTE COVERAGE OF THE VASCULATURE

Pericytes are involved in vascular stability and provide survival signals to endothelial cells. Inhibition of VEGF signaling may spare endothelial cells that are in strict contact with pericytes in "mature vessels" (Benjamin *et al.*, 1999). Conversely, anti-VEGF therapy not only may lead to endothelial cell apoptosis and pruning of immature tumor vasculature (without pericyte coverage) but also may increase angiopoietin 1 that enhances pericyte recruitment to the vessels, thereby reversing the effect of anti-VEGF therapy (Winkler *et al.*, 2004). Indeed, a number of studies have shown that targeting both pericytes and endothelial cells (PDGFR and VEGFR inhibitors) may lead to synergistic inhibition of tumor growth (Bergers *et al.*, 2003). Conversely, recent evidence suggests that targeting pericytes in the tumor vasculature may lead to disruption of vessel integrity, enabling tumor cells to transit into the circulation system and metastasize (Xian *et al.*, 2006). Moreover, a negative rather than a positive effect of VEGF on pericyte function and vessel maturation has also been recently suggested, adding complexity to the potential effects of VEGF/PDGF modulation (Greenberg *et al.*, 2008). Due to the similarities between VEGFRs and PDGFRs, many RTK inhibitors that target VEGFRs also inhibit PDGFRs functions. The clinical benefit of targeting both endothelial cells and pericytes remains to be determined.

C. Role of the Hypoxic Tumor Microenvironment in the Resistance to Antiangiogenic Therapies

The functional consequences of antiangiogenic therapies on the tumor microenvironment are still poorly understood and controversial. Indeed, at least two hypotheses have been proposed: (1) "normalization" of the vasculature, with a consequent decrease in intratumor hypoxia and interstitial pressure, which would be associated with a better delivery of chemotherapy; (2) vascular "regression," resulting in an increase of intratumor hypoxia, selection of more metastatic clones, and resistance to therapy (Jain, 2005a;

Kerbel and Folkman, 2002). Several lines of evidence in preclinical models support the hypothesis that antiangiogenic therapy might be associated with an increase in intratumor hypoxia and selection of a more malignant phenotype (Bergers and Hanahan, 2008; Bottaro and Liotta, 2003; Casanovas *et al.*, 2005; Ebos *et al.*, 2009; Franco *et al.*, 2006; Keunen *et al.*, 2011; Paez-Ribes *et al.*, 2009; Pennacchietti *et al.*, 2003; Steeg, 2003). Moreover, these preclinical data appear to be consistent with clinical findings demonstrating increased intratumor hypoxia in patients with nonsmall cell lung cancer and primary liver following treatment with bevacizumab (Smit *et al.*, 2011; Yopp *et al.*, 2011). Notably, it has been recently shown that administration of antiangiogenic agents, such as sunitinib and bevacizumab, increases the cancer stem cell (CSC) population in breast cancer xenografts as a consequence of the generation of tumor hypoxia (Conley *et al.*, 2012). This study strongly indicates that hypoxia-driven CSC stimulation limits the effectiveness of antiangiogenic agents and suggests that, to improve patient outcome, antiangiogenic therapies might have to be combined with CSC-targeting drugs. Interestingly, several studies have demonstrated the acquisition of an invasive phenotype in glioblastoma patients who have developed multifocal recurrence of tumors during the course of antiangiogenic therapy (Narayana *et al.*, 2009, 2011; Norden *et al.*, 2008). This data strongly suggests that reduction of tumor vasculature and increase in intratumor hypoxia might result in enhanced tumor cell invasiveness. In addition, intratumor hypoxia has been implicated not only in the increased metastatic phenotype of tumors in response to antiangiogenic agents but also in a number of mechanisms of resistance that have been described so far (Rapisarda and Melillo, 2009). Indeed, hypoxia plays an important role in the regulation of angiogenic factors (FGFs, PDGFs, PIGF) (Fischer *et al.*, 2007; Relf *et al.*, 1997), such as regulation of Notch/Dll-4 signaling (Diez *et al.*, 2007), recruitment of BMDCs (Ceradini *et al.*, 2004) (that have the capacity to elicit tumor growth and angiogenesis; Kerbel, 2008), recruitment of $CD11b^{+}Gr1^{+}$ myeloid cells (triggered by G-CSF, IL6, and SDF-1 secreted by tumor and stromal cells) (Shojaei and Ferrara, 2008a), recruitment of $CD11b^{+}$ myeloid cells at the premetastatic sites (in response to SDF-1 and LOX gradients) (Erler *et al.*, 2009; Yang *et al.*, 2008) and pericyte recruitment to vessels (Winkler *et al.*, 2004) (in response to the HIF-1 regulated genes PDGF, PAI-1, angiopoietin 1, and Tie-2; Hirota and Semenza, 2006).

The hypoxic tumor microenvironment may also be an important predictive factor to identify tumors that may be more sensitive or resistant to anti-VEGF therapy (Dang *et al.*, 2008). For example, treatment with antiangiogenic agents has been shown to increase plasma levels of VEGF in cancer patients, and such an increase has been proposed to be a potential predictive biomarker for tumor response (Bertolini *et al.*, 2006, 2007; Bocci *et al.*, 2004). These observations underline the complexity of the relationship between antiangiogenic therapies and the tumor microenvironment and they emphasize the need to identify biomarkers that may guide the selection of patients in which combined targeting of tumor hypoxia and angiogenesis may be more beneficial.

IV. IMPROVING THE THERAPEUTIC OUTCOME OF VEGF-TARGETING AGENTS BY COMBINATION STRATEGIES

Considering the complexity of pathways regulating tumor angiogenesis and the limited activity observed by targeting VEGF-dependent responses, combination strategies that target multiple pathways involved in angiogenesis might be beneficial. Hence, combining VEGFR2 inhibitors with a blockade of PDGFR-β (Bergers *et al.*, 2003), VEGFR1 (Gille *et al.*, 2007), MMPs (Mancuso *et al.*, 2006), and other growth factors (e.g., EGF) shows additive antitumor activity in preclinical models (Ciardiello *et al.*, 2004; Wedge *et al.*, 2002). In addition, combinatorial therapies are being conducted that target VEGFA and stroma-derived growth factors, such as EGF or FGF. A preclinical study by Cascone *et al.* showed that dual targeting of VEGFR and EGFR increased progression-free survival and delayed the appearance of resistance associated with antiangiogenic therapy (Cascone *et al.*, 2011). Brivanib, a dual inhibitor of VEGFR and fibroblast growth factor receptor-1 (FGFR1) is already being evaluated in about 20 clinical trials, including hepatocellular carcinoma and colorectal carcinoma (Diaz-Padilla and Siu, 2011).

A. Can Intratumor Hypoxia be Exploited in Combination Strategies with AntiAngiogenic Agents?

The potential therapeutic relevance of hypoxia in the development of resistance to antiangiogenic agents argues in favor of the development of combination strategies aimed to thwart adaptive hypoxia-dependent responses during anti-VEGF treatment. Indeed, a number of therapeutic strategies have been devised to target the hypoxic microenvironment: (1) targeting hypoxic cells by using bioreductive prodrugs that are converted to cytotoxins under hypoxic conditions (Wilson and Hay, 2011), (2) development of inhibitors of HIF-1 activity (Melillo, 2006; Onnis *et al.*, 2009), (3) inhibition of downstream pathways activated by hypoxia such as metabolism (Denko, 2008; Papandreou *et al.*, 2011), (4) pH homeostasis (Chiche

Table III Examples of Pharmacological Strategies to Target Hypoxic Cells

Pathway	Target	Agents
Hypoxia	Hypoxia-activated cytotoxin	Tirapazamine
HIF-1 inhibitors	HIF-1α mRNA expression	EZN-2968, Aminoflavone
	HIF-1α protein synthesis	Topotecan, EZN-2208, Cardiac glycosides, PX-478, Temsirolimus, Everolimus
	HIF-1α degradation	17AAG/17DMAG, HDAC inhibitors
	HIF-1-DNA binding	Anthracyclines
	HIF-1α transcriptional activity	Bortezomib
Metabolism	Hexokinase 2	2DG, Lonidamine
	PDK1-4	DCA
Invasion and migration	Met/ALK	Crizotinib
	MET/VEGF	XL-880/XL-184
UPR and autophagy	HSP90	17AAG/17DMAG
	IRE1	Salicaldehydes
	Proteasome	Bortezomib
	Autophagy	Chloroquine

et al., 2009, 2010), (5) invasion/migration, (6) unfolded protein response (UPR) (Wouters and Koritzinsky, 2008), (7) authopagy (Rouschop and Wouters, 2009), and (8) DNA damage response and repair pathways (Olcina *et al.*, 2010; Table III).

Several studies have already addressed the question of whether combining inhibition of hypoxic targets with anti-VEGF agents might result in a therapeutic advantage. In this regard, evidence has been provided that combination of bevazizumab with low-dose daily topotecan, a camptothecin analog Top1 poison that inhibits HIF-1α protein synthesis *in vitro* and *in vivo* (Rapisarda *et al.*, 2004a,b), results in increased antitumor activity relative to either agent alone in xenografts models (Rapisarda *et al.*, 2009). Consistent with these findings, combination of bevacizumab with irinotecan (a topoisomerase I inhibitor that also inhibits HIF-1) has shown clinical benefit in glioblastoma patients with a 6-month OS of 62–77% (Chen *et al.*, 2007; Vredenburgh *et al.*, 2007). Given that HIF-1-dependent genes may play key roles in multiple mechanisms implicated in the resistance to anti-VEGF therapies, a combination of these agents with HIF-1 inhibitors might result in inhibition of adaptive pathways and increased therapeutic efficacy. Likewise, activity of HIF-1 inhibitors might be maximized in the presence of therapy-induced intratumor hypoxia.

Recent work from the McDonald laboratory has combined a blockade of VEGFR with that of c-Met, an RTK that binds hepatocyte growth factor and has been shown to play an important role in angiogenesis, epithelial–mesenchymal transformation, drug resistance, invasion, and metastasis. This combinatorial blockade improves antitumor activity in the RIP-Tag2 pancreatic islet cancer model when compared to an agent that targets only VEGFR. VEGFR and c-Met inhibition reduced pericyte vascular coverage, induced intratumor hypoxia and tumor cell apoptosis, slowed tumor vasculature regrowth after treatment, and reduced invasiveness of primary tumors and metastasis. These results suggest that combining VEGFRs and c-Met inhibition is a viable option to achieve a better therapeutic outcome (You *et al.*, 2011).

V. THE IMPORTANCE OF BIOMARKERS FOR PATIENTS' SELECTION

Profiling tumors from individual patients has the potential to radically change therapeutic strategies by identifying patients that will most likely benefit from a particular agent or combination. Despite the obvious benefits potentially provided by this approach, identification of predictive biomarkers to efficiently select patients remains elusive at this time. Several biomarkers that might predict sensitivity to antiangiogenic therapies have been evaluated, including VEGF levels and polymorphisms, VEGFR expression and imaging parameters, but with mixed results (Murukesh *et al.*, 2010).

A. VEGF/VEGFRs Expression and Polymorphisms

One of the first biomarkers to be evaluated has been the plasma concentration of VEGFA. Of the many trials, only results with E4599 indicated that the pretreatment plasma concentration of VEGF was of prognostic significance in nonsmall cell lung cancer patients (Dowlati *et al.*, 2008). Intuitively, one would predict that the pretreatment plasma concentration of VEGF would be most helpful in diseases that respond to single-agent VEGF inhibitors (e.g., renal, ovarian, and hepatic cancer), however, this hypothesis hasn't been fully investigated. The increase in plasma VEGF concentration in patients treated with anti-VEGF antibodies has also been seen in those receiving low-molecular-weight RTKIs. A VEGFR inhibitor biomarker signature has emerged in which the drugs induce an increase in plasma VEGF and PIGF, as well as reductions in soluble VEGFR2 and VEGFR3. Presumably, this biomarker signature reflects the larger repertoire of receptors targeted by RTKIs compared with anti-VEGF antibodies. If true, one might not expect to see an increase in VEGFR3 concentrations in patients receiving bevacizumab, although this has not been formally reported.

Interestingly, in patients with upper gastrointestinal cancers, VEGFA and VEGFR2 appear to be potential predictive biomarkers to identify responders to a combination therapy of bevacizumab and erlotinib (Rohrberg *et al.*, 2011). Moreover, in renal cell cancer (RCC), the ratio of VEGFA121/VEGFA165 mRNA levels seems to predict responsiveness to sunitinib (Paule *et al.*, 2010).

Few studies have reported a potential association between clinical outcome and single-nucleotide polymorphisms (SNPs) in VEGF genes. When patients with metastatic breast cancer were treated with paclitaxel and bevacizumab (E2100 trial), SNP analysis demonstrated that VEGF-2578 AA and VEGF 1154-A genotypes were associated with better OS, but not response rate (RR) or PFS (Schneider *et al.*, 2008). In contrast, those patients who received bevacizumab alone had a better RR and PFS but not OS, thereby challenging the pathophysiological role of these SNPs with regard to bevacizumab efficacy. Moreover, in patients with metastatic clear cell renal cell carcinoma treated with sunitinib, VEGF SNP-634 is associated with hypertension and a combination of VEGF SNP 936 and VEGFR2 SNP 889 genotypes is associated with OS (Kim *et al.*, 2012).

Perhaps the most attractive tissue biomarker that could be used to predict sensitivity is phosphorylated VEGFR2. In patients with inflammatory breast carcinoma, administration of bevacizumab resulted in a significant reduction of phospho-VEGFR2, which was coupled with a marked increase in tumor cell apoptosis, but no significant change in proliferation (Wedam *et al.*, 2006). In a phase I trial of a VEGFR2-binding di-Fab fragment, biopsy data were compatible with the proposed mechanism of action (Ton *et al.*, 2007). However, such reports are very infrequent for at least two reasons: (a) detection of phosphorylated proteins requires extremely rapid tissue preservation to avoid dephosphorylation of receptors and (b) limited choice of antibodies that bind with sufficient specificity to phosphorylated VEGFR2. Whether a validated biomarker assay of antiphosphorylated VEGFR2 could be used successfully in a multisite study remains to be established.

B. Imaging as a Biomarker

Early clinical trials of VEGF inhibitors sought pharmacological proof of concept by examining changes in the tumor vasculature, predominantly through the use of MRI, which is a technology that is noninvasive, sensitive, and avoids ionizing radiation. Of all the biomarkers that have been tested in trials of VEGF inhibitors, the most consistent findings have been achieved with dynamic contrast-enhanced MRI (DCE-MRI). Although many of these studies were small and confounded by interpatient heterogeneity, overall data show that patients whose tumors undergo at least a 50% reduction in

DCE-MRI parameters attain stable disease or a better response (Murukesh *et al.*, 2010). Thus, DCE-MRI perhaps holds the greatest promise as a biomarker associated with responses to VEGF inhibitors.

Recent interest in MRI techniques that do not require contrast has highlighted blood oxygenation level-dependent (BOLD) imaging and arterial spin labeling (ASL). ASL is a technique in which protons entering the zone of interest are magnetized and was developed for imaging the vasculature of the brain. Although initial results with ASL in patients treated with VEGF inhibitors have shown promise as a potential biomarker (de Bazelaire *et al.*, 2008), ASL is technically challenging and usually requires 3T MRI machines. BOLD imaging, a technique that relies on the paramagnetic effects of deoxyhemoglobin, can be used to provide information on the oxygenation status of the patient's tumor and in particular the oxygen status in tumor vessels.

VI. CONCLUSION AND PERSPECTIVES

The identification of the VEGF/VEGFRs pathway as an important regulator of the angiogenesis process has prompted considerable research into its role in the pathogenesis of cancer. Continued progress has been made in the identification and characterization of new VEGF ligands and receptors, as well as their respective function, roles, and regulatory mechanisms. Clinical trials with anti-VEGF agents have initially generated great enthusiasm for the potential universal application of this novel therapeutic approach to human cancers. However, the premise that the efficacy of antiangiogenic agents would not be limited by the inevitable occurrence of drug resistance has turned out to be a hopeful but incorrect prediction. Clearly, a better understanding of the VEGF/VEGFR family and their role in tumor angiogenesis is necessary to improve treatment outcome and design appropriate combination strategies. Identification of biomarkers predictive of response is essential to select patients that might respond to therapy. The rapid translation of promising and validated hypothesis from preclinical models to the clinical setting may be another way to expedite the development of more effective and desperately needed therapeutic strategies.

ACKNOWLEDGMENTS

The authors would like to thank Nicole Fer and Monica Mancini for helpful discussion. This project has been funded in whole or in part with Federal funds from the National Cancer Institute, National Institutes of Health, under Contract No. N01-CO-12400. The content of this publication does not necessarily reflect the views or policies of the Department of Health and Human Services, nor does mention of trade names, commercial products, or organizations imply endorsement by the U.S. Government. This research was supported (in part) by the Developmental Therapeutics Program, DCTD, of the National Cancer Institute, NIH.

REFERENCES

Achen, M. G., and Stacker, S. A. (2008). Targeting tumor stroma. *Curr. Cancer Drug Targets* **8**, 446.

Albuquerque, R. J. C., Hayashi, T., Cho, W. G., Kleinman, M. E., Dridi, S., Takeda, A., Baffi, J. Z., Yamada, K., Kaneko, H., Green, M. G., Chappell, J., Wilting, J., *et al.* (2009). Alternatively spliced vascular endothelial growth factor receptor-2 is an essential endogenous inhibitor of lymphatic vessel growth. *Nat. Med.* **15**, 1023–1030.

Batchelor, T. T., Sorensen, A. G., di, T. E., Zhang, W. T., Duda, D. G., Cohen, K. S., Kozak, K. R., Cahill, D. P., Chen, P. J., Zhu, M., Ancukiewicz, M., Mrugala, M. M., *et al.* (2007). AZD2171, a pan-VEGF receptor tyrosine kinase inhibitor, normalizes tumor vasculature and alleviates edema in glioblastoma patients. *Cancer Cell* **11**, 83–95.

Becker, J., Pavlakovic, H., Ludewig, F., Wilting, F., Weich, H. A., Albuquerque, R., Ambati, J., and Wilting, J. (2010). Neuroblastoma progression correlates with downregulation of the lymphangiogenesis inhibitor sVEGFR-2. *Clin. Cancer Res.* **16**, 1431–1441.

Benjamin, L. E., Golijanin, D., Itin, A., Pode, D., and Keshet, E. (1999). Selective ablation of immature blood vessels in established human tumors follows vascular endothelial growth factor withdrawal. *J. Clin. Invest.* **103**, 159–165.

Bergers, G., and Hanahan, D. (2008). Modes of resistance to anti-angiogenic therapy. *Nat. Rev. Cancer* **8**, 592–603.

Bergers, G., Song, S., Meyer-Morse, N., Bergsland, E., and Hanahan, D. (2003). Benefits of targeting both pericytes and endothelial cells in the tumor vasculature with kinase inhibitors. *J. Clin. Invest.* **111**, 1287–1295.

Bertolini, F., Shaked, Y., Mancuso, P., and Kerbel, R. S. (2006). The multifaceted circulating endothelial cell in cancer: towards marker and target identification. *Nat. Rev. Cancer* **6**, 835–845.

Bertolini, F., Mancuso, P., Shaked, Y., and Kerbel, R. S. (2007). Molecular and cellular biomarkers for angiogenesis in clinical oncology. *Drug Discov. Today* **12**, 806–812.

Bocci, G., Man, S., Green, S. K., Francia, G., Ebos, J. M., du Manoir, J. M., Weinerman, A., Emmenegger, U., Ma, L., Thorpe, P., Davidoff, A., Huber, J., *et al.* (2004). Increased plasma vascular endothelial growth factor (VEGF) as a surrogate marker for optimal therapeutic dosing of VEGF receptor-2 monoclonal antibodies. *Cancer Res.* **64**, 6616–6625.

Bottaro, D. P., and Liotta, L. A. (2003). Cancer: out of air is not out of action. *Nature* **423**, 593–595.

Bruns, C. J., Shrader, M., Harbison, M. T., Portera, C., Solorzano, C. C., Jauch, K. W., Hicklin, D. J., Radinsky, R., and Ellis, L. M. (2002). Effect of the vascular endothelial growth factor receptor-2 antibody DC101 plus gemcitabine on growth, metastasis and angiogenesis of human pancreatic cancer growing orthotopically in nude mice. *Int. J. Cancer* **102**, 101–108.

Burris, H., III, and Rocha-Lima, C. (2008). New therapeutic directions for advanced pancreatic cancer: targeting the epidermal growth factor and vascular endothelial growth factor pathways. *Oncologist* **13**, 289–298.

Calvani, M., Rapisarda, A., Uranchimeg, B., Shoemaker, R. H., and Melillo, G. (2006). Hypoxic induction of an HIF-1alpha-dependent bFGF autocrine loop drives angiogenesis in human endothelial cells. *Blood* **107**, 2705–2712.

Carmeliet, P., Ferreira, V., Breier, G., Pollefeyt, S., Kieckens, L., Gertsenstein, M., Fahrig, M., Vandenhoeck, A., Harpal, K., Eberhardt, C., Declercq, C., Pawling, J., *et al.* (1996). Abnormal blood vessel development and lethality in embryos lacking a single VEGF allele. *Nature* **380**, 435–439.
Carmeliet, P., Moons, L., Luttun, A., Vincenti, V., Compernolle, V., De Mol, M., Wu, Y., Bono, F., Devy, L., Beck, H., Scholz, D., Acker, T., *et al.* (2001). Synergism between vascular endothelial growth factor and placental growth factor contributes to angiogenesis and plasma extravasation in pathological conditions. *Nat. Med.* **7**, 575–583.
Carmeliet, P., De Smet, F., Loges, S., and Mazzone, M. (2009). Branching morphogenesis and antiangiogenesis candidates: tip cells lead the way. *Nat. Rev. Clin. Oncol.* **6**, 315–326.
Casanovas, O., Hicklin, D. J., Bergers, G., and Hanahan, D. (2005). Drug resistance by evasion of antiangiogenic targeting of VEGF signaling in late-stage pancreatic islet tumors. *Cancer Cell* **8**, 299–309.
Cascone, T., Herynk, M. H., Xu, L., Du, Z., Kadara, H., Nilsson, M. B., Oborn, C. J., Park, Y. Y., Erez, B., Jacoby, J., Jr., Lee, J. S., Lin, H. Y., *et al.* (2011). Upregulated stromal EGFR and vascular remodeling in mouse xenograft models of angiogenesis inhibitor-resistant human lung adenocarcinoma. *J. Clin. Invest.* **121**, 1313–1328.
Ceradini, D. J., Kulkarni, A. R., Callaghan, M. J., Tepper, O. M., Bastidas, N., Kleinman, M. E., Capla, J. M., Galiano, R. D., Levine, J. P., and Gurtner, G. C. (2004). Progenitor cell trafficking is regulated by hypoxic gradients through HIF-1 induction of SDF-1. *Nat. Med.* **10**, 858–864.
Chen, W., Delaloye, S., Silverman, D. H., Geist, C., Czernin, J., Sayre, J., Satyamurthy, N., Pope, W., Lai, A., Phelps, M. E., and Cloughesy, T. (2007). Predicting treatment response of malignant gliomas to bevacizumab and irinotecan by imaging proliferation with [18F] fluorothymidine positron emission tomography: a pilot study. *J. Clin. Oncol.* **25**, 4714–4721.
Chiche, J., Ilc, K., Laferriere, J., Trottier, E., Dayan, F., Mazure, N. M., Brahimi-Horn, M. C., and Pouyssegur, J. (2009). Hypoxia-inducible carbonic anhydrase IX and XII promote tumor cell growth by counteracting acidosis through the regulation of the intracellular pH. *Cancer Res.* **69**, 358–368.
Chiche, J., Brahimi-Horn, M. C., and Pouyssegur, J. (2010). Tumour hypoxia induces a metabolic shift causing acidosis: a common feature in cancer. *J. Cell. Mol. Med.* **14**, 771–794.
Ciardiello, F., Bianco, R., Caputo, R., Caputo, R., Damiano, V., Troiani, T., Melisi, D., De Vita, F., De Placido, S., Bianco, A. R., and Tortora, G. (2004). Antitumor activity of ZD6474, a vascular endothelial growth factor receptor tyrosine kinase inhibitor, in human cancer cells with acquired resistance to antiepidermal growth factor receptor therapy. *Clin. Cancer Res.* **10**, 784–793.
Cohen, T., Herzog, Y., Brodzky, A., Greenson, J. K., Eldar, S., Gluzman-Poltorak, Z., Neufeld, G., and Resnick, M. B. (2002). Neuropilin-2 is a novel marker expressed in pancreatic islet cells and endocrine pancreatic tumours. *J. Pathol.* **198**, 77–82.
Conley, S. J., Gheordunescu, E., Kakarala, P., Newman, B., Korkaya, H., Heath, A. N., Clouthier, S. G., and Wicha, M. S. (2012). Antiangiogenic agents increase breast cancer stem cells via the generation of tumor hypoxia. *Proc. Natl. Acad. Sci. USA***109**(8), 2784–2789.
Crawford, Y., Kasman, I., Yu, L., Zhong, C., Wu, X., Modrusan, Z., Kaminker, J., and Ferrara, N. (2009). PDGF-C mediates the angiogenic and tumorigenic properties of fibroblasts associated with tumors refractory to anti-VEGF treatment. *Cancer Cell* **15**, 21–34.
Dang, D. T., Chun, S. Y., Burkitt, K., Abe, M., Chen, S., Havre, P., Mabjeesh, N. J., Heath, E. I., Vogelzang, N. J., Cruz-Correa, M., Blayney, D. W., Ensminger, W. D., *et al.* (2008). Hypoxia-inducible factor-1 target genes as indicators of tumor vessel response to vascular endothelial growth factor inhibition. *Cancer Res.* **68**, 1872–1880.
de Bazelaire, C., Alsop, D. C., George, D., Pedrosa, I., Wang, Y., Michaelson, M. D., and Rofsky, N. M. (2008). Magnetic resonance imaging-measured blood flow change after antiangiogenic therapy with PTK787/ZK 222584 correlates with clinical outcome in metastatic renal cell carcinoma. *Clin. Cancer Res.* **14**, 5548–5554.

De Luca, A., and Normanno, N. (2010). Tivozanib, a pan-VEGFR tyrosine kinase inhibitor for the potential treatment of solid tumors. *IDrugs* **13**, 636–645.

Demetri, G. D., van Oosterom, A. T., Garrett, C. R., Blackstein, M. E., Shah, M. H., Verweij, J., McArthur, G., Judson, I. R., Heinrich, M. C., Morgan, J. A., Desai, J., Fletcher, C. D., *et al.* (2006). Efficacy and safety of sunitinib in patients with advanced gastrointestinal stromal tumour after failure of imatinib: a randomised controlled trial. *Lancet* **368**, 1329–1338.

Denko, N. C. (2008). Hypoxia, HIF1 and glucose metabolism in the solid tumour. *Nat. Rev. Cancer* **8**, 705–713.

Diaz-Padilla, I., and Siu, L. L. (2011). Brivanib alaninate for cancer. *Expert Opin. Investig. Drugs* **20**, 577–586.

Diez, H., Fischer, A., Winkler, A., Hu, C. J., Hatzopoulos, A. K., Breier, G., and Gessler, M. (2007). Hypoxia-mediated activation of Dll4-Notch-Hey2 signaling in endothelial progenitor cells and adoption of arterial cell fate. *Exp. Cell Res.* **313**, 1–9.

DiSalvo, J., Bayne, M. L., Conn, G., Kwok, P. W., Trivedi, P. G., Soderman, D. D., Palisi, T. M., Sullivan, K. A., and Thomas, K. A. (1995). Purification and characterization of a naturally occurring vascular endothelial growth factor.placenta growth factor heterodimer. *J. Biol. Chem.* **270**, 7717–7723.

Dowlati, A., Gray, R., Sandler, A. B., Schiller, J. H., and Johnson, D. H. (2008). Cell adhesion molecules, vascular endothelial growth factor, and basic fibroblast growth factor in patients with non-small cell lung cancer treated with chemotherapy with or without bevacizumab—an eastern cooperative oncology group study. *Clin. Cancer Res.* **14**, 1407–1412.

Ebos, J. M., Lee, C. R., Cruz-Munoz, W., Bjarnason, G. A., Christensen, J. G., and Kerbel, R. S. (2009). Accelerated metastasis after short-term treatment with a potent inhibitor of tumor angiogenesis. *Cancer Cell* **15**, 232–239.

Ellis, L. M., and Hicklin, D. J. (2008a). Pathways mediating resistance to vascular endothelial growth factor-targeted therapy. *Clin. Cancer Res.* **14**, 6371–6375.

Ellis, L. M., and Hicklin, D. J. (2008b). VEGF-targeted therapy: mechanisms of anti-tumour activity. *Nat. Rev. Cancer* **8**, 579–591.

Ellis, L. M., Liu, W., and Wilson, M. (1996). Down-regulation of vascular endothelial growth factor in human colon carcinoma cell lines by antisense transfection decreases endothelial cell proliferation. *Surgery* **120**, 871–878.

Erler, J. T., Bennewith, K. L., Cox, T. R., Lang, G., Bird, D., Koong, A., Le, Q. T., and Giaccia, A. J. (2009). Hypoxia-induced lysyl oxidase is a critical mediator of bone marrow cell recruitment to form the premetastatic niche. *Cancer Cell* **15**, 35–44.

Escudier, B., Eisen, T., Stadler, W. M., Szczylik, C., Oudard, S.p., Siebels, M., Negrier, S., Chevreau, C., Solska, E., Desai, A. A., Rolland, F.d.r., Demkow, T., *et al.* (2007). Sorafenib in advanced clear-cell renal-cell carcinoma. *N. Engl. J. Med.* **356**, 125–134.

Escudier, B., Bellmunt, J., Negrier, S., Bajetta, E., Melichar, B., Bracarda, S., Ravaud, A., Golding, S., Jethwa, S., and Sneller, V. (2010). Phase III trial of bevacizumab plus interferon alfa-2a in patients with metastatic renal cell carcinoma (AVOREN): final analysis of overall survival. *J. Clin. Oncol.* **28**, 2144–2150.

Failla, C. M., Odorisio, T., Cianfarani, F., Schietroma, C., Puddu, P., and Zambruno, G. (2000). Placenta growth factor is induced in human keratinocytes during wound healing. *J. Invest. Dermatol.* **115**, 388–395.

Fan, F., Wey, J. S., McCarty, M. F., Belcheva, A., Liu, W., Bauer, T. W., Somcio, R. J., Wu, Y., Hooper, A., Hicklin, D. J., and Ellis, L. M. (2005). Expression and function of vascular endothelial growth factor receptor-1 on human colorectal cancer cells. *Oncogene* **24**, 2647–2653.

Ferrara, N. (2005). VEGF as a therapeutic target in cancer. *Oncology* **69**(Suppl 3), 11–16.

Ferrara, N., and Davis-Smyth, T. (1997). The biology of vascular endothelial growth factor. *Endocr. Rev.* **18**, 4–25.

Ferrara, N., Carver-Moore, K., Chen, H., Dowd, M., Lu, L., O'Shea, K. S., Powell-Braxton, L., Hillan, K. J., and Moore, M. W. (1996). Heterozygous embryonic lethality induced by targeted inactivation of the VEGF gene. *Nature* **380**, 439–442.

Fischer, C., Jonckx, B., Mazzone, M., Zacchigna, S., Loges, S., Pattarini, L., Chorianopoulos, E., Liesenborghs, L., Koch, M., De, M. M., Autiero, M., Wyns, S., *et al.* (2007). Anti-PlGF inhibits growth of VEGF(R)-inhibitor-resistant tumors without affecting healthy vessels. *Cell* **131**, 463–475.

Franco, M., Man, S., Chen, L., Emmenegger, U., Shaked, Y., Cheung, A. M., Brown, A. S., Hicklin, D. J., Foster, F. S., and Kerbel, R. S. (2006). Targeted anti-vascular endothelial growth factor receptor-2 therapy leads to short-term and long-term impairment of vascular function and increase in tumor hypoxia. *Cancer Res.* **66**, 3639–3648.

Friedman, H. S., Prados, M. D., Wen, P. Y., Mikkelsen, T., Schiff, D., Abrey, L. E., Yung, W. K. A., Paleologos, N., Nicholas, M. K., Jensen, R., Vredenburgh, J., Huang, J., *et al.* (2009). Bevacizumab alone and in combination with irinotecan in recurrent glioblastoma. *J. Clin. Oncol.* **27**, 4733–4740.

Gerber, H. P., Kowalski, J., Sherman, D., Eberhard, D. A., and Ferrara, N. (2000). Complete inhibition of rhabdomyosarcoma xenograft growth and neovascularization requires blockade of both tumor and host vascular endothelial growth factor. *Cancer Res.* **60**, 6253–6258.

Gille, H., Kowalski, J., Yu, L., Chen, H., Pisabarro, M. T., Davis-Smyth, T., and Ferrara, N. (2000). A repressor sequence in the juxtamembrane domain of Flt-1 (VEGFR-1) constitutively inhibits vascular endothelial growth factor-dependent phosphatidylinositol 3′-kinase activation and endothelial cell migration. *EMBO J.* **19**, 4064–4073.

Gille, J., Heidenreich, R., Pinter, A., Schmitz, J., Boehme, B., Hicklin, D. J., Henschler, R., and Breier, G. (2007). Simultaneous blockade of VEGFR-1 and VEGFR-2 activation is necessary to efficiently inhibit experimental melanoma growth and metastasis formation. *Int. J. Cancer* **120**, 1899–1908.

Green, C. J., Lichtlen, P., Huynh, N. T., Yanovsky, M., Laderoute, K. R., Schaffner, W., and Murphy, B. J. (2001). Placenta growth factor gene expression is induced by hypoxia in fibroblasts: a central role for metal transcription factor-1. *Cancer Res.* **61**, 2696–2703.

Greenberg, J. I., Shields, D. J., Barillas, S. G., Acevedo, L. M., Murphy, E., Huang, J., Scheppke, L., Stockmann, C., Johnson, R. S., Angle, N., and Cheresh, D. A. (2008). A role for VEGF as a negative regulator of pericyte function and vessel maturation. *Nature* **456**, 809–813.

Haiko, P., Makinen, T., Keskitalo, S., Taipale, J., Karkkainen, M. J., Baldwin, M. E., Stacker, S. A., Achen, M. G., and Alitalo, K. (2008). Deletion of vascular endothelial growth factor C (VEGF-C) and VEGF-D is not equivalent to VEGF receptor 3 deletion in mouse embryos. *Mol. Cell. Biol.* **28**, 4843–4850.

Hansel, D. E., Wilentz, R. E., Yeo, C. J., Schulick, R. D., Montgomery, E., and Maitra, A. (2004). Expression of neuropilin-1 in high-grade dysplasia, invasive cancer, and metastases of the human gastrointestinal tract. *Am. J. Surg. Pathol.* **28**, 347–356.

He, Y., Rajantie, I., Ilmonen, M., Makinen, T., Karkkainen, M. J., Haiko, P., Salven, P., and Alitalo, K. (2004). Preexisting lymphatic endothelium but not endothelial progenitor cells are essential for tumor lymphangiogenesis and lymphatic metastasis. *Cancer Res.* **64**, 3737–3740.

He, Y., Rajantie, I., Pajusola, K., Jeltsch, M., Holopainen, T., Yla-Herttuala, S., Harding, T., Jooss, K., Takahashi, T., and Alitalo, K. (2005). Vascular endothelial cell growth factor receptor 3-mediated activation of lymphatic endothelium is crucial for tumor cell entry and spread via lymphatic vessels. *Cancer Res.* **65**, 4739–4746.

Hicklin, D. J., and Ellis, L. M. (2005). Role of the vascular endothelial growth factor pathway in tumor growth and angiogenesis. *J. Clin. Oncol.* **23**, 1011–1027.

Hirota, K., and Semenza, G. L. (2006). Regulation of angiogenesis by hypoxia-inducible factor 1. *Crit. Rev. Oncol. Hematol.* **59**, 15–26.

Hurwitz, H., Fehrenbacher, L., Novotny, W., Cartwright, T., Hainsworth, J., Heim, W., Berlin, J., Baron, A., Griffing, S., Holmgren, E., Ferrara, N., Fyfe, G., *et al.* (2004). Bevacizumab plus irinotecan, fluorouracil, and leucovorin for metastatic colorectal cancer. *N. Engl. J. Med.* **350**, 2335–2342.

Jain, R. K. (2005a). Antiangiogenic therapy for cancer: current and emerging concepts. *Oncology* **19**, 7–16.

Jain, R. K. (2005b). Normalization of tumor vasculature: an emerging concept in antiangiogenic therapy. *Science* **307**, 58–62.

Jain, R. K., Duda, D. G., Clark, J. W., and Loeffler, J. S. (2006). Lessons from phase III clinical trials on anti-VEGF therapy for cancer. *Nat. Clin. Pract. Oncol.* **3**, 24–40.

Karjalainen, K., Jaalouk, D. E., Bueso-Ramos, C. E., Zurita, A. J., Kuniyasu, A., Eckhardt, B. L., Marini, F. C., Lichtiger, B., O'Brien, S., Kantarjian, H. M., Cortes, J. E., Koivunen, E., *et al.* (2011). Targeting neuropilin-1 in human leukemia and lymphoma. *Blood* **117**, 920–927.

Kawakami, T., Tokunaga, T., Hatanaka, H., Kijima, H., Yamazaki, H., Abe, Y., Osamura, Y., Inoue, H., Ueyama, Y., and Nakamura, M. (2002). Neuropilin 1 and neuropilin 2 co-expression is significantly correlated with increased vascularity and poor prognosis in non-small cell lung carcinoma. *Cancer* **95**, 2196–2201.

Kelly, R. J., Darnell, C., and Rixe, O. (2010). Target inhibition in antiangiogenic therapy a wide spectrum of selectivity and specificity. *Cancer J.* **16**, 635–642.

Kendall, R. L., Wang, G., DiSalvo, J., and Thomas, K. A. (1994). Specificity of vascular endothelial cell growth factor receptor ligand binding domains. *Biochem. Biophys. Res. Commun.* **201**, 326–330.

Kerbel, R. S. (2008). Tumor angiogenesis. *N. Engl. J. Med.* **358**, 2039–2049.

Kerbel, R., and Folkman, J. (2002). Clinical translation of angiogenesis inhibitors. *Nat. Rev. Cancer* **2**, 727–739.

Keunen, O., Johansson, M., Oudin, A., Sanzey, M., Rahim, S. A., Fack, F., Thorsen, F., Taxt, T., Bartos, M., Jirik, R., Miletic, H., Wang, J., *et al.* (2011). Anti-VEGF treatment reduces blood supply and increases tumor cell invasion in glioblastoma. *Proc. Natl. Acad. Sci. USA* **108**, 3749–3754.

Kim, K. J., Li, B., Winer, J., Armanini, M., Gillett, N., Phillips, H. S., and Ferrara, N. (1993). Inhibition of vascular endothelial growth factor-induced angiogenesis suppresses tumour growth in vivo. *Nature* **362**, 841–844.

Kim, J. J., Vaziri, S. A. J., Rini, B. I., Elson, P., Garcia, J. A., Wirka, R., Dreicer, R., Ganapathi, M. K., and Ganapathi, R. (2012). Association of VEGF and VEGFR2 single nucleotide polymorphisms with hypertension and clinical outcome in metastatic clear cell renal cell carcinoma patients treated with sunitinib. *Cancer* **118**(7), 1946–1954.

Klohs, W. D., and Hamby, J. M. (1999). Antiangiogenic agents. *Curr. Opin. Biotechnol.* **10**, 544–549.

Kreisl, T. N., Kim, L., Moore, K., Duic, P., Royce, C., Stroud, I., Garren, N., Mackey, M., Butman, J. A., Camphausen, K., Park, J., Albert, P. S., *et al.* (2009). Phase II trial of single-agent bevacizumab followed by bevacizumab plus irinotecan at tumor progression in recurrent glioblastoma. *J. Clin. Oncol.* **27**, 740–745.

Laakkonen, P., Waltari, M., Holopainen, T., Takahashi, T., Pytowski, B., Steiner, P., Hicklin, D., Persaud, K., Tonra, J. R., Witte, L., and Alitalo, K. (2007). Vascular endothelial growth factor receptor 3 is involved in tumor angiogenesis and growth. *Cancer Res.* **67**, 593–599.

Lantuejoul, S., Constantin, B., Drabkin, H., Brambilla, C., Roche, J., and Brambilla, E. (2003). Expression of VEGF, semaphorin SEMA3F, and their common receptors neuropilins NP1 and NP2 in preinvasive bronchial lesions, lung tumours, and cell lines. *J. Pathol.* **200**, 336–347.

Larcher, F., Franco, M., Bolontrade, M., Rodriguez-Puebla, M., Casanova, L., Navarro, M., Yancopoulos, G., Jorcano, J. L., and Conti, C. J. (2003). Modulation of the angiogenesis response through Ha-ras control, placenta growth factor, and angiopoietin expression in mouse skin carcinogenesis. *Mol. Carcinog.* **37,** 83–90.

Latil, A., Bieche, I., Pesche, S., Valeri, A., Fournier, G., Cussenot, O., and Lidereau, R. (2000). VEGF overexpression in clinically localized prostate tumors and neuropilin-1 overexpression in metastatic forms. *Int. J. Cancer* **89,** 167–171.

Lesslie, D. P., Summy, J. M., Parikh, N. U., Fan, F., Trevino, J. G., Sawyer, T. K., Metcalf, C. A., Shakespeare, W. C., Hicklin, D. J., Ellis, L. M., and Gallick, G. E. (2006). Vascular endothelial growth factor receptor-1 mediates migration of human colorectal carcinoma cells by activation of Src family kinases. *Br. J. Cancer* **94,** 1710–1717.

Levine, A. M., Tulpule, A., Quinn, D. I., Gorospe, G., Smith, D. L., Hornor, L., Boswell, W. D., Espina, B. M., Groshen, S. G., Masood, R., and Gill, P. S. (2006). Phase I study of antisense oligonucleotide against vascular endothelial growth factor: decrease in plasma vascular endothelial growth factor with potential clinical efficacy. *J. Clin. Oncol.* **24,** 1712–1719.

Li, B., Sharpe, E. E., Maupin, A. B., Teleron, A. A., Pyle, A. L., Carmeliet, P., and Young, P. P. (2006). VEGF and PlGF promote adult vasculogenesis by enhancing EPC recruitment and vessel formation at the site of tumor neovascularization. *FASEB J.* **20,** 1495–1497.

Li, J. L., Sainson, R. C., Oon, C. E., Turley, H., Leek, R., Sheldon, H., Bridges, E., Shi, W., Snell, C., Bowden, E. T., Wu, H., Chowdhury, P. S., *et al.* (2011). DLL4-notch signaling mediates tumor resistance to anti-VEGF therapy in vivo. *Cancer Res.* **71,** 6073–6083.

Liang, W. C., Wu, X., Peale, F. V., Lee, C. V., Meng, Y. G., Gutierrez, J., Fu, L., Malik, A. K., Gerber, H. P., Ferrara, N., and Fuh, G. (2006). Cross-species vascular endothelial growth factor (VEGF)-blocking antibodies completely inhibit the growth of human tumor xenografts and measure the contribution of stromal VEGF. *J. Biol. Chem.* **281,** 951–961.

Lindsay, C. R., MacPherson, I. R., and Cassidy, J. (2009). Current status of cediranib: the rapid development of a novel anti-angiogenic therapy. *Future Oncol.* **5,** 421–432.

Llovet, J. M., Ricci, S., Mazzaferro, V., Hilgard, P., Gane, E., Blanc, J. F., de Oliveira, A. C., Santoro, A., Raoul, J. L., Forner, A., Schwartz, M., Porta, C., *et al.* (2008). Sorafenib in advanced hepatocellular carcinoma. *N. Engl. J. Med.* **359,** 378–390.

Mancuso, M. R., Davis, R., Norberg, S. M., O'Brien, S., Sennino, B., Nakahara, T., Yao, V. J., Inai, T., Brooks, P., Freimark, B., Shalinsky, D. R., Hu-Lowe, D. D., *et al.* (2006). Rapid vascular regrowth in tumors after reversal of VEGF inhibition. *J. Clin. Invest.* **116,** 2610–2621.

McColl, B. K., Baldwin, M. E., Roufail, S., Freeman, C., Moritz, R. L., Simpson, R. J., Alitalo, K., Stacker, S. A., and Achen, M. G. (2003). Plasmin activates the lymphangiogenic growth factors VEGF-C and VEGF-D. *J. Exp. Med.* **198,** 863–868.

Melillo, G. (2006). Inhibiting hypoxia-inducible factor 1 for cancer therapy. *Mol. Cancer Res.* **4,** 601–605.

Morabito, A., Piccirillo, M. C., Falasconi, F., De Feo, G., Del Giudice, A., Bryce, J., Di Maio, M., De Maio, E., Normanno, N., and Perrone, F. (2009). Vandetanib (ZD6474), a dual inhibitor of vascular endothelial growth factor receptor (VEGFR) and epidermal growth factor receptor (EGFR) tyrosine kinases: current status and future directions. *Oncologist* **14,** 378–390.

Motzer, R. J., Michaelson, M. D., Rosenberg, J., Bukowski, R. M., Curti, B. D., George, D. J., Hudes, G. R., Redman, B. G., Margolin, K. A., and Wilding, G. (2007). Sunitinib efficacy against advanced renal cell carcinoma. *J. Urol.* **178,** 1883–1887.

Muramatsu, M., Yamamoto, S., Osawa, T., and Shibuya, M. (2010). Vascular endothelial growth factor receptor-1 signaling promotes mobilization of macrophage lineage cells from bone marrow and stimulates solid tumor growth. *Cancer Res.* **70,** 8211–8221.

Murukesh, N., Dive, C., and Jayson, G. C. (2010). Biomarkers of angiogenesis and their role in the development of VEGF inhibitors. *Br. J. Cancer* **102,** 8–18.

Mylona, E., Alexandrou, P., Giannopoulou, I., Liapis, G., Sofia, M., Keramopoulos, A., and Nakopoulou, L. (2007). The prognostic value of vascular endothelial growth factors (VEGFs)-A and -B and their receptor, VEGFR-1, in invasive breast carcinoma. *Gynecol. Oncol.* **104,** 557–563.

Narayana, A., Kelly, P., Golfinos, J., Parker, E., Johnson, G., Knopp, E., Zagzag, D., Fischer, I., Raza, S., Medabalmi, P., Eagan, P., and Gruber, M. L. (2009). Antiangiogenic therapy using bevacizumab in recurrent high-grade glioma: impact on local control and patient survival. *J. Neurosurg.* **110,** 173–180.

Narayana, A., Gruber, D., Kunnakkat, S., Golfinos, J. G., Parker, E., Raza, S., Zagzag, D., Eagan, P., and Gruber, M. L. (2011). A clinical trial of bevacizumab, temozolomide, and radiation for newly diagnosed glioblastoma. *J. Neurosurg.* **116,** 341–345.

Norden, A. D., Young, G. S., Setayesh, K., Muzikansky, A., Klufas, R., Ross, G. L., Ciampa, A. S., Ebbeling, L. G., Levy, B., Drappatz, J., Kesari, S., and Wen, P. Y. (2008). Bevacizumab for recurrent malignant gliomas: efficacy, toxicity, and patterns of recurrence. *Neurology* **70,** 779–787.

Nowak, D. G., Woolard, J., Amin, E. M., Konopatskaya, O., Saleem, M. A., Churchill, A. J., Ladomery, M. R., Harper, S. J., and Bates, D. O. (2008). Expression of pro- and anti-angiogenic isoforms of VEGF is differentially regulated by splicing and growth factors. *J. Cell Sci.* **121,** 3487–3495.

Olcina, M., Lecane, P. S., and Hammond, E. M. (2010). Targeting hypoxic cells through the DNA damage response. *Clin. Cancer Res.* **16,** 5624–5629.

Onnis, B., Rapisarda, A., and Melillo, G. (2009). Development of HIF-1 inhibitors for cancer therapy. *J. Cell. Mol. Med.* **13,** 2780–2786.

Paez-Ribes, M., Allen, E., Hudock, J., Takeda, T., Okuyama, H., Vinals, F., Inoue, M., Bergers, G., Hanahan, D., and Casanovas, O. (2009). Antiangiogenic therapy elicits malignant progression of tumors to increased local invasion and distant metastasis. *Cancer Cell* **15,** 220–231.

Pan, Q., Chathery, Y., Wu, Y., Rathore, N., Tong, R. K., Peale, F., Bagri, A., Tessier-Lavigne, M., Koch, A. W., and Watts, R. J. (2007). Neuropilin-1 binds to VEGF121 and regulates endothelial cell migration and sprouting. *J. Biol. Chem.* **282,** 24049–24056.

Papandreou, I., Goliasova, T., and Denko, N. C. (2011). Anticancer drugs that target metabolism: is dichloroacetate the new paradigm? *Int. J. Cancer* **128,** 1001–1008.

Paule, B., Bastien, L., Deslandes, E., Cussenot, O., Podgorniak, M. P., Allory, Y., Nami, B., Porcher, R., de La Taille, A., Menashi, S., Calvo, F., and Mourah, S. (2010). Soluble isoforms of vascular endothelial growth factor are predictors of response to sunitinib in metastatic renal cell carcinomas. *PLoS One* **5,** e10715.

Pennacchietti, S., Michieli, P., Galluzzo, M., Mazzone, M., Giordano, S., and Comoglio, P. M. (2003). Hypoxia promotes invasive growth by transcriptional activation of the met protooncogene. *Cancer Cell* **3,** 347–361.

Petrova, T. V., Bono, P., Holnthoner, W., Chesnes, J., Pytowski, B., Sihto, H., Laakkonen, P., Heikkila, P., Joensuu, H., and Alitalo, K. (2008). VEGFR-3 expression is restricted to blood and lymphatic vessels in solid tumors. *Cancer Cell* **13,** 554–556.

Podar, K., and Anderson, K. C. (2005). The pathophysiologic role of VEGF in hematologic malignancies: therapeutic implications. *Blood* **105,** 1383–1395.

Prewett, M., Huber, J., Li, Y., Santiago, A., O'Connor, W., King, K., Overholser, J., Hooper, A., Pytowski, B., Witte, L., Bohlen, P., and Hicklin, D. J. (1999). Antivascular endothelial growth factor receptor (fetal liver kinase 1) monoclonal antibody inhibits tumor angiogenesis and growth of several mouse and human tumors. *Cancer Res.* **59,** 5209–5218.

Rapisarda, A., and Melillo, G. (2009). Role of the hypoxic tumor microenvironment in the resistance to anti-angiogenic therapies. *Drug Resist. Updat.* **12,** 74–80.

Rapisarda, A., Uranchimeg, B., Sordet, O., Pommier, Y., Shoemaker, R. H., and Melillo, G. (2004a). Topoisomerase I-mediated inhibition of hypoxia-inducible factor 1: mechanism and therapeutic implications. *Cancer Res.* **64,** 1475–1482.

Rapisarda, A., Zalek, J., Hollingshead, M., Braunschweig, T., Uranchimeg, B., Bonomi, C. A., Borgel, S. D., Carter, J. P., Hewitt, S. M., Shoemaker, R. H., and Melillo, G. (2004b). Schedule-dependent inhibition of hypoxia-inducible factor-1alpha protein accumulation, angiogenesis, and tumor growth by topotecan in U251-HRE glioblastoma xenografts. *Cancer Res.* **64,** 6845–6848.

Rapisarda, A., Hollingshead, M., Uranchimeg, B., Bonomi, C. A., Borgel, S. D., Carter, J. P., Gehrs, B., Raffeld, M., Kinders, R. J., Parchment, R., Anver, M. R., Shoemaker, R. H., *et al.* (2009). Increased antitumor activity of bevacizumab in combination with hypoxia inducible factor-1 inhibition. *Mol. Cancer Ther.* **8,** 1867–1877.

Raymond, E., Dahan, L., Raoul, J. L., Bang, Y. J., Borbath, I., Lombard-Bohas, C., Valle, J., Metrakos, P., Smith, D., Vinik, A., Chen, J. S., Horsch, D., *et al.* (2011). Sunitinib malate for the treatment of pancreatic neuroendocrine tumors. *N. Engl. J. Med.* **364,** 501–513.

Relf, M., LeJeune, S., Scott, P. A., Fox, S., Smith, K., Leek, R., Moghaddam, A., Whitehouse, R., Bicknell, R., and Harris, A. L. (1997). Expression of the angiogenic factors vascular endothelial cell growth factor, acidic and basic fibroblast growth factor, tumor growth factor beta-1, platelet-derived endothelial cell growth factor, placenta growth factor, and pleiotrophin in human primary breast cancer and its relation to angiogenesis. *Cancer Res.* **57,** 963–969.

Rini, B. I., Halabi, S., Rosenberg, J. E., Stadler, W. M., Vaena, D. A., Ou, S. S., Archer, L., Atkins, J. N., Picus, J., Czaykowski, P., Dutcher, J., and Small, E. J. (2008). Bevacizumab plus interferon alfa compared with interferon alfa monotherapy in patients with metastatic renal cell carcinoma: CALGB 90206. *J. Clin. Oncol.* **26,** 5422–5428.

Rohrberg, K. S., Pappot, H., Lassen, U., Westman, M., Olesen, R. K., Pfeiffer, P., Ladekarl, M., Srensen, M., Christensen, I. J., and Skov, B. G. (2011). Biomarkers in tissue from patients with upper gastrointestinal cancers treated with erlotinib and bevacizumab. *Cancer Biol. Ther.* **11,** 732–739.

Rouschop, K. M., and Wouters, B. G. (2009). Regulation of autophagy through multiple independent hypoxic signaling pathways. *Curr. Mol. Med.* **9,** 417–424.

Saif, M. W. (2007). Pancreatic cancer: are we moving forward yet? Highlights from the Gastrointestinal Cancers Symposium. Orlando, FL, USA. January 20th, 2007. *JOP* **8,** 166–176.

Sandler, A., Gray, R., Perry, M. C., Brahmer, J., Schiller, J. H., Dowlati, A., Lilenbaum, R., and Johnson, D. H. (2006). Paclitaxel-carboplatin alone or with bevacizumab for non-small-cell lung cancer. *N. Engl. J. Med.* **355,** 2542–2550.

Sathornsumetee, S., Cao, Y., Marcello, J. E., Herndon, J. E., McLendon, R. E., Desjardins, A., Friedman, H. S., Dewhirst, M. W., Vredenburgh, J. J., and Rich, J. N. (2008). Tumor angiogenic and hypoxic profiles predict radiographic response and survival in malignant astrocytoma patients treated with bevacizumab and irinotecan. *J. Clin. Oncol.* **26,** 271–278.

Scheufler, K. M., Drevs, J., van, V. V., Reusch, P., Klisch, J., Augustin, H. G., Zentner, J., and Marme, D. (2003). Implications of vascular endothelial growth factor, sFlt-1, and sTie-2 in plasma, serum and cerebrospinal fluid during cerebral ischemia in man. *J. Cereb. Blood Flow Metab.* **23,** 99–110.

Schneider, B. P., Wang, M., Radovich, M., Sledge, G. W., Badve, S., Thor, A., Flockhart, D. A., Hancock, B., Davidson, N., Gralow, J., Dickler, M., Perez, E. A., *et al.* (2008). Association of vascular endothelial growth factor and vascular endothelial growth factor receptor-2 genetic polymorphisms with outcome in a trial of paclitaxel compared with paclitaxel plus bevacizumab in advanced breast cancer: ECOG 2100. *J. Clin. Oncol.* **26,** 4672–4678.

Scott, E. N., Meinhardt, G., Jacques, C., Laurent, D., and Thomas, A. L. (2007). Vatalanib: the clinical development of a tyrosine kinase inhibitor of angiogenesis in solid tumours. *Expert Opin. Investig. Drugs* **16,** 367–379.

Seto, T., Higashiyama, M., Funai, H., Imamura, F., Uematsu, K., Seki, N., Eguchi, K., Yamanaka, T., and Ichinose, Y. (2006). Prognostic value of expression of vascular endothelial growth factor and its flt-1 and KDR receptors in stage I non-small-cell lung cancer. *Lung Cancer* **53,** 91–96.

Shaheen, R. M., Tseng, W. W., Vellagas, R., Liu, W., Ahmad, S. A., Jung, Y. D., Reinmuth, N., Drazan, K. E., Bucana, C. D., Hicklin, D. J., and Ellis, L. M. (2001). Effects of an antibody to vascular endothelial growth factor receptor-2 on survival, tumor vascularity, and apoptosis in a murine model of colon carcinomatosis. *Int. J. Oncol.* **18,** 221–226.

Shaked, Y., Ciarrocchi, A., Franco, M., Lee, C. R., Man, S., Cheung, A. M., Hicklin, D. J., Chaplin, D., Foster, F. S., Benezra, R., and Kerbel, R. S. (2006). Therapy-induced acute recruitment of circulating endothelial progenitor cells to tumors. *Science* **313,** 1785–1787.

Shibuya, M., and Claesson-Welsh, L. (2006). Signal transduction by VEGF receptors in regulation of angiogenesis and lymphangiogenesis. *Exp. Cell Res.* **312,** 549–560.

Shojaei, F., and Ferrara, N. (2008a). Refractoriness to antivascular endothelial growth factor treatment: role of myeloid cells. *Cancer Res.* **68,** 5501–5504.

Shojaei, F., and Ferrara, N. (2008b). Role of the microenvironment in tumor growth and in refractoriness/resistance to anti-angiogenic therapies. *Drug Resist. Updat.* **11,** 219–230.

Shojaei, F., Wu, X., Malik, A. K., Zhong, C., Baldwin, M. E., Schanz, S., Fuh, G., Gerber, H. P., and Ferrara, N. (2007). Tumor refractoriness to anti-VEGF treatment is mediated by CD11b +Gr1+ myeloid cells. *Nat. Biotechnol.* **25,** 911–920.

Smit, E. F., Lubberink, M., Bahce, I., Walraven, M., de Boer, M. P., Greuter, H. N., Hendrikse, N. H., Eriksson, J., Windhorst, A. D., Postmus, P. E., Verheul, H. M., Serne, E. H., *et al.* (2011). Effects of the antiangiogenic drug bevacizumab on tumor perfusion and drug delivery of 11C-labeled docetaxel in patients with non-small cell lung cancer (NSCLC): implications for scheduling of antiangiogenic agents. *ASCO Meeting Abstr.* **29,** 3059.

Smith, N. R., Baker, D., James, N. H., Ratcliffe, K., Jenkins, M., Ashton, S. E., Sproat, G., Swann, R., Gray, N., Ryan, A., Jrgensmeier, J. M., and Womack, C. (2010). Vascular endothelial growth factor receptors VEGFR-2 and VEGFR-3 are localized primarily to the vasculature in human primary solid cancers. *Clin. Cancer Res.* **16,** 3548–3561.

Spratlin, J. (2011). Ramucirumab (IMC-1121B): monoclonal antibody inhibition of vascular endothelial growth factor receptor-2. *Curr. Oncol. Rep.* **13,** 97–102.

Steeg, P. S. (2003). Angiogenesis inhibitors: motivators of metastasis? *Nat. Med.* **9,** 822–823.

Stephenson, J. M., Banerjee, S., Saxena, N. K., Cherian, R., and Banerjee, S. K. (2002). Neuropilin-1 is differentially expressed in myoepithelial cells and vascular smooth muscle cells in preneoplastic and neoplastic human breast: a possible marker for the progression of breast cancer. *Int. J. Cancer* **101,** 409–414.

Sternberg, C. N., Davis, I. D., Mardiak, J., Szczylik, C., Lee, E., Wagstaff, J., Barrios, C. H., Salman, P., Gladkov, O. A., Kavina, A., Zarba, J. J., Chen, M., *et al.* (2010). Pazopanib in locally advanced or metastatic renal cell carcinoma: results of a randomized phase III trial. *J. Clin. Oncol.* **28,** 1061–1068.

Sulkes, A. (2010). Novel multitargeted anticancer oral therapies: sunitinib and sorafenib as paradigm. *Israel Med. Assoc. J.* **12,** 628–632.

Teng, L. S., Jin, K. T., He, K. F., Zhang, J., Wang, H. H., and Cao, J. (2010). Clinical applications of VEGF-Trap (Aflibercept) in cancer treatment. *J. Chin. Med. Assoc.* **73,** 449–456.

Toi, M., Bando, H., Ogawa, T., Muta, M., Hornig, C., and Weich, H. A. (2002). Significance of vascular endothelial growth factor (VEGF)/soluble VEGF receptor-1 relationship in breast cancer. *Int. J. Cancer* **98,** 14–18.

Ton, N. C., Parker, G. J. M., Jackson, A., Mullamitha, S., Buonaccorsi, G. A., Roberts, C., Watson, Y., Davies, K., Cheung, S., Hope, L., Power, F., Lawrance, J., *et al.* (2007). Phase I evaluation of CDP791, a PEGylated Di-Fab conjugate that binds vascular endothelial growth factor receptor 2. *Clin. Cancer Res.* **13,** 7113–7118.

Van Meter, M. E., and Kim, E. S. (2010). Bevacizumab: current updates in treatment. *Curr. Opin. Oncol.* **22,** 586–591.

Vanveldhuizen, P. J., Zulfiqar, M., Banerjee, S., Cherian, R., Saxena, N. K., Rabe, A., Thrasher, J. B., and Banerjee, S. K. (2003). Differential expression of neuropilin-1 in malignant and benign prostatic stromal tissue. *Oncol. Rep.* **10,** 1067–1071.

Vredenburgh, J. J., Desjardins, A., Herndon, J. E., Marcello, J., Reardon, D. A., Quinn, J. A., Rich, J. N., Sathornsumetee, S., Gururangan, S., Sampson, J., Wagner, M., Bailey, L., *et al.* (2007). Bevacizumab plus irinotecan in recurrent glioblastoma multiforme. *J. Clin. Oncol.* **25,** 4722–4729.

Waldner, M. J., Wirtz, S., Jefremow, A., Warntjen, M., Neufert, C., Atreya, R., Becker, C., Weigmann, B., Vieth, M., Rose-John, S., and Neurath, M. F. (2010). VEGF receptor signaling links inflammation and tumorigenesis in colitis-associated cancer. *J. Exp. Med.* **207,** 2855–2868.

Wedam, S. B., Low, J. A., Yang, S. X., Chow, C. K., Choyke, P., Danforth, D., Hewitt, S. M., Berman, A., Steinberg, S. M., Liewehr, D. J., Plehn, J., Doshi, A., *et al.* (2006). Antiangiogenic and antitumor effects of bevacizumab in patients with inflammatory and locally advanced breast cancer. *J. Clin. Oncol.* **24,** 769–777.

Wedge, S. R., Ogilvie, D. J., Dukes, M., Kendrew, J., Chester, R., Jackson, J. A., Boffey, S. J., Valentine, P. J., Curwen, J. O., Musgrove, H. L., Graham, G. A., Hughes, G. D., *et al.* (2002). ZD6474 inhibits vascular endothelial growth factor signaling, angiogenesis, and tumor growth following oral administration. *Cancer Res.* **62,** 4645–4655.

White, J. (2010). The challenge of rational development of complex natural products as cancer therapeutics. *J. Natl. Cancer Inst.* **102,** 834–835.

Wilson, W. R., and Hay, M. P. (2011). Targeting hypoxia in cancer therapy. *Nat. Rev. Cancer* **11,** 393–410.

Winkler, F., Kozin, S. V., Tong, R. T., Chae, S. S., Booth, M. F., Garkavtsev, I., Xu, L., Hicklin, D. J., Fukumura, D., di, T. E., Munn, L. L., and Jain, R. K. (2004). Kinetics of vascular normalization by VEGFR2 blockade governs brain tumor response to radiation: role of oxygenation, angiopoietin-1, and matrix metalloproteinases. *Cancer Cell* **6,** 553–563.

Wouters, B. G., and Koritzinsky, M. (2008). Hypoxia signalling through mTOR and the unfolded protein response in cancer. *Nat. Rev. Cancer* **8,** 851–864.

Wu, Y., Hooper, A. T., Zhong, Z., Witte, L., Bohlen, P., Rafii, S., and Hicklin, D. J. (2006a). The vascular endothelial growth factor receptor (VEGFR-1) supports growth and survival of human breast carcinoma. *Int. J. Cancer* **119,** 1519–1529.

Wu, Y., Zhong, Z., Huber, J., Bassi, R., Finnerty, B., Corcoran, E., Li, H., Navarro, E., Balderes, P., Jimenez, X., Koo, H., Mangalampalli, V. R. M., *et al.* (2006b). Anti-vascular endothelial growth factor receptor-1 antagonist antibody as a therapeutic agent for cancer. *Clin. Cancer Res.* **12,** 6573–6584.

Xian, X., Hakansson, J., Stahlberg, A., Lindblom, P., Betsholtz, C., Gerhardt, H., and Semb, H. (2006). Pericytes limit tumor cell metastasis. *J. Clin. Invest.* **116,** 642–651.

Yan, M., and Plowman, G. D. (2007). Delta-like 4/Notch signaling and its therapeutic implications. *Clin. Cancer Res.* **13,** 7243–7246.

Yang, J. C., Haworth, L., Sherry, R. M., Hwu, P., Schwartzentruber, D. J., Topalian, S. L., Steinberg, S. M., Chen, H. X., and Rosenberg, S. A. (2003). A randomized trial of bevacizumab, an anti-vascular endothelial growth factor antibody, for metastatic renal cancer. *N. Engl. J. Med.* **349,** 427–434.

Yang, L., Huang, J., Ren, X., Gorska, A. E., Chytil, A., Aakre, M., Carbone, D. P., Matrisian, L. M., Richmond, A., Lin, P. C., and Moses, H. L. (2008). Abrogation of TGF beta signaling in mammary carcinomas recruits Gr-1+CD11b+ myeloid cells that promote metastasis. *Cancer Cell* **13**, 23–35.

Yopp, A. C., Schwartz, L. H., Kemeny, N., Gultekin, D. H., Gonen, M., Bamboat, Z., Shia, J., Haviland, D., D'Angelica, M. I., Fong, Y., DeMatteo, R. P., Allen, P. J., *et al.* (2011). Antiangiogenic therapy for primary liver cancer: correlation of changes in dynamic contrast-enhanced magnetic resonance imaging with tissue hypoxia markers and clinical response. *Ann. Surg. Oncol.* **18**, 2192–2199.

You, W. K., Sennino, B., Williamson, C. W., Falcon, B., Hashizume, H., Yao, L. C., Aftab, D. T., and McDonald, D. M. (2011). VEGF and c-Met blockade amplify angiogenesis inhibition in pancreatic islet cancer. *Cancer Res.* **71**, 4758–4768.

Youssoufian, H., Hicklin, D. J., and Rowinsky, E. K. (2007). Review: monoclonal antibodies to the vascular endothelial growth factor receptor-2 in cancer therapy. *Clin. Cancer Res.* **13**, 5544s–5548s.

Zhu, Z., Lu, D., Kotanides, H., Santiago, A., Jimenez, X., Simcox, T., Hicklin, D. J., Bohlen, P., and Witte, L. (1999). Inhibition of vascular endothelial growth factor induced mitogenesis of human endothelial cells by a chimeric anti-kinase insert domain-containing receptor antibody. *Cancer Lett.* **136**, 203–213.

Index

Note: Page numbers followed by "*f*" indicate figures, and "*t*" indicate tables.

D

E

T

V

Y

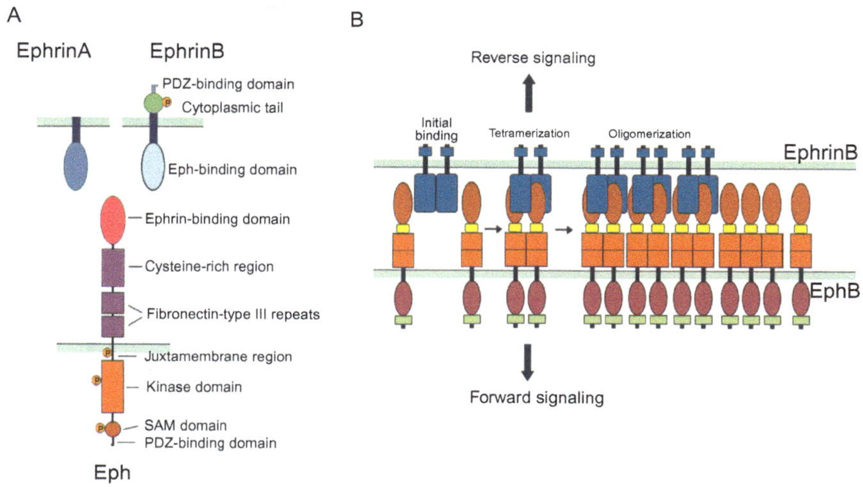

Fig. 1, Ombretta Salvucci and Giovanna Tosato (See Page 24 of this volume.)

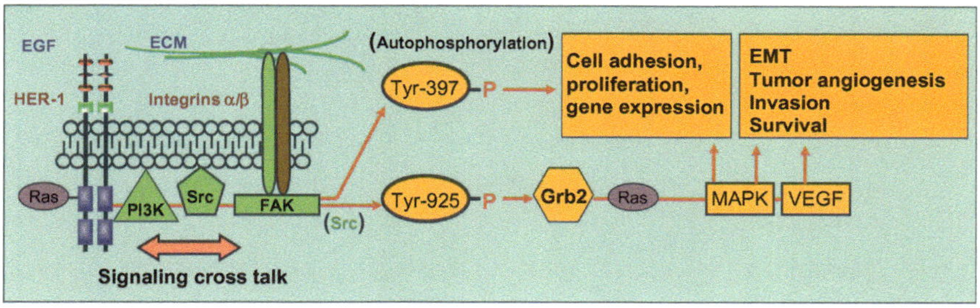

Fig. 1, Christian Gespach (See Page 90 of this volume.)

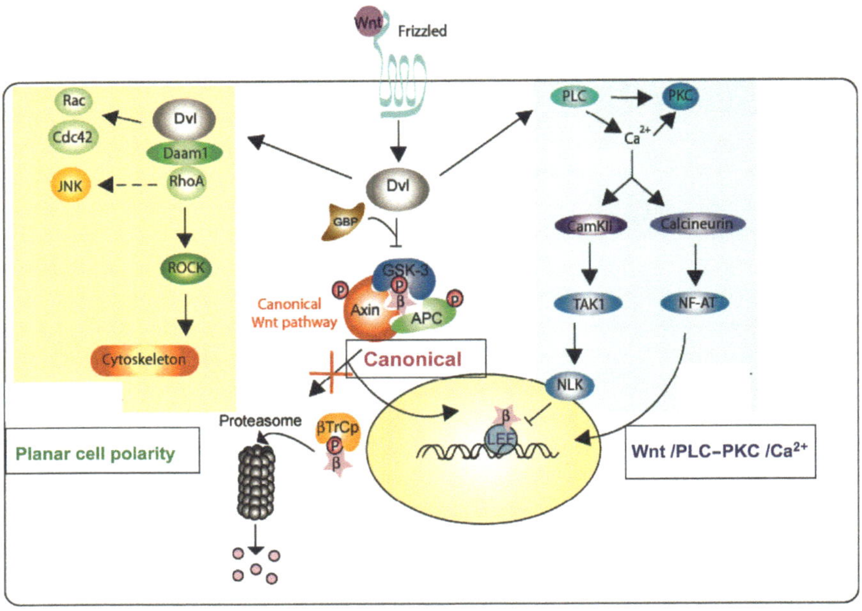

Fig. 3, Christian Gespach (See Page 97 of this volume.)

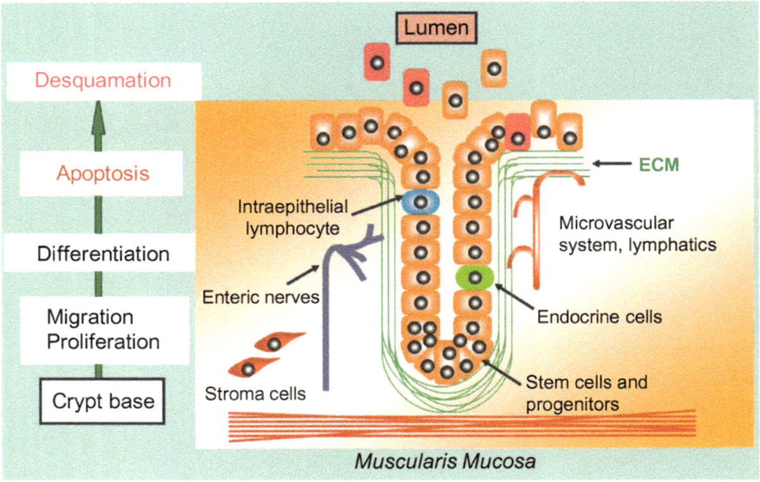

Fig. 4, Christian Gespach (See Page 109 of this volume.)

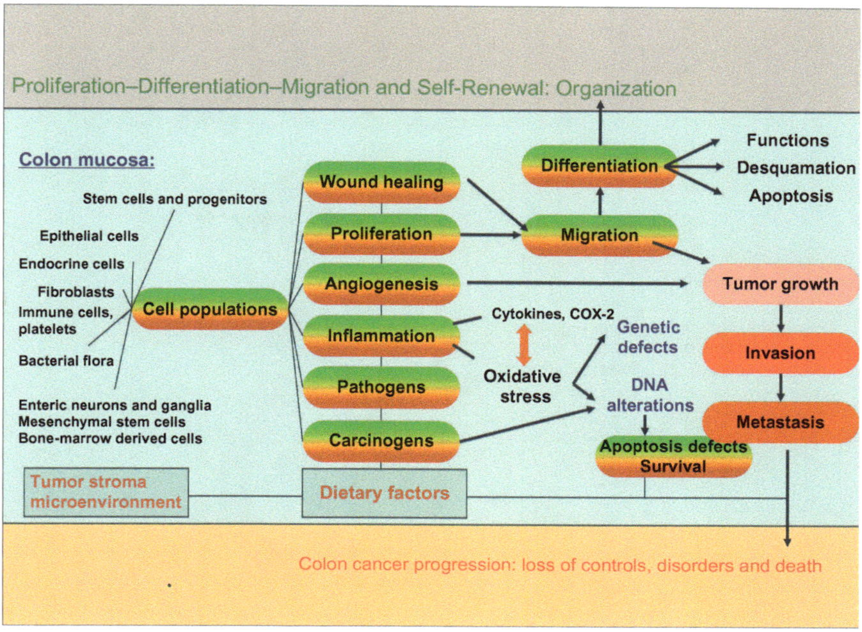

Fig. 5, Christian Gespach (See Page 111 of this volume.)

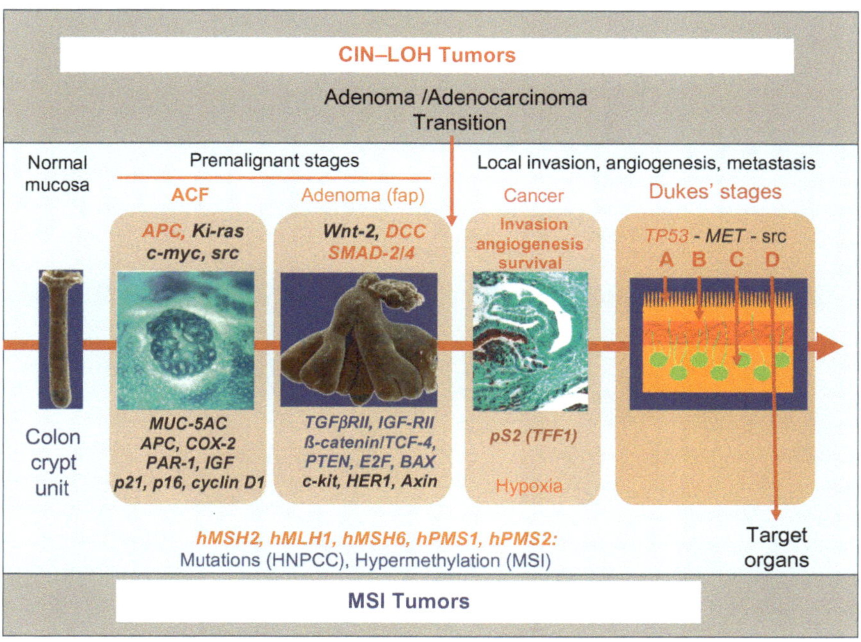

Fig. 6, Christian Gespach (See Page 114 of this volume.)

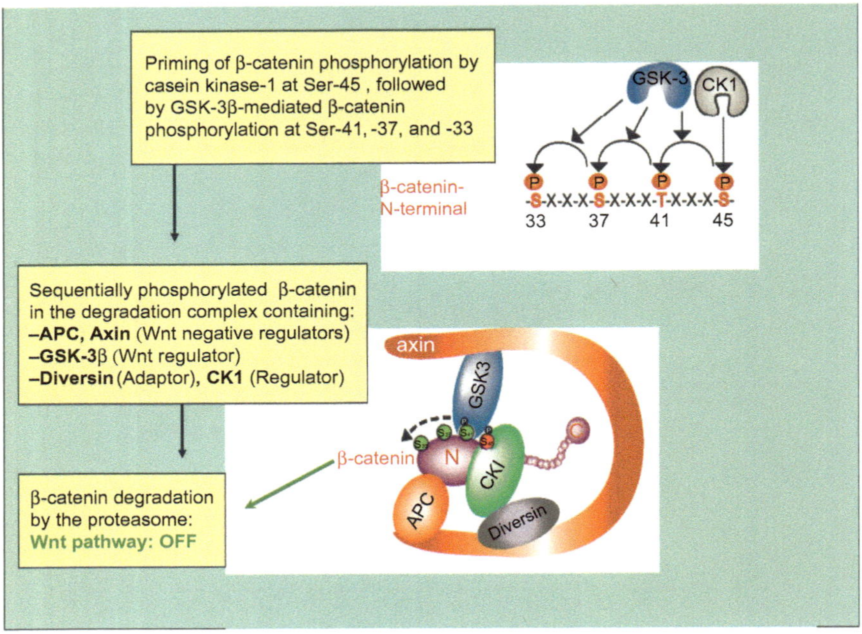

Fig. 7, Christian Gespach (See Page 123 of this volume.)

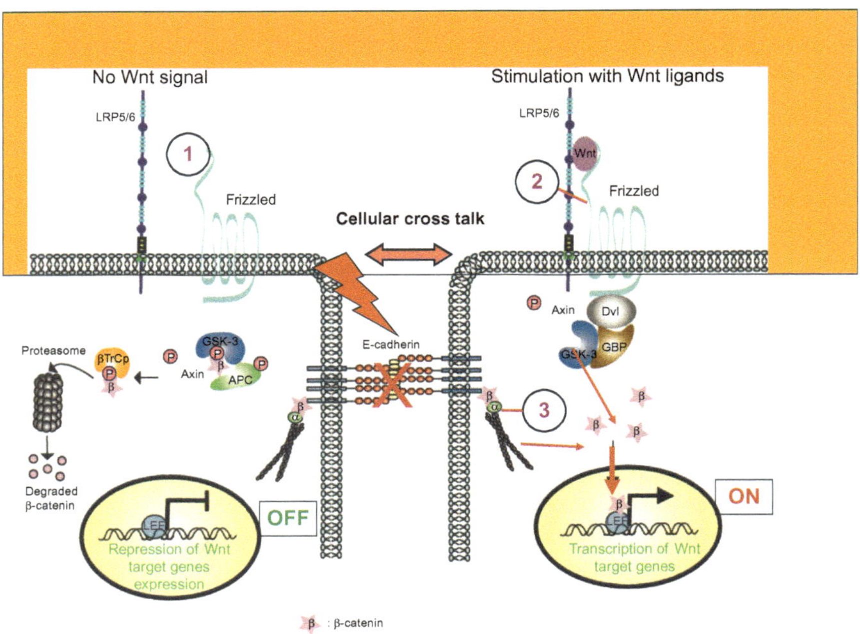

Fig. 8, Christian Gespach (See Page 124 of this volume.)

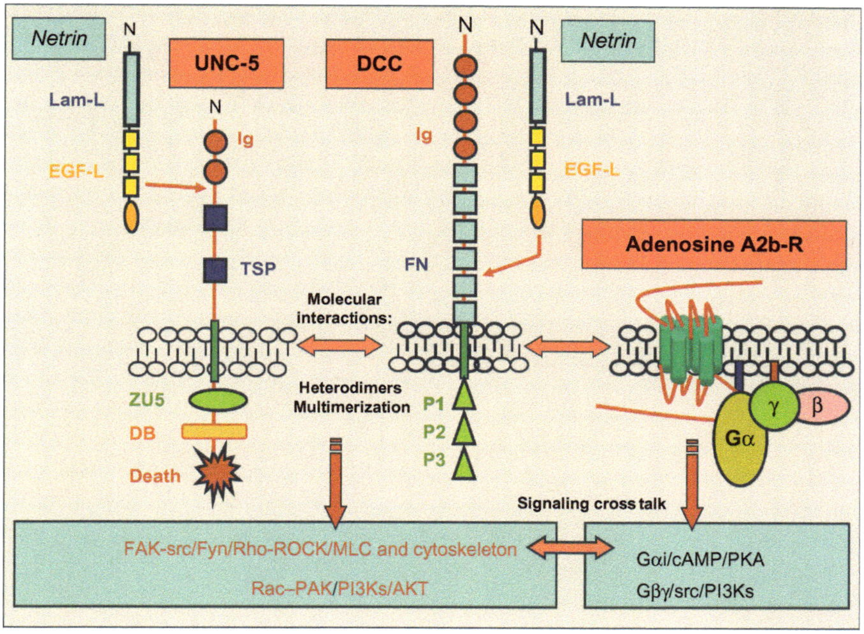

Fig. 9, Christian Gespach (See Page 133 of this volume.)

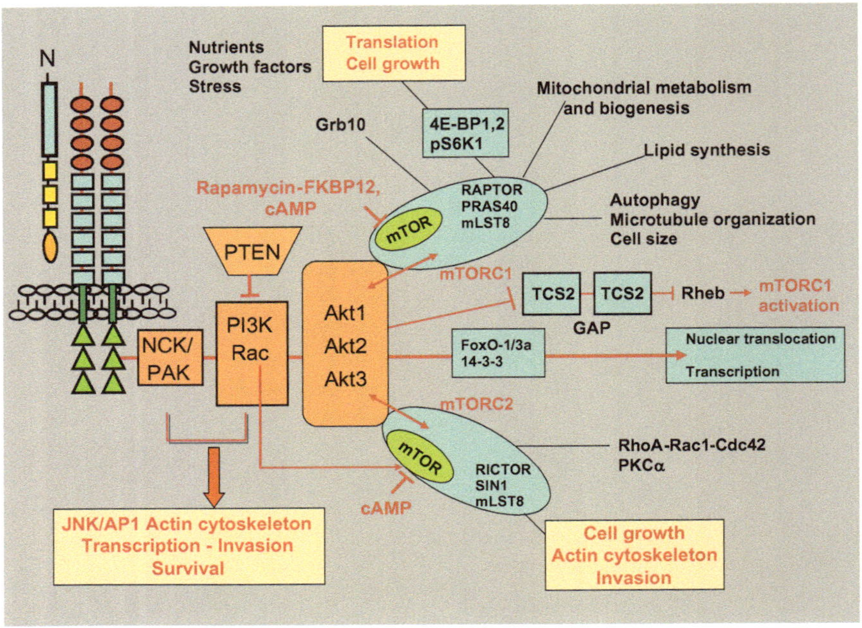

Fig. 10, Christian Gespach (See Page 138 of this volume.)

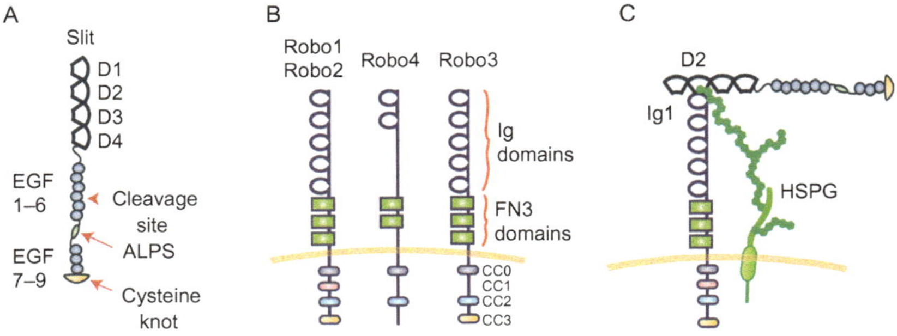

Fig. 1, Mimmi S. Ballard and Lindsay Hinck (See Page 190 of this volume.)

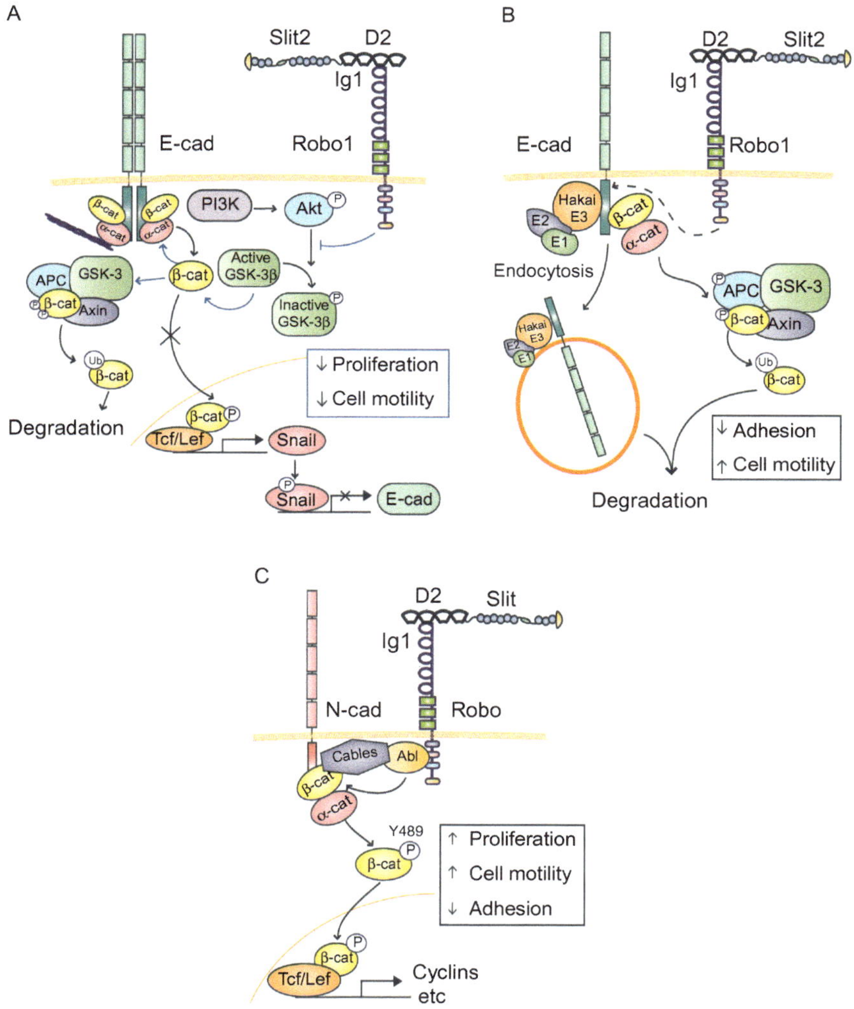

Fig. 2, Mimmi S. Ballard and Lindsay Hinck (See Page 207 of this volume.)

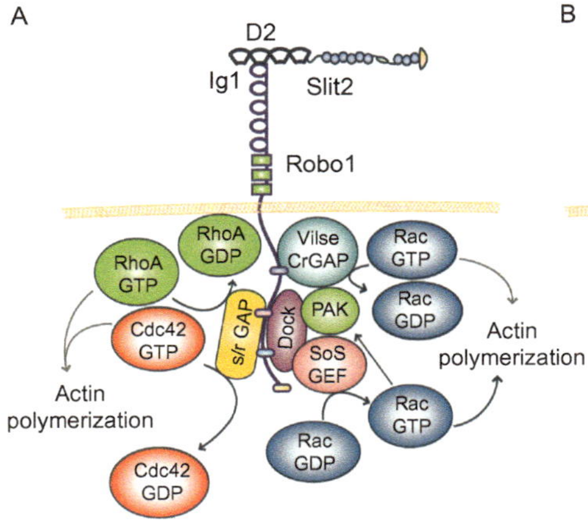

Fig. 3, Mimmi S. Ballard and Lindsay Hinck (See Page 212 of this volume.)

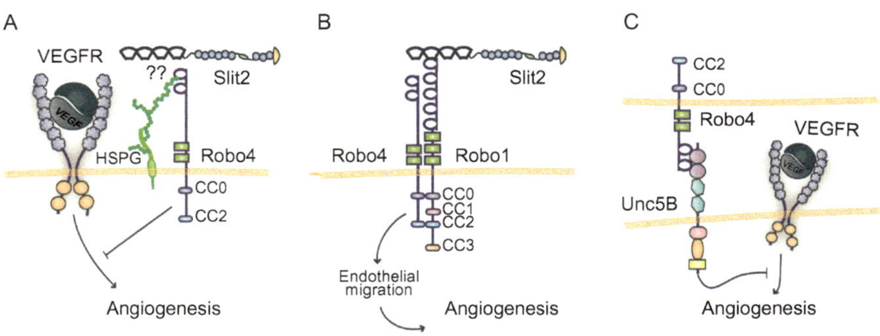

Fig. 4, Mimmi S. Ballard and Lindsay Hinck (See Page 221 of this volume.)

Kolibri 360 GmbH
Ferdinand-Jühlke-Straße 7
99095 Erfurt, Deutschland
produktsicherheit@kolibri360.de